HILBERT-HUANG TRANSFORM ANALYSIS OF HYDROLOGICAL AND ENVIRONMENTAL TIME SERIES

Water Science and Technology Library

VOLUME 60

The titles published in this series are listed at the end of this volume.

HILBERT-HUANG TRANSFORM ANALYSIS OF HYDROLOGICAL AND ENVIRONMENTAL TIME SERIES

by

A. RAMACHANDRA RAO
School of Civil Engineering, Purdue University, West Lafayette, IN, U.S.A.

and

EN-CHING HSU
School of Civil Engineering, Purdue University, West Lafayette, IN, U.S.A.

Springer

Library of Congress Control Number: 2007936514

ISBN 978-1-4020-6453-1 ISBN 978-1-4020-6454-8 (eBook)

Published by Springer,
P.O. Box 17, 3300 AA Dordrecht, The Netherlands.

www.springer.com

Cover Image: Time-frequency distribution of monthly streamflows in the Warta river (Fig 5.3.5 (b))

Printed on acid-free paper

DEDICATION

This book is respectfully dedicated to
the unique yogini of the twentieth century
Maatha Jayalakshmi
and to her son the great Siddha Purusha
Sri Sri Sri Ganapathi Sachchidananda Swamiji
of
Avadhootha Datta Peetham
Sri Ganapathi Sachchidananda Ashrama, Mysore 570 025, India
with namaskarams

CONTENTS

PREFACE

To accommodate the inherent non-linearity and non-stationarity of many natural time series, empirical mode decomposition (EMD) and Hilbert-Huang transform (HHT) provide an adaptive and efficient method. The HHT is based on the local characteristic time scale of the data. The HHT method provides not only a precise definition in time-frequency representation than the other conventional signal processing methods, but also more physically meaningful interpretation of the underlying dynamic processes. The EMD also works as a filter to extract the variability of signals with different scales and is applicable to non-linear and non-stationary processes. This promising algorithm has been applied in many fields since it was developed, but it has not been applied to hydrological and climatic time series.

The discussion in this book starts with several simulated data sets in order to investigate the capability of this method and to compare it to other conventional frequency-domain analysis methods that assume stationarity. Rainfall, streamflow, temperature, wind speed time series and lake temperature data are investigated in this study. The aim of the work is to investigate periodicity, long term oscillations and trends embedded in these data by using HHT. The analysis is performed in both the time and frequency domains. The results from HHT are compared to those from the multi-taper method (MTM) which is based on Fourier Transform of the data. The results indicate that the HHT is clearly superior to MTM in delineating the stochastic structure of the data. Details about the data which cannot be investigated by traditional methods are clearly seen with HHT. The nonstationarities of climatic and hydrologic data are also brought out. The HHT is seen to be an excellent tool to investigate the characteristics of environmental and hydrologic time series.

The details regarding the definition and application of Hilbert-Huang transform (Huang et al. (1998, 2005)) are discussed. It includes the sifting process used for empirical mode decomposition, Hilbert transform spectral analysis, some in the time-frequency domain (degree of stationarity, volatility, and instantaneous energy), and the trend tests (the modified Mann-Kendall test). Simulated data are first analyzed to investigate the performance of HHT analysis. Different types of synthetic data are discussed. One of the innovations based on HHT is the generation of nonstationary data. This aspect is of interest in time series analysis. Generation of data makes it possible to determine the confidence limits of the spectrum and furthermore to identify the significant peaks in HHT spectra.

The material herein puts much emphasis on the analysis of climatic, hydrological and environmental time series. Rainfall, temperature, streamflow and wind speed data in the state of Indiana, U.S.A., are studied. Also, long historical temperature records in Europe are investigated. The trends in European temperature data are clearly brought about by the results of EMD which compare well with the results of parametric trend tests. The other issue which may be brought to readers' attention is that the HHT spectra are often characterized by power law equations. The detection of periodicities, long-term oscillations, trends, nonstationarities embedded in the data by using HHT technique is a promising approach in time series analysis.

We would like to thank Dr. Miki Hondzo of the St. Anthony Falls Laboratory for sending us the lake temperature data of Chapter 8. He also reviewed Chapter 8 where the data acquisition is discussed. Dr. Tim Whalen of Purdue University contributed the wind data discussed in Chapter 7. Dr. Whalen wrote a draft paper on Chapter 7 based on which Chapter 7 has been written.

Professor Rao would like to thank the numbers of his family, Mamatha Rao his wife, Dr. Malini Rao Prasad his daughter, Dr. Sathya Prasad his son-in-law, Karthik A. Rao and Siddhartha S. Rao his sons and especially Shambhavi N. Prasad, his delightful grand daughter for their support. Dr. Hsu would like to give thanks to her family for their continuous love and support.

We would like to thank a number of people for both direct and indirect support during the period that we worked on this book project. We would like to thank Dr. V.P. Singh for his support. We thank the Publishing Editorial and Production staff at Springer Publishers (Dordrecht, The Netherlands) who helped to bring this book project to a successful conclusion. Our special thanks to Petra D. van Steenbergen (Publishing Editor).

A. Ramachandra Rao, Bangalore, India (April, 2007)
En-Ching Hsu, West Lafayette, Indiana, USA (April 2007)

CHAPTER 1

INTRODUCTION

In earlier studies of climatic time series, long-term time series have been assumed to be either periodic or stationary to apply the time domain or frequency domain analysis methods. Traditional frequency analysis techniques based on Fourier Transforms tend to spread the energy of the signal into several frequencies, which sometimes leads to misinterpretations of the characteristics of the data. In particular, trends in data would seriously distort the low frequency characteristics of the data.

There are several methods used for analyzing the non-stationary processes, such as spectrogram (short-time Fourier transform), Wagner-Ville distribution (Loutridis, 2005), empirical orthogonal function (EOF) expansion for metrological and oceanographic data and wavelet transforms. The spectrogram or the fixed window Fourier spectral analysis is widely used for musical and speech signal analysis. To calculate a spectrogram, the Fourier transform is applied by sliding a window along the time axis and repeatedly calculating the Fourier transform to obtain a time-frequency distribution. The disadvantage of this approach is that we have to ensure that the data within the time window is stationary. Even if it is stationary, the spectrogram method has an additional problem of having a trade-off in time and frequency resolution. The Wagner-Ville distribution is a quadratic-form time-frequency distribution with optimized resolution in both time and frequency domain, but it is not always nonnegative. There are also miscellaneous other methods such as the least squared estimation of the trend (Brockwell and Davis, 1991), which have problems and disadvantages.

Huang et al. (2001) and Flandrin (1999) point out that wavelets, though a good tool to investigate features of data, is a poor method to analyze time-energy-frequency distributions. This lack of frequency resolution is also addressed in great detail in Huang et al. (1998). Loutridis (2005) points out that the time and frequency resolution leads to compromises, as large scale wavelets are chosen for determining general signal features and small scale wavelets for extracting the signal details. Consequently, time localization is poor for low frequency signals and frequency resolution is poor for high frequency signals. Peng et al. (2005) demonstrate that wavelet transforms may generate many small undesirable spikes over all frequency scales and make the results confusing and difficult to interpret.

None of these methods can simultaneously provide a good resolution in both time and frequency domain. Huang et al. (1998, 2003a) proposed a new technique to efficiently extract the information in both time and frequency domains directly from the data. It is adaptive, efficient and without any prior assumptions. This scheme is called as Hilbert-Huang Transform (HHT), which is the combination of empirical mode decomposition (EMD) and Hilbert spectral analysis (HSA). It offers a different approach to processing time-series data. The signal is decomposed into several oscillation modes by extracting the characteristic scales embedded in the data. Traditionally, filtering is carried out in frequency domain; however, frequency domain filtering is difficult when the data are either nonstationary or nonlinear, or both. Filtering can eliminate some of the harmonics. Empirical mode decomposition can be treated as a time-frequency filtering method through the representation of the intrinsic mode function (IMF) components. Therefore the low-pass, high-pass and band-pass filters can be designed from the IMF components.

This technique is widely applied in science, engineering and financial analysis. In mechanical system analysis, it has been used for gear fault detection (Loutridis, 2004), fault diagnosis of roller bearings (Yu et al., 2005; Peng et al., 2005), and the processing of rotor startup signals (Gai, 2006). In biomedical science and health monitoring, it is applied for analyzing neural data (Liang et al., 2005), indicial responses of pulmonary blood pressure to step change of oxygen tension in the breathing gas (Huang et al., 1999a, 1999b), deriving the respiratory sinus arrhythmia from the heartbeat time series (Balocchi et al., 2004), and deriving main rhythms of the human cardiovascular system from the heartbeat time series and detecting their synchronization (Ponomarenko, 2005). Huang et al. (2003b) applied the empirical mode decomposition to financial market data analysis; they used the HHT algorithm to examine the changeability of the market, as a measure of volatility of the market. Montesinos et al. (2003) analyze the BWR neutron detector signals by using empirical mode decomposition and compare the result to those based on autoregressive models.

In testing structures, the HHT has been applied to detecting anomalies in beams and plates (Quek et al., 2003), vibration signal analysis (Peng et al., 2004), time-frequency analysis of the free vibration response of a beam with a breathing crack (Douka and Hadjileontiadis, 2005), and investigating the dynamic response of bridges to controlled pile damage (Zhang et al., 2005). Huang et al. (1998) conclude that HHT is a potential tool for cost-effective, efficient structural damage diagnosis procedures and health-monitoring systems. In coastal engineering applications, A.D. Veltcheva (2002) discusses the wave and group transformation by the HHT. Hwang et al. (2002) compare the energy flux computation of shoaling waves by using Hilbert and wavelet spectral analysis techniques.

HHT has also been applied to analyze earthquake signals. Huang et al. (2001) apply the HHT spectral analysis to the earthquake data of 21 September 1999 from Chi-Chi. Zhang et al. (2004) estimate the damping factor of non-linear soils and their role in estimating seismic wave responses at soil sites from earthquake recordings. Chen et al. (2004) tried to identify the natural frequencies and modal damping ratios of the Tsing Ma suspension bridge during Typhoon Victor using the HHT algorithm. In atmospheric and geophysical sciences, Pan et al. (2002) use the intrinsic mode functions to interpret the scattermeter ocean surface wind vector EOFs over the Northwestern Pacific. Gloersen and Huang (2003) compare interannual intrinsic modes in hemispheric sea ice cover and other geophysical parameters.

There are several extended studies and theoretical discussions of this method. Flandrin and Gonçalvès (2003, 2004) apply the empirical mode decomposition as an equivalent filter bank structure to analyze the fractional Gaussian noise and further rationalize the method as an alternative way to estimate the Hurst exponent. Yang et al. (2003) identified general linear structures with complex modes using free vibration response data polluted by noise.

As far as we can ascertain, results of analysis of climatic and hydrologic time series by using HHT have not been reported. The objective of the research discussed here is to investigate climatic and hydrologic time series, such as those of rainfall, runoff, and temperature by using HHT analysis, and discuss the results. These time series may be nonstationary and nonlinear. The results obtained by conventional spectral analysis cannot be well interpreted in such cases. The properties of these data are investigated in time, frequency and in time-frequency domains.

In this study, the rainfall (HCN and NCDC), streamflow (USGS data and three other cases from Warta (Poland), Godavari (India) and Krishna (India) rivers), temperature (HCN, NCDC and long-term measurements in Europe), wind speed (in the state of Indiana), and lake temperature data (four stations in the state of Minnesota) are analyzed. The results obtained by HHT analysis are compared to those from Fourier and multi-taper methods. The trend and periodicity in the data are studied by performing empirical mode decomposition to obtain the intrinsic mode functions. This procedure decomposes the data into several components representing different frequencies which helps in the interpretation of the data more efficiently and adaptively. The degree of stationarity is a statistic used to investigate the variation in power spectral density in time. In addition, another measure, that of volatility, provides information of how the intrinsic mode functions are related to the signal. When similar data are analyzed, common characteristics are of interest and are investigated.

The material herein is presented as follows. In Chapter 2, the details regarding the definition and application of Hilbert-Huang transform are discussed. It includes the sifting process used for empirical mode decomposition, Hilbert transform spectral analysis, some statistics to evaluate the results in the time-frequency domain (degree of stationarity, volatility, and instantaneous energy), and the trend tests (the modified Mann-Kendall test) are discussed.

Simulated data are analyzed by using HHT in Chapter 3. The performance and results of HHT analysis for different types of data are discussed first. One of the innovations based on HHT is the synthetic generation of nonstationary data. This aspect is discussed in Chapter 3. These generated data are used to identify the significance of peaks in HHT spectra. They may also be used to generate synthetic data commonly used in stochastic hydrology.

The analysis and discussion of climatic and environmental data are discussed in Chapters 4–8. Rainfall data from the state of Indiana are analyzed by using HHT and the results are discussed in Chapter 4. Both the NCDC and HCN data are analyzed. The NCDC data, because it is averaged over a region, are more consistent than the HCN data and the results reflect this characteristic. The commonly occurring periodicities are identified. Long term (greater than about 20 years) oscillations are not present in these data.

Monthly streamflow data from Indiana are analyzed and reported in Chapter 5. These data show greater variability than the rainfall data, but the common periods of oscillation correspond to those found in rainfall data. One other series, a very long daily flow series from Warta River in Poland, has been analyzed and the results discussed. The spectrum of this series is of considerable interest as this series is one of the longest streamflow sequences available. The Krishna and Godavari River data from India are analyzed. These data which give strong spurious peaks in spectra when Multi-taper and other methods are used do not give such peaks with HHT.

The monthly temperature data from Indiana and some long historical data from Europe are analyzed and discussed in Chapter 6. The monthly data from Indiana indicate variability corresponding to 1, 2, 4 and 11 years. The trends in European temperature data are clearly brought about by the results of EMD which compare well with the results of parametric trend tests.

Daily wind speed data from four stations in Indiana are analyzed in Chapter 7. The HHT spectra of these data exhibit considerable similarity, thereby indicating the potential of HHT spectra to characterize similar regions of wind velocity. The spectra of wind speed data are also characterized by power law equations. There are some significant periodicities in wind speed data also.

The lake temperature and PAR data are obviously nonstationary. Previous attempts to analyze these data used the technique of segmenting the data. These segments were approximately stationary. The Fourier spectra of these segments were computed. Naturally there are considerable variations in these spectra and in the results based on HHT. These nonstationary data are analyzed by the HHT and the results are discussed.

CHAPTER 2

HILBERT-HUANG TRANSFORM (HHT) SPECTRAL ANALYSIS

2.1. INTRODUCTION

Huang et al. (1998) introduced a general signal-analysis technique, called Hilbert-Huang Transform (HHT). It is a two-step algorithm, combining empirical mode decomposition (EMD) and Hilbert spectral analysis, to accommodate the non-linear and non-stationary processes. This method is not based on a priori selection of kernel functions, but instead it decomposes the signal into intrinsic oscillation modes derived from the succession of extrema.

Before discussing the Hilbert-Huang Transform algorithm, traditional spectral analysis methods which are used for comparison are reviewed. They are the Fourier transform and Multi-Taper methods. In addition, statistical measures to investigate the time series and spectral properties are discussed. These include the time-frequency representation known as spectrogram, degree of stationarity, volatility, and trend tests.

2.2. CONVENTIONAL SPECTRAL ANALYSIS METHODS

2.2.1 Fourier Transform Analysis

The common definition for Fourier transform of a continuous-time signal $x(t)$ is given in Eq. (2.2.1).

$$X(\omega) = \int_{-\infty}^{\infty} x(t)e^{-j\omega t}dt, \ \omega \in (-\infty, \infty) \tag{2.2.1}$$

Almost all data analysis is carried out not with functions in continuous time but with discrete-time data. Hence discrete-time Fourier transform (DFT) is used in data analysis. The DFT replaces the infinite integral in Eq. (2.2.1) with a finite summation representation,

$$X(\omega_k) \cong \sum_{n=0}^{N-1} x(t_n)e^{-j\omega_k t_n}, \quad k = 0, 1, 2, \ldots, N-1 \tag{2.2.2}$$

where N is the number of time samples, and ω_k is the kth frequency. This formula has finite summation limits. DFT is implemented by using the Fast Fourier Transform (FFT) algorithm when possible. The FFT yields an efficient way to calculate the DFT.

Fourier amplitude spectrum defines harmonic components globally and thus yields average characteristics over the entire duration of the data. In order to investigate nonstationary data and investigate the time-frequency characteristics, the Fourier transform can be utilized with segments of data to produce the so-called spectrogram. In using this technique, each segment should be stationary so that we can minimize the non-stationarity in the signal caused by different types of propagating waves. However, the frequency resolution is reduced when the length of the window is shortened. A trade-off situation is faced in this approach: the shorter the window, the better the temporal localization of Fourier amplitude spectrum, but the poorer the resolution in frequency.

2.2.2 Multi-Taper Method (MTM) of Spectral Analysis

The MTM method (Thomson, 1982) makes use of an extended version of conventional spectral representation. The process $x(t)$ may include a number of periodic components in addition to an underlying stationary process,

$$x(t) = \sum_j c_j \cos(2\pi f_j t + \phi_j) + \zeta_t = \sum_j \mu_j e^{i2\pi f_j t} + \mu_j^* e^{-i2\pi f_j t} + \zeta_t \tag{2.2.3}$$

where ζ_t is a zero mean stationary process with spectral density $S(f)$, c_j and f_j are the amplitude and frequency of periodic or line components j, $\mu_j = (c_j/2)e^{i\phi_j}$ is the complex amplitude corresponding to the real amplitude c_j. These types of processes are known as centered stationary or conditional stationary processes and often have mixed spectra.

The basic idea of MTM spectral analysis is using multiple data windows known as "discrete prolate sheroidal sequences" or "Slepian sequences", which are defined as the solution of the symmetric Toeplitz matrix eigenvalue problem in Eq. (2.2.4),

$$\lambda_k v_n^{(k)}(N, W) = \sum_{m=0}^{N-1} \frac{\sin\{2\pi W(n-m)\}}{\pi(n-m)} v_m^{(k)}(N, W) \tag{2.2.4}$$

where N is the number of data points, W is the spectral band width and λ_k are the eigenvalues associated with the Slepian sequences $v_n^{(k)}(N, W)$. The values of Slepian sequences can be calculated numerically by using methods given by Percival and Walden (1993) and Thomson (1982). The Fourier transform of these sequences are given in Eq. (2.2.5),

$$V_k(f) = \sum_{n=0}^{N-1} v_n^{(k)}(N, W) e^{-i2\pi fn} \tag{2.2.5}$$

The Slepian functions have the maximum energy concentration within the interval $(f - W, f + W)$. In this method, the bias from all frequencies remote from the frequency range of interest decreases exponentially as a function of NW; thus this method very effectively eliminates window leakage. The first step in MTM spectral estimation is the expansion of the time series $x(t)$ as Eq. (2.2.6),

$$y_k(f) = \sum_{t=0}^{N-1} x(t) v_t^{(k)}(N, W) e^{-i2\pi ft} \tag{2.2.6}$$

where $k = 0, 1, \ldots, K-1$ and K is usually taken as $2NW-1$. The band width W is usually chosen between $2/N$ and $20/N$, with $4/N$ a good initial choice. If W is too small, the resulting spectral estimate is unstable, but if W is too large, it results in poor resolution. The spectrum can be estimated by Eq. (2.2.7).

$$S(f) = \frac{1}{K} \sum_{k=0}^{K-1} |y_k(f)|^2 \tag{2.2.7}$$

Priestley (1965) suggests a method for calculating the evolutionary spectra $S(t,f)$ of nonstationary time series based on a double windowing technique, which can reduce the variance of the estimate of the evolutionary spectrum. This technique is similar to that applied in the multitaper method for spectral analysis. The difference is that in MTM, the variance is reduced by averaging the spectra from the same data segment using multiple data tapers. MTM is used as an alternative to the double window technique to evaluate the evolutionary spectra. To apply multi-taper method to study the time-frequency spectra, the signal is divided into a number of segments (possibly overlapping as a sliding window) each of length T and MTM spectra are calculated for each segment to obtain $S(t,f)$.

2.2.3 Spectrogram

Spectrograms are usually created in one of two ways; either with a series of bandpass filters or they are calculated from time signals by using the short-time Fourier transform (STFT). Piece-wise stationarity is assumed and sliding a window across the time series and performing Fourier analysis to construct the spectrograms. STFT is simply described in a continuous case. A window function, which is nonzero for a short period of time, is convolved with the function to be transformed and Fourier transformed. The resulting signal is taken as the window sliding along the time axis and written as

$$STFT(t, \omega) = \int_{-\infty}^{\infty} x(\tau) w(t - \tau) e^{-j\omega\tau} d\tau \tag{2.2.8}$$

where $w(t)$ is the window function and $x(t)$ is the signal to be transformed. $STFT(t, \omega)$ is then a complex function representing the phase and magnitude of the signal over time and frequency. The spectrogram, $SP(t, \omega)$, is given by the magnitude of the STFT function:

$$SP(t, \omega) = |STFT(t, \omega)|^2 \tag{2.2.9}$$

To calculate the spectrogram, the digital sampled data in time domain is broken up into several segments, which usually overlap, and Fourier transformed to calculate the magnitude of the power spectrum of each segment. Each segment then corresponds to a vertical line in the image of time-frequency representation- a representation of magnitude versus frequency at a specific moment in time.

2.3. EMPIRICAL MODE DECOMPOSITION

In the traditional Fourier analysis, the frequency is defined by using the sine and cosine functions spanning the entire length of data. Such a definition would not make sense for non-stationary data in which changes occur with time. This difficulty is overcome by the introduction of the approaches based on the Hilbert transform. For an arbitrary time series, $x(t)$, its Hilbert transform, $y(t)$, is obtained by

$$y(t) = \frac{1}{\pi} P \int_{-\infty}^{\infty} \frac{x(t')}{t - t'} dt' \tag{2.3.1}$$

where P indicates the Cauchy principal value. It is the convolution of $x(t)$ with $1/t$; hence, the transform emphasizes the local properties of $x(t)$. $x(t)$ and $y(t)$ form the complex conjugate pair by definition, so we can have an analytical signal, $z(t)$ as shown in Eq. (2.3.2),

$$z(t) = x(t) + iy(t) = a(t)e^{i\theta(t)} \tag{2.3.2}$$

in which

$$a(t) = [x^2(t) + y^2(t)]^{1/2}, \theta(t) = \arctan\left(\frac{y(t)}{x(t)}\right) \tag{2.3.3}$$

The polar coordinate expression is the local fit of an amplitude and phase varying trigonometric function to $x(t)$. Based on Hilbert transform, the instantaneous frequency is defined as

$$\omega(t) = \frac{d\theta(t)}{dt} \tag{2.3.4}$$

In practice, at any time, it is quite possible that the signal may involve more than one oscillation mode, and consequently the signal has more than one local instantaneous frequency at a time. There is still considerable controversy in defining the instantaneous frequency with Hilbert transform. A detailed discussion is found in Huang et al. (1998). Restrictive conditions have to be imposed on the data in order to obtain meaningful instantaneous frequency. For this purpose, Huang et al. (1998) suggest modifying the restrictive condition from a global to a local one so that we can translate the requirement into physically implementable steps. Furthermore, this local restriction also suggests a method to decompose the data into components for which the instantaneous frequency can be defined. Hence, EMD is needed. Otherwise negative amplitudes may appear in Hilbert transform. Intrinsic mode function (IMF) is thus designated as a class of functions so that the instantaneous frequency can be defined everywhere based on the local properties. As a result, the limitation of interest here is not on the existence of the Hilbert transform which is general and global, but on the existence of a meaningful instantaneous frequency which is restrictive and local.

Physically, the required conditions to define a meaningful instantaneous frequency are that the functions are symmetric with respect to the local zero mean and have the same number of zero crossings and extrema. An intrinsic mode function (IMF) is defined as a function that satisfies two conditions: (1) the number of extrema and the number of zero crossings must either equal or differ at most by one in the whole data set, and (2) the mean value of the envelope defined by the local maxima and the envelope defined by the local minima is zero everywhere. An illustration of local mean, and envelopes of local maxima and minima are shown in Figure 2.3.1.

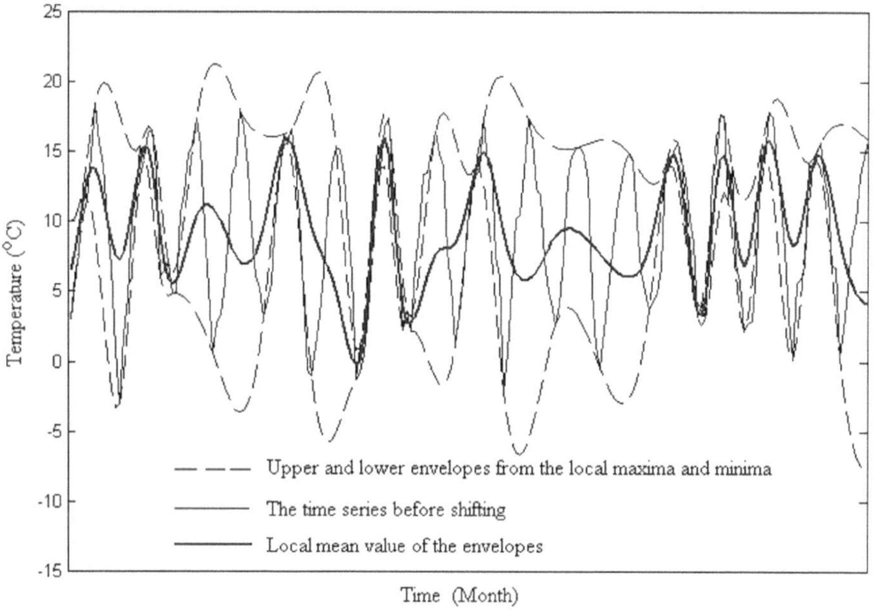

Figure 2.3.1. Definition of sifting

Knowing the well-behaved Hilbert transform of the IMF components is only a starting point. In most cases, we have to decompose the data into several IMFs since most time series involve more than one oscillatory mode. A systematic way to extract the IMFs, designated as a sifting process, is based on the following assumptions: (1) the signal has at least two extrema, and (2) the characteristic time scale is defined by the time lapse between the extrema. The sifting process is described as follows.

(1) Identify all extrema (maxima and minima) of the signal $x(t)$.
(2) Connect these maxima with the cubic spline lines to construct an upper envelope, $e_{max}(t)$; use the same procedure for minima and to construct a lower envelope, $e_{min}(t)$.
(3) Compute the mean of the upper and lower envelope: $m(t) = [e_{min}(t) + e_{max}(t)]/\ 2$.
(4) Calculate $d(t) = x(t) - m(t)$.
(5) Let $d(t)$ be the new signal $x(t)$. Follow the previous procedure again until $d(t)$ becomes a zero-mean process according to a stopping criterion. These iterations are shown in Figure 2.3.2.
(6) Once we have the zero-mean $d(t)$, it is designated as the first intrinsic mode function(IMF 1), c_1.
(7) The IMF 1 is subtracted from the original signal and the residual is used as a new signal $x(t)$. The sifting process is repeated to get IMF 2.
(8) Continuing like this, we obtain c_3, c_4, and so on. This process is stopped when the residual is a monotonic function having only one minimum or one maximum.

In practice, after a certain number of iterations, the resulting signal does not carry significant physical information. The sifting process is stopped by limiting the standard deviation, which is computed from the two consecutive sifting results. The threshold is usually set as 0.2 and 0.3. Also, the number of extrema decreases while moving to the higher order IMF, and this guarantees that the sifting process ends with a finite number of intrinsic mode functions. Basically, the sifting process eliminates the riding waves and makes the IMF profiles mode symmetrical in order to obtain meaningful results for instantaneous frequency.

IMF components represent simple oscillatory modes embedded in the signal and is much more general compared to the simple harmonic functions. As a check of the completeness of using Eq. (2.3.5), we can reconstruct the data by adding all the IMF components and the residual trend. Assume that we have n IMF components $(c_1, c_2, \ldots, c_n)$ and one residual (r_n), which follow the order from the shortest to the longest period. Hence it implies that they range from the highest frequency to the lowest frequency. The characteristic scale is physical which helps us to examine the physical meaning of each IMF component.

$$x(t) = \sum_{j=1}^{n} c_j + r_n \tag{2.3.5}$$

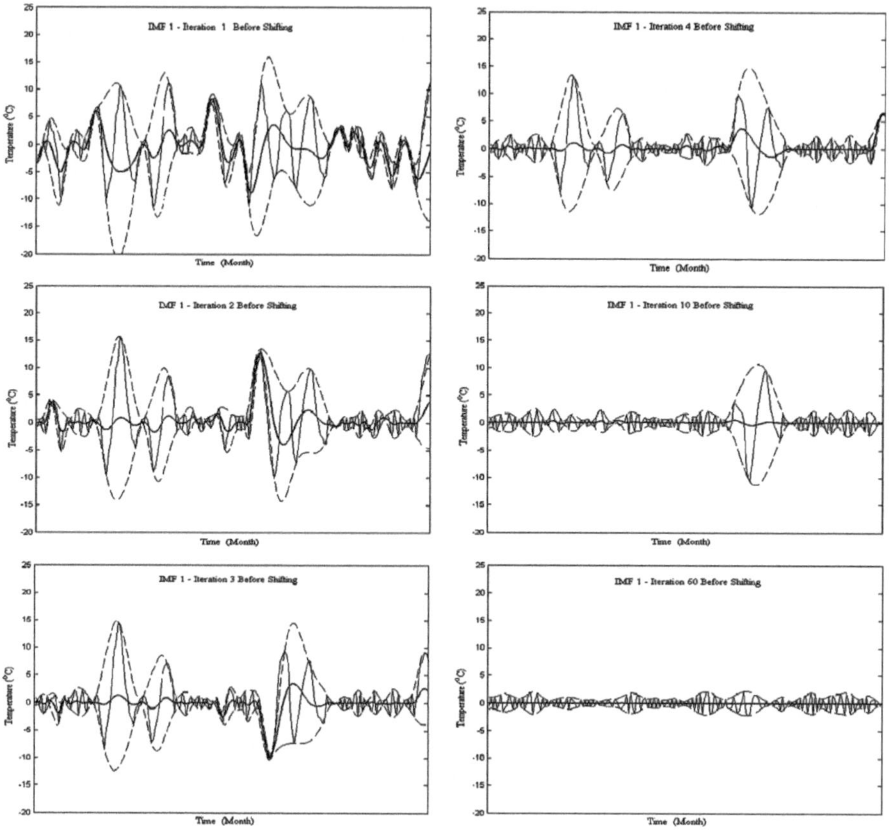

Figure 2.3.2. A demonstration of the iteration process to obtain a zero-mean process, i.e. an intrinsic mode function. The thin-solid line is the time series before sifting, the dot-dashed lines are the upper and lower envelopes from the local maxima and minima, and the thick solid line is the local mean value of the envelopes

A new identified use of the IMF component is filtering. For example, a low pass filtered result of a signal having n IMFs can be expressed by

$$x_{LP}(t) = \sum_{j=k}^{n} c_j + r_n \tag{2.3.6}$$

and high pass filtered results can be expressed as

$$x_{HP}(t) = \sum_{j=1}^{k} c_j \tag{2.3.7}$$

Further, a band pass filtered result can be expressed as

$$x_{BP}(t) = \sum_{j=b}^{k} c_j \tag{2.3.8}$$

In order words, we can add the long period components to get a lowpass filter result, or we can add all components with selected omissions to get the band-pass result.

The orthogonality of the IMF components should be checked *a posteriori* in order to investigate the goodness of the decomposition process. Let the residue $r(t)$ be the last IMF, i.e., $c_n + 1 = r(t)$. Then Eq. (2.3.9) can be rewritten as

$$x(t) = \sum_{j=1}^{n+1} c_j(t) \tag{2.3.9}$$

Then taking square of this signal $x(t)$ we have

$$x^2(t) = \sum_{j=1}^{n+1} c_j^2(t) + 2\sum_{j=1}^{n+1}\sum_{k=1}^{n+1} c_j(t)c_k(t) \tag{2.3.10}$$

If the decomposition is orthogonal, the cross terms given in the second part of the right-hand side should be zero when they are integrated along time. Therefore, an overall index of the orthogonality, IO, is defined as

$$IO = \sum_{t=0}^{T} \left(\sum_{j=1}^{n+1}\sum_{k=1}^{n+1} c_j(t)c_k(t)/x^2(t) \right) \tag{2.3.11}$$

T is the time interval under consideration. The index IO should be very small in order to have a good decomposition of IMF. Typically, values between 0.01 and 0.001 are acceptable.

2.4. HILBERT-HUANG SPECTRA

Once we have these intrinsic mode function components, Hilbert transform can be applied to each component to get the amplitudes, and meanwhile the instantaneous frequency is calculated using Eq. (2.3.4). Therefore, Eq. (2.3.5) is rewritten in the following expression,

$$x(t) = \Re \sum_{j=1}^{n} a_j(t) \exp\left(i \int \omega_j(\tau) d\tau\right) \tag{2.4.1}$$

where $\Re$ is the real part of the complex number.

The time-frequency distribution of the amplitude is designated as the Hilbert amplitude spectrum, $H(\omega, t)$, or simply the Hilbert spectrum. At a given time t, the instantaneous frequency ω and the amplitude are calculated simultaneously so that these values are assigned to Hilbert spectrum, $H(\omega, t)$. $a_j(t)$ is a time-dependent expansion coefficient similar to the constant in the Fourier expansion and ω_j is the instantaneous frequency at a time τ which differs from the constant frequency ω_j in Fourier transform Eq. (2.4.2). It represents a generalized form of Fourier expansion.

$$x(t) = \Re \sum_{j=1}^{n} a_j(t) \exp(i\omega_j t) \tag{2.4.2}$$

With the Hilbert spectrum defined, the marginal Hilbert spectrum, $h(\omega)$, is defined in Eq. (2.4.3). It is a measure of total energy contribution from each frequency over the entire data span in a probabilistic sense. It provides a quantitative way to describe the time-frequency-energy representation by integrating the Hilbert spectrum over the entire time span,

$$h(\omega) = \int_0^T H(\omega, t)dt \tag{2.4.3}$$

where T is the total data length.

Another integration over the frequency span is the instantaneous energy $IE(t)$, which is defined as Eq. (2.4.4). It provides information about the time variation of the energy.

$$IE(t) = \int_\omega H(\omega, t)d\omega \tag{2.4.4}$$

The raw Hilbert spectrum presentation gives desirable and quantitative results. But, the higher resolution representation and small scattered points in time-frequency-energy plot are not easy to interpret in raw Hilbert spectrum. Hence, a Gaussian weighted Laplacian filter is applied to the Hilbert spectrum. The schematic of this filter is shown in Figure 2.4.1. A "fuzzy" or "smoothed" view thus can be derived from the original presentation by using two-dimensional filtering. The properties of four spectral analysis methods based on the capability of handling the nonlinear and nonstationary time series are listed in Table 2.4.1.

A flowchart in Figure 2.4.2 summarizes the calculation procedure of this two-step HHT algorithm. The left-hand side of Figure 2.4.2 is basically the procedure for using sifting process to define the intrinsic mode functions or the empirical mode decomposition while the right-hand side is the procedure to construct the Hilbert spectrum.

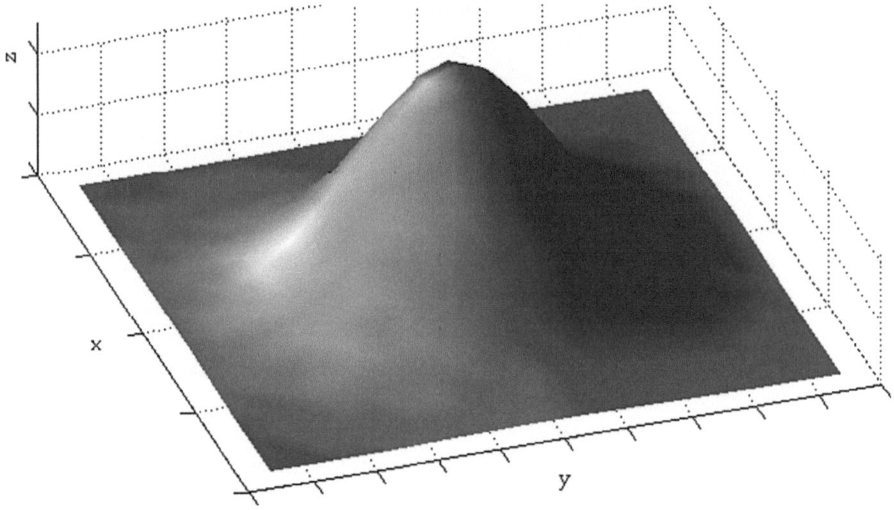

Figure 2.4.1. Schematic of a 2D Gaussian weighted Laplacian filter

Table 2.4.1. Comparison of different spectral analysis methods (Huang et al., 1998)

	Fourier	MTM	Wavelet	HHT
Basic	A priori	A priori	A priori	Adaptive
Frequency	Convolution: Global	Convolution: Global	Convolution: Regional	Differentiation: Local
Presentation	Energy-frequency	Energy-frequency	Energy-time-frequency	Energy-time-frequency
Nonlinear	No	No	No	Yes
Nonstationary	No	Yes	Yes	Yes
Feature extraction	No	No	No (discrete) Yes (Continuous)	Yes

2.5. RELATIONSHIP BETWEEN HHT AND FOURIER SPECTRA

As mentioned previously, the representations of a signal by using Fourier series or Hilbert transform are given in Eq. (2.5.1). It is clear that Hilbert transform is a more general representation than Fourier transform.

$$
\begin{aligned}
x(t) &= \Re \sum_{j=1}^{n} a_j(t) \exp(i\omega_j t) && \text{(Fourier transform)} \\
&= \Re \sum_{j=1}^{n} a_j(t) \exp\left(i \int \omega_j(\tau) d\tau\right) && \text{(Hilbert transform)}
\end{aligned}
\tag{2.5.1}
$$

A way to investigate the relationship between HHT and Fourier transform is by calculating the energy of HHT and Fourier spectra. Fourier transform of a time

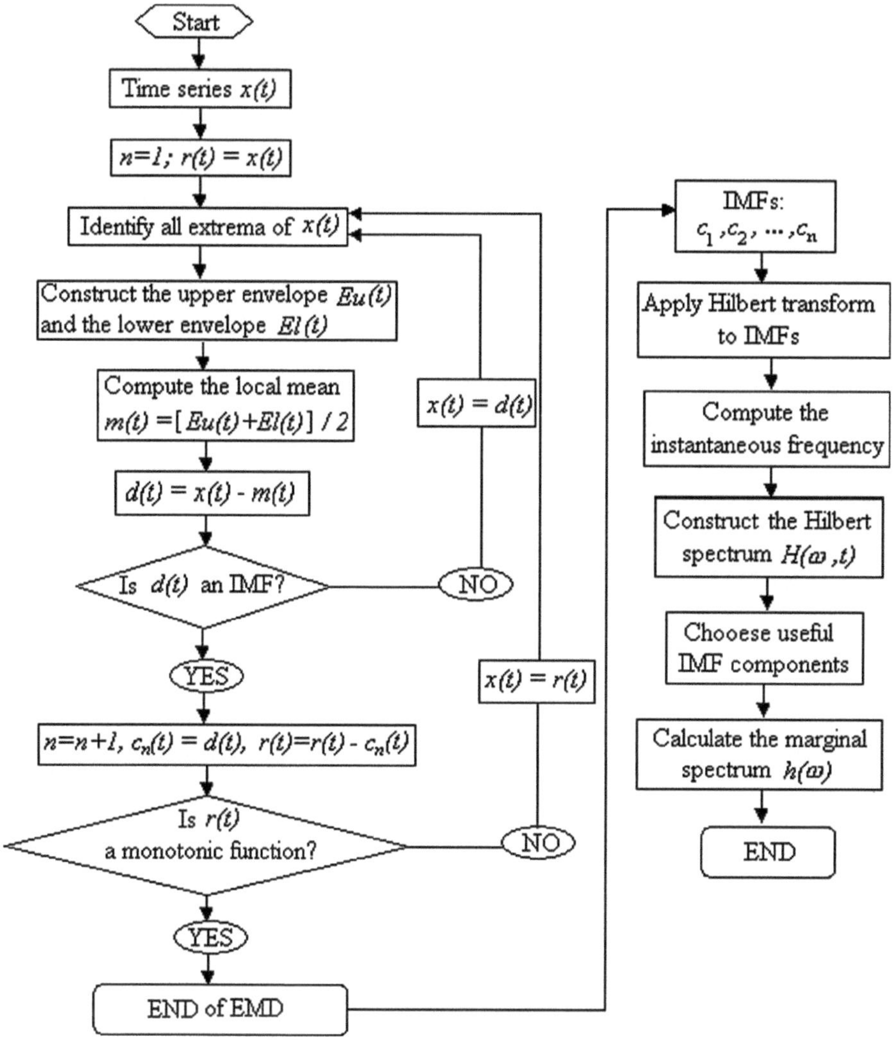

Figure 2.4.2. Flowchart of the empirical mode decomposition and Hilbert spectrum analysis

series $x(t)$ is:

$$X(\omega) = \int_{-\infty}^{\infty} x(t)e^{-i\omega t}dt \tag{2.5.2}$$

Its complex conjugate is

$$X^*(\omega) = \int_{-\infty}^{\infty} x(t)e^{i\omega t}dt \tag{2.5.3}$$

The area under the Fourier energy spectrum for all frequencies is then given by

$$\begin{aligned} E &= \int_{-\infty}^{\infty} X(\omega)X^*(\omega)d\omega = \int_{-\infty}^{\infty}\int_{-\infty}^{\infty}\int_{-\infty}^{\infty} x(t_1)x(t_2)e^{-i\omega t_1}e^{i\omega t_2}d\omega dt_1 dt_2 \\ &= \int_{-\infty}^{\infty}\int_{-\infty}^{\infty} x(t_1)x(t_2)2\pi\delta(t_1 - t_2)dt_1 dt_2 = 2\pi\int_{-\infty}^{\infty} x^2(t)dt \end{aligned} \tag{2.5.4}$$

If $x(t)$ is defined for $0 < t < T$ and zero otherwise, the relationship is simplified to

$$E = |X(\omega)|^2 = 2\pi \int_0^T x^2(t)dt \tag{2.5.5}$$

The area under the marginal Hilbert spectrum can be illustrated by the general case of a time series which is a summation of n cosine functions defined for a given time period 0 to T (Wen and Gu, 2004).

$$x(t) = \sum_{j=1}^{n} a_j(t) \cdot \cos[\theta_j(t)] \tag{2.5.6}$$

if the residual term from Eq. (2.3.6) is neglected. The total area under the marginal Hilbert spectrum is

$$\sum_{j=1}^{n}\int_0^T a_j^2(t)dt \tag{2.5.7}$$

Its total energy over time is

$$\int_0^T x^2(t)dt = \sum_{i=1}^{n}\sum_{j=1}^{n}\int_0^T a_i(t)a_j(t)\cos[\theta_i(t)]\cos[\theta_j(t)]dt \tag{2.5.8}$$

Due to the orthogonality of IMFs, the cross terms ($i \neq j$) can be neglected. Thus, Eq. (2.5.8) reduces to Eq. (2.5.9).

$$\int_0^T x^2(t)dt = \sum_{j=1}^{n}\int_0^T a_j^2(t)\cos^2[\theta_j(t)]dt \tag{2.5.9}$$

When the IMFs are relatively smooth and sinusoidal, an approximation for the integer number of quarter-waves is written as

$$\int_0^T a_j^2(t)\cos^2[\theta_j(t)]dt \approx \frac{1}{2}\int_0^T a_j^2(t)dt \tag{2.5.10}$$

Thus

$$\int_0^T x^2(t)dt \approx \frac{1}{2}\sum_{j=1}^{n}\int_0^T a_j^2(t)dt \tag{2.5.11}$$

This is an approximation. For high frequency IMFs, the number of quarter-waves is usually large so the portion which does not contain complete quarter-waves is small and their contribution can be neglected. For low frequency IMFs, this portion cannot be neglected, but these IMFs usually have small amplitudes and contribute insignificantly to the total energy. (Wen and Gu, 2004). Hence, by Eq. (2.5.4) and Eq. (2.5.11), the relationship between marginal Hilbert spectra and Fourier spectra are approximately connected with a factor of π. Thus there is a linear relationship between these two spectra. This relationship may be used in interpreting the HHT spectra.

2.6. VOLATILITY OF TIME SERIES

To characterize the variation in time of time series, Huang et al. (2003) designated a measure, $V(t, T)$, to indicate the volatility of the signal. The volatility is defined as the ratio of the absolute value of IMF components to the signal at any time,

$$V(t, T) = \frac{S_h(t)}{S(t)} \tag{2.6.1}$$

where T corresponds to the period at the Hilbert spectrum peak of the high pass signal up to h terms.

$$S_h(t) = \sum_{j=1}^{h} c_j(t) \tag{2.6.2}$$

By definition, it is similar to the idea of check of completeness, which is performed by adding IMFs to make sure the summation is the same as the original data. As for volatility, it focuses on the ratio of variation to the signal. The h value is flexible. For example, the volatility of HCN120177 monthly rainfall data is shown in Figure 2.6.1. The volatilities are computed for $h = 2, 3, 4$ and 5. The measure of instantaneous energy, IE, which is calculated by integrating over frequency bins in Hilbert spectrum Eq. (2.4.4), is also plotted for comparison. For the subplot of $IE(t)$, the bold gray line is the mean value of the instantaneous energy and the dotted line is the normalized standard deviation of the instantaneous energy. Both are shifted with respect to the mean values. The mean IE is quite smooth while the standard deviation presents more information about the energy variation in time. Comparing the peaks of $IE(t)$ to the original data, they are consistent in pointing out the portion which has more variations in time. Therefore, the standard deviation of instantaneous energy is an alternative way to investigate the characteristics of time series.

Other subplots of volatility in Figure 2.6.1 indicate only small differences among them. This result is observed not only in the rainfall data but also in other types of data analyzed. It is found that by using Eq. (2.6.1), these volatilities are controlled by the first IMF component, which has positive and negative values alternating on the x axis. When it is divided by the data, it results in several extreme values and

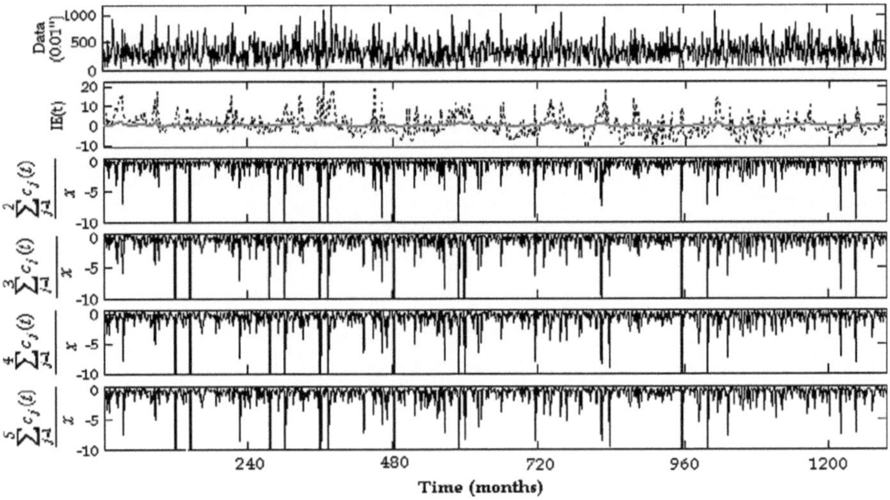

Figure 2.6.1. Volatility and IE for the HCN120177 monthly rainfall data

makes the results difficult to interpret. These extreme values are mostly from the unreasonable numerical values, such as an IMF point value divided by a nearly zero value.

By definition, the mean of an IMF is nearly zero. To have a stable division result from Eq. (2.6.1), it is necessary to put the residual into Eq. (2.6.2) especially when the residuals are large. In this case, the residual is around 300 (0.01 inches). Therefore, to fix the problem caused by the numerical division by small values and the shifting of data, a modification by adding residual is made for the calculation of $S_h(t)$ in Eq. (2.6.2) and we arrive at Eq. (2.6.3).

$$S_h(t) = r(t) + \sum_{j=1}^{h} c_j(t) \tag{2.6.3}$$

The results are shown in Figure 2.6.2. These results are a stable and recognizable along with more IMF components considered.

There is another way to utilize the volatility measure as shown in Eq. (2.6.4). It is the summation computed from the higher order IMFs. Without adding the first IMF, the volatility is shown in Figure 2.6.3. It yields a consistent result to detect the variation in IE and data.

$$S_h(t) = r(t) + \sum_{j=1}^{h} c_{nimf-j}(t) \tag{2.6.4}$$

where *nimf* is the total number of intrinsic mode functions decomposed.

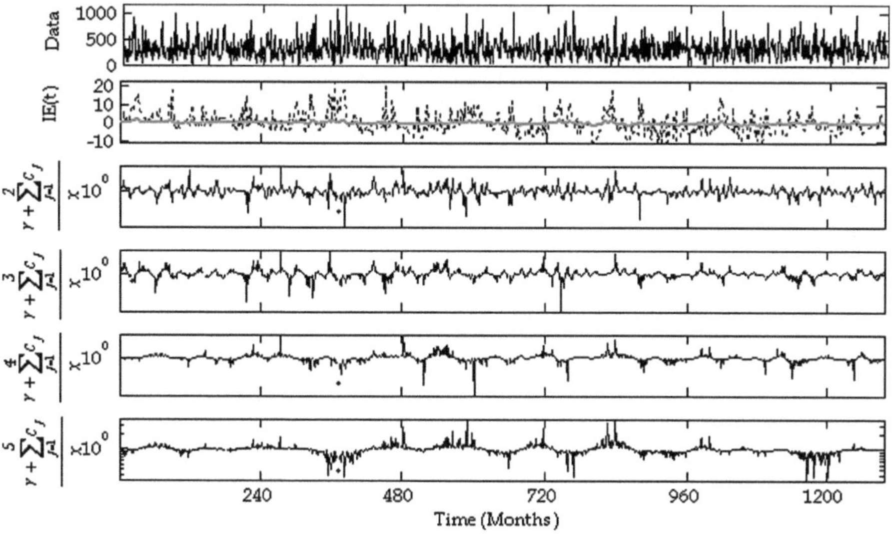

Figure 2.6.2. Volatility and *IE* for the HCN120177 monthly rainfall data

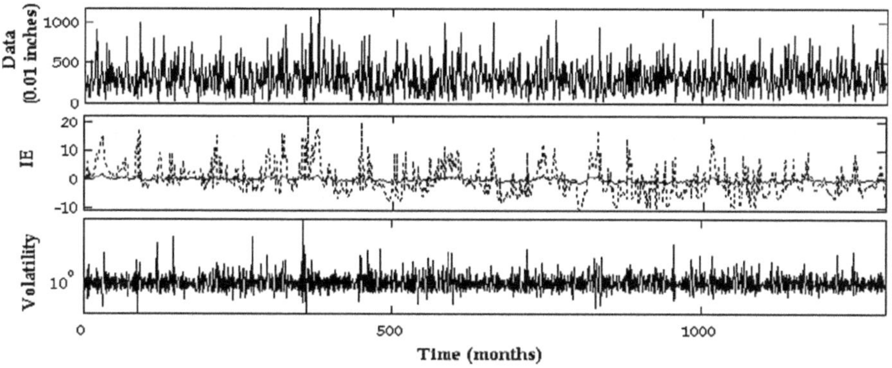

Figure 2.6.3. Volatility and IE for the HCN120177 monthly rainfall data

2.7. DEGREE OF STATIONARITY OF TIME SERIES

There are several measures used for examining the spectral properties of time series. One of them, the degree of stationarity, $DS(\omega)$, is defined as (Huang et al., 1998)

$$DS(\omega) = \frac{1}{T}\int_0^T \left(1 - \frac{H(\omega, t)}{n(\omega)}\right)^2 dt \tag{2.7.1}$$

where $n(\omega)$ is the mean marginal spectrum and calculated by $h(\omega)/T$. $DS(\omega)$ gives a quantitative measure of the entire dataset. It is a function of frequency.

For certain frequency components $DS(\omega)$ can be nonstationary while other components may remain stationary. Obviously, for a stationary process, the Hilbert spectrum cannot be a function of time; in such a case the Hilbert spectrum only contains horizontal lines when plotted against ω. For a pure stationary case, the $DS(\omega)$ will then be identically zero. Only under this condition, marginal Hilbert spectrum will be identical to Fourier spectrum and then Fourier spectrum makes physical sense.

The degree of stationarity can be modified slightly to include the statistically stationary signals, for which the degree of statistical stationarity, $DSS(\omega,\Delta T)$, is defined as

$$DSS(\omega, \Delta T) = \frac{1}{T}\int_0^T \left(1 - \frac{\overline{H(\omega, t)}}{n(\omega)}\right)^2 dt \tag{2.7.2}$$

where the overline indicates averaging over a definite but shorter time span, ΔT, than the overall time duration of the data, T. The definition for DSS could be useful in characterizing random variables from natural phenomena. For example, the degree of stationarity could be calculated over the piecewise span, ΔT, such as 10, 50, and 100 time steps. Degree of stationarity is used to investigate the variation in frequency bins.

The monthly temperature data in Europe is used as an example. Their degree of stationarity $DS(\omega)$, Eq. (2.7.1) and degree of statistical stationarity $DSS(\omega, \Delta T)$ (for $\Delta T = 10, 50, 100$ and 300 in Eq. (2.7.2) are shown in Figure 2.7.1. $DS(\omega)$ is the darkest of the lines (series (5)) in Figure 2.7.1 while the others are $DSS(\omega, \Delta T)$.

Overall, $DS(\omega)$ has the higher value than $DSS(\omega, \Delta T)$. $DSS(\omega, \Delta T)$ decreases and approaches zero with decreasing length of ΔT, especially in the high-frequency range. For a stationary case, the $DS(\omega)$ is identically zero. Hence, the results show that while the high-frequency components are nonstationary, they can still be statistically stationary with shorter time spans. The other result we can observe from Figure 2.7.1 is that the nonstationary components occur around the significant periodicity, such as 12-month cycle, and a valley is formed in some cases.

2.8. STATIONARITY TESTS

2.8.1 Modified Mann-Kendall Test

The Mann-Kendall test (Mann, 1945; Kendall, 1975) is a commonly used non-parametric trend test. The null hypothesis is that the data are independent and random. However, the existence of positive autocorrelation in the data increases the probability of detecting trend when actually it does not exist, and vice versa. The effect of autocorrelation in the data is considered in the modified Mann-Kendall test. Hamed and Rao (1998) derived a theoretical relationship to modify the Mann-Kendall test statistics so that it can be used with correlated data. The algorithm of the modified Mann-Kendall test is as follows.

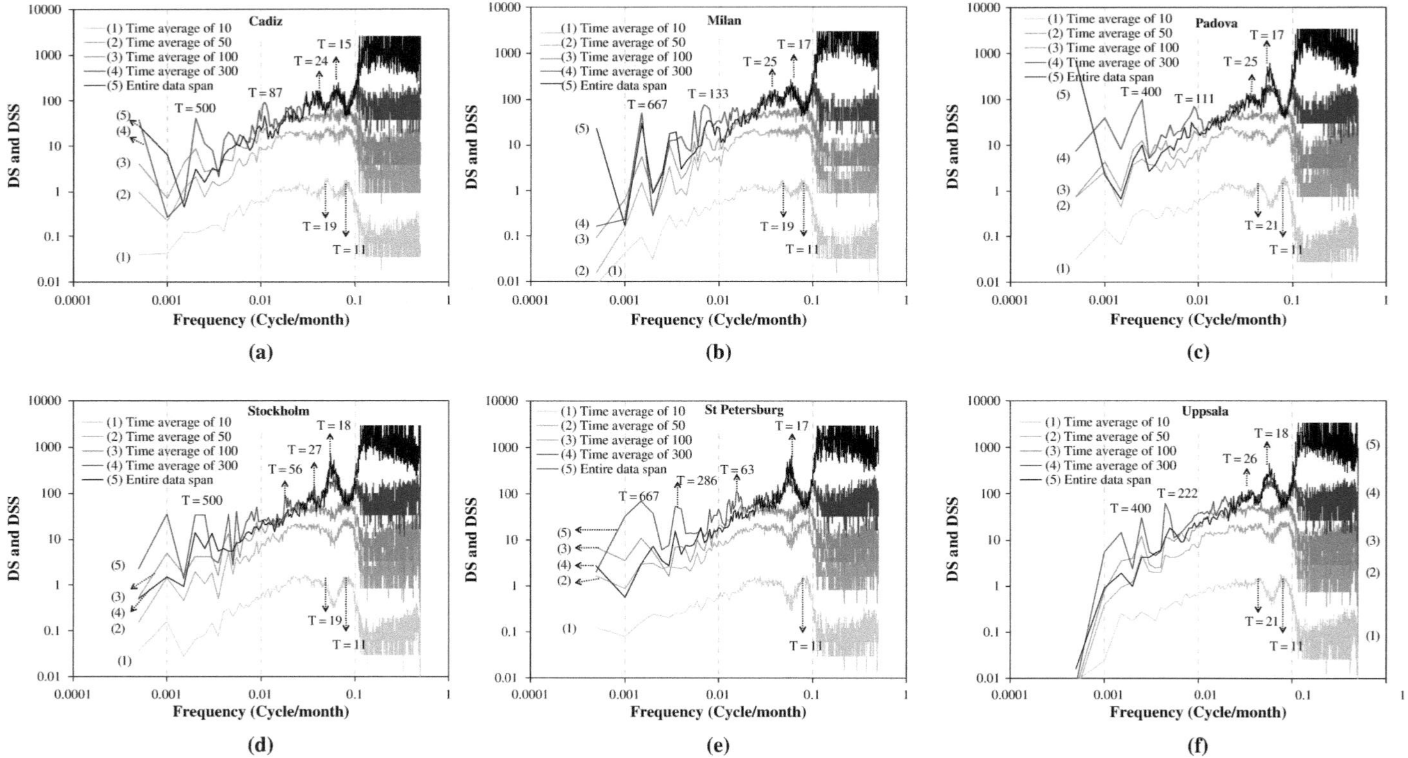

Figure 2.7.1. The degree of stationarity DS (Series 5 by using Eq. (2.7.1)), and the degree of statistical stationarity DSS, (Series 1, 2, 3 and 4 by using Eq. (2.7.2)) for the European monthly temperature data. In computing the DSS, time average of 10, 50, 100 and 300 have been used

The rank correlation test for two sets of observations $X = x_1, x_2, \ldots, x_n$ and $Y = y_1, y_2, \ldots, y_n$ is used in the test as follows. The statistic S is calculated as Eq. (2.8.1).

$$S = \sum_{i<j} a_{ij} b_{ij} \tag{2.8.1}$$

where

$$a_{ij} = sign(x_j - x_i) = \begin{cases} 1 & x_i < x_j \\ 0 & x_i = x_j \\ -1 & x_i > x_j \end{cases} \tag{2.8.2}$$

The significance of trends is tested by comparing the standardized test statistic $Z = S/\sqrt{V(S)}$ with the standard normal variate at the desired significance level. The modified $V(S)$ is calculated by

$$V(S) = \frac{n(n-1)(2n+5)}{18} \cdot \frac{n}{n_S^*} \tag{2.8.3}$$

The n/n_S^* is obtained by an approximation to the theoretical values.

$$\frac{n}{n_S^*} = 1 + \frac{2}{n(n-1)(n-2)} \cdot \sum_{i=1}^{n-1} (n-i)(n-i-1)(n-i-2)\rho_S(i) \tag{2.8.4}$$

where n is the actual number of observations and $\rho_S(i)$ is the autocorrelation function of the ranks of the observations. The test result, which is the standardized test statistic Z, gives not only the information about stationarity or nonstationarity but also whether the trend is upward or downward.

2.8.2 Trend Test of Segments Derived from IMFs

Sometimes the system or signal exhibits changes during the time when the data are collected. It might be important in certain applications to find the time when the changes occur and to develop models for the different segments at which the series does change. This is like a segmentation problem. However, detection of trend changes in time series is not easy. In the past, some prior assumptions and computational trials are made to detect the trends. The highest order IMF components may be treated as long term oscillations or as stepwise trends. These give us a technique to perform segmentation or to detect trend changes based on the empirical mode decomposition. The EMD simply extracts the IMF components from the data without any assumptions. The breaking points of each segment are determined by using the highest order IMF. It yields a unique set of segments which may be used for trend tests while common approaches need several trials to reduce the time series to segments. This method was tested with several series as discussed below.

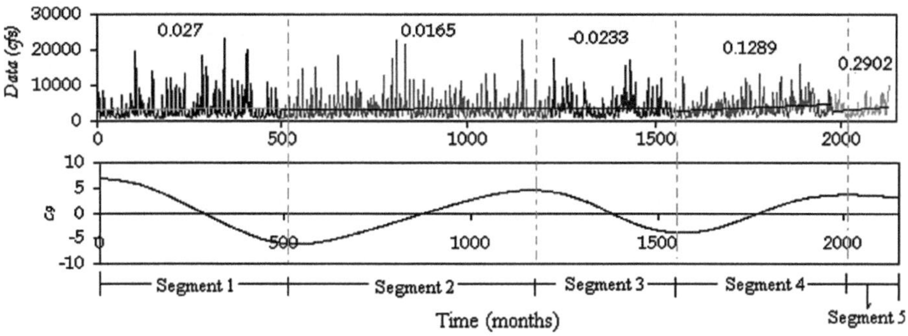

Figure 2.8.1. Segmentation by the highest order IMF for the streamflow in Warta River

Several time series are used for these tests discussed herein. They are monthly streamflow data in Warta River at Poznan, monthly temperature data in Cadiz, and daily peak wind speed data in South Bend, IN. The results are discussed separately as follows.

1) The monthly flows of the Warta at Poznan from 1920–2000

The highest order IMF component, in this case, is the 9th component. The entire time series is divided into five series as in Figure 2.8.1. For each segment, the modified Mann-Kendall test is used to test the significance of trend. The results for 90% confidence intervals are listed in Table 2.8.1. The modified Mann-Kendall test provides the information that the time series is stationary or nonstationary by giving a measure, z. The z value is an indication for an upward trend if positive or a downward trend if negative. The slopes observed from the c_9 component in Figure 2.8.1 are [−, +, −, +, −] ("+" means positive trend and "−" means negative trend), which are consistent with the z values in Table 2.8.1 except for the fifth segment. The first and last segments defined by IMF component do not correspond to the trend well because there is instability introduced through the sifting process. These are called as end effects. These two segments are not recommended for

Table 2.8.1. Results of trend test for Warta River streamflow data

Series	Modified Mann-Kendall
1	No significant trend ($z = -1.37$)
2	No significant trend ($z = 1.63$)
3	No significant trend ($z = -0.41$)
4	Upward trend ($z = 3.49$)
5	Upward trend ($z = 2.82$)

These statistics are based on 90% significant level

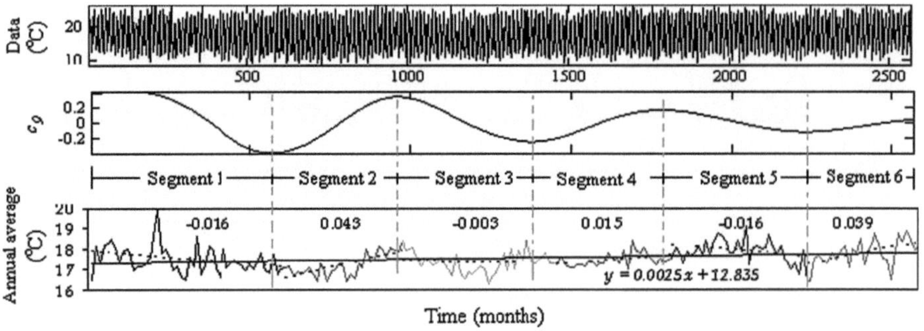

Figure 2.8.2. Segmentation by the highest order IMF of Cadiz monthly temperature data

further discussion. The slopes regressed for each segment by simple linear regression is also shown in the subplot of data in Figure 2.8.1. They are [0.027, 0.0165, −0.0233, 0.1289, 0.2902] for each segment. Similarly, except for the first and last segments, the others are consistent with the slope tendencies of c_9 or z values in Mann-Kendell test.

2) Monthly temperature data in Cadiz

The slopes for each segment observed from the c_9 component in Figure 2.8.2 are [−, +, −, +, −, +], which are consistent with the z values in Table 2.8.2. As in the previous analysis, the linear slopes for each segment of data are computed and shown in the third subplot. For Cadiz, they are [−0.016, 0.043, −0.003, 0.015, −0.016, 0.039] and they are consistent with the tendencies from c_9 component and z values. The results from the modified Mann-Kendall test indicate the possible upward or downward trend with a quantified z values.

3) Daily peak wind speed data in South Bend, Indiana

The slopes for each segment observed from the c_{10} component in Figure 2.8.3 are [−, +, −, +, −, +, −], which are consistent with the z values in Table 2.8.3. The linear regressed slopes are [−0.0024, 0.0029, −0.003, 0.0012, −0.0044, 0.0003, −0.0016] and again they are consistent with the tendencies from c_9 component and z values.

Table 2.8.2. Results of trend test for Cadiz monthly temperature data

Series	Modified Mann-Kendall
1	No significant trend (z = −1.29)
2	Upward trend (z = 1.75)
3	No significant trend (z = −0.78)
4	No significant trend (z = 0.57)
5	Downward trend (z = −1.71)
6	No significant trend (z = 1.50)

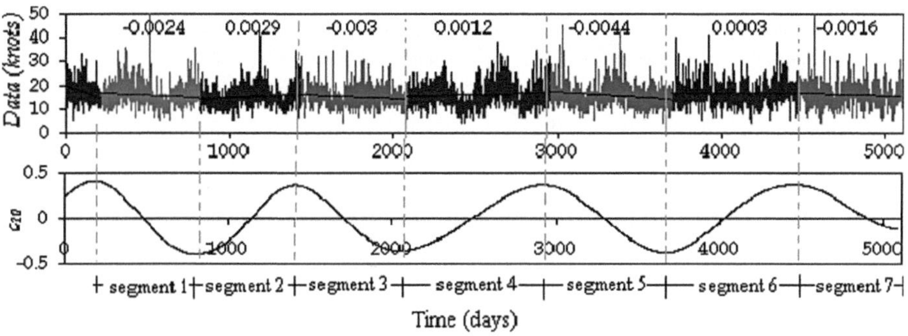

Figure 2.8.3. Segmentation by the highest order IMF for the daily peak wind speed data in South Bend, Indiana

Table 2.8.3. Results of trend test for daily peak wind speed data from South Bend, Indiana

Series	Modified Mann-Kendall
1	Downward trend ($z = -1.71$)
2	No significant trend ($z = 0.45$)
3	Downward trend ($z = -1.97$)
4	No significant trend ($z = 0.69$)
5	Downward trend ($z = -3.16$)
6	No significant trend ($z = 1.02$)
7	No significant trend ($z = -0.32$)

These statistics are based on 90% significance level

2.9. CONCLUDING COMMENTS

The Hilbert-Huang Transform provides local, adaptive and efficient information directly from the data without stationarity assumption. These advantages make it a powerful tool to investigate hydrologic and climatic time series. There is a trade-off limitation of using spectrogram which is obtained from Fourier transform. The gain of resolution in time corresponds to a loss of resolution in frequency. However, the HHT time-frequency representation, Hilbert spectrum, addresses the variations of frequency in time very well. The marginal Hilbert spectrum is a quantitative way to express the Hilbert spectrum in frequency domain. The empirical mode decomposition is used to produce components with zero local mean and ensures accurate estimates of the Hilbert transform. However, the selection of fitting spline lines and the end effects may lead to some erroneous results. Spline fitting is unsteady in high frequency region. The end effect may be ignored if the time series is long enough.

There are several measures in time and frequency domains used to investigate the Hilbert spectra. In time domain, the volatility, instantaneous energy (IE), standard deviation of instantaneous energy, and trend tests may be used. To make the plots easier to interpret, the mean IE and standard deviation of IE are shifted with respect to mean IE. From the results in this study, standard deviation of instantaneous energy is the most sensitive measure of the variation in time. In frequency domain, degree of stationarity (DS) is used to investigate the variation of energy in frequency. Higher DS means that the time series is likely nonstationary at that frequency while lower DS indicates that the time series is likely to be stationary at that frequency.

The modified Mann-Kendall test is a good technique to examine the trends in the data. It considers the effect of autocorrelation and gives a positive or negative z value, which refers a possible upward or downward trend. The residual obtained from empirical mode decomposition indicates the overall trend of the data. The trend test is also applied to the segments defined by the maxima and minima of the last IMF component and the results indicate that most segments have consistent trends.

CHAPTER 3

HILBERT-HUANG SPECTRA OF SIMULATED DATA

3.1. INTRODUCTION

In this chapter, the performance of Hilbert-Huang spectra is examined by using some synthetic time series. These series are a simple harmonic signal with well-separated fundamental frequencies and with close fundamental frequencies, a decaying signal, and series generated from a second-order autoregressive model. Fourier, multitaper and Hilbert-Huang spectra are used for comparison. The results of spectrogram and spectra in frequency domain are investigated.

The other issue addressed in this chapter is the capabilities of the Hilbert-Huang transform based methods to simulate nonstationary time series. Several methods are suggested for generation of data. These are random phase generation, random phase and amplitude generation and Wen-Yeh method. With these simulated series, the confidence limits can be calculated for the marginal Hilbert spectra. These confidence intervals can be used to systematically identify significant peaks in the marginal Hilbert spectra.

3.2. SYNTHETIC DATA ANALYSIS

3.2.1 Introduction

Before natural hydrological and climatic time series are investigated, properties of several spectral analysis methods are examined by using several synthetic data series. The first set of data is the sum of harmonic cosine waves composed of several fundamental frequencies, second set is a decaying signal, third is a signal with three close frequencies and the last case is generated from a second order autoregressive time series.

3.2.2 Simple Harmonic Data

A simple harmonic wave, Eq. (3.2.1), with three well separated fundamental frequencies: 0.02, 0.06 and 0.12 cycle/second is considered here. Hence, the f value in Eq. (3.2.1) is 0.02 cycles/second.

$$x(t) = \cos(2\pi ft) - \cos(2\pi \cdot 3ft) + \cos(2\pi \cdot 6ft) \qquad (3.2.1)$$

The empirical mode decomposition is performed for this signal and it yields five IMF components (c_1, c_2, c_3, c_4 and c_5) and a residual series (called r or c_6) as shown in Figure 3.2.1a. From the top to the bottom are original data, IMF 1 through IMF 5 and residual, r. In this case, IMF components have clear characteristics over the frequency spans. IMF 1 (c_1) extracts the highest frequency component, which refers to frequency at 0.12 cycles/second. Similarly, IMF 2 extracts the frequency 0.06 cycle/second and IMF 3 refers to frequency at 0.02 cycle/second. The last two IMFs (c_4 and c_5) are the processed components in order to achieve a monotonic residual. The right hand side of Figure 3.2.1 is used for checking the completeness of decomposition in which we add the IMF components from the highest to the lowest. For instance, the first solid line is obtained by adding c_5 and c_6 together. The last one is the sum from c_1 to c_6 and it reconstructs the signal. The solid line represents the reconstructed signal and the dotted line is the original signal.

These IMF components are Hilbert transformed so that we can obtain the instantaneous frequency to construct the time-frequency-energy relationship. The Hilbert spectrum is shown in Figure 3.2.2a and the Fourier (DFT) and Multi-taper spectra are shown in Figure 3.2.2b, c respectively. We can observe from Figure 3.2.2a that there is a ripple phenomenon in the high frequency band and it looks as if frequency changes periodically with time. But as this is a case with given fixed frequency, the frequency should be constant. This phenomenon of variation in high frequency occurs quite frequently. This is because Hilbert transform is computed on the signal that cannot satisfy the monocomponent strictly, especially in high frequencies. But, even with this oscillation, the bandwidth of the spectrogram obtained from HHT is narrower than those obtained from Fourier spectrum or Multi-taper spectrum. As a result, HHT has better resolution in frequency domain than the others.

The Fourier and Multi-taper spectrograms, with their limited temporal and frequency resolutions, seem to be a two-dimensional smoothed version of the Hilbert spectrum. The Hilbert-Huang spectrum represents the instantaneous frequency and does not involve the concept of time and frequency resolution. Hence, HHT does not consider the entire data for its resolution. In addition, Figure 3.2.2d is a "smeared" result because of applying a Gaussian 5 $\times$ 5 filter to Figure 3.2.2a. Sometimes it is much more distinguishable than the skeleton one (Figure 3.2.2a). The marginal Hilbert spectrum, Fourier spectrum and Multi-taper spectrum are given in Figure 3.2.3. All these methods can capture the fundamental frequencies, 0.02, 0.06 and 0.12 cycles/second. HHT and MTM have sharper spectra and HHT

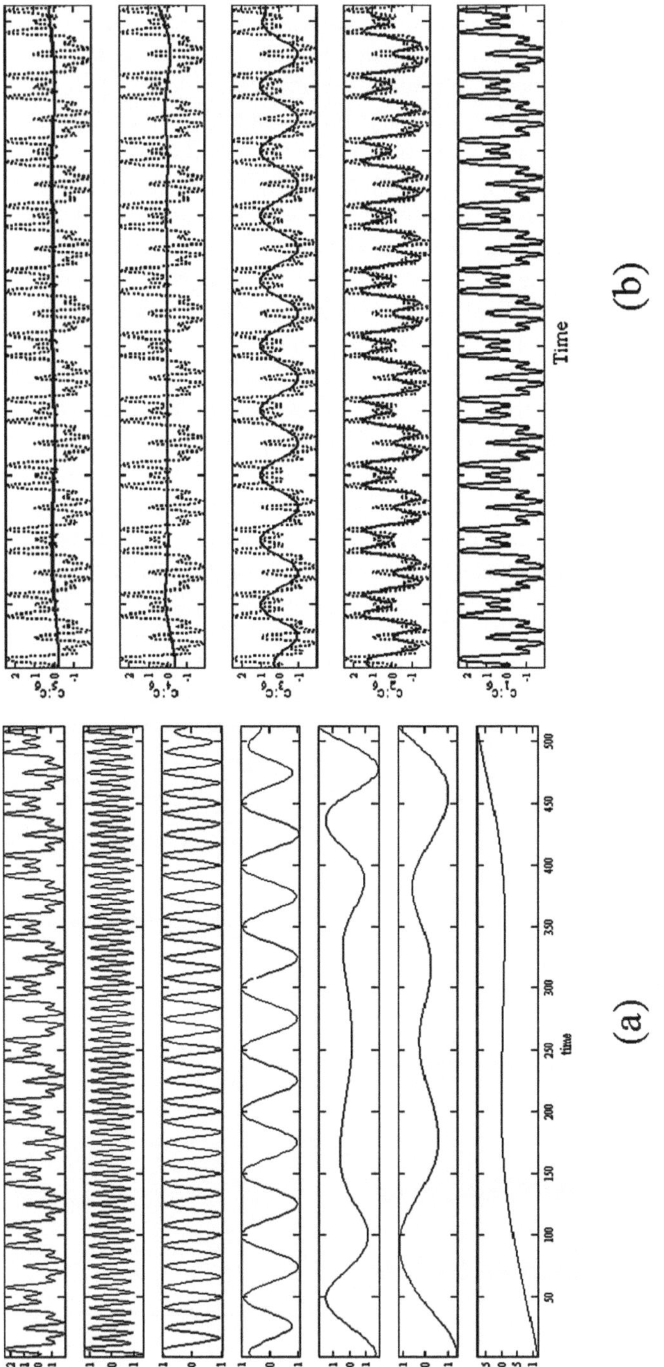

Figure 3.2.1. (a) Intrinsic mode functions of case 1. (b) Signal reconstruction of case 1

has frequency splitting. It is also noticed that there is a small peak around frequency 0.01 cycle/second, which is caused by the IMF 4 and IMF 5, but its energy is very small compared to the energy in three main frequencies.

The lengths of the series are different for HHT and DFT/MTM in time-frequency representation (Figure 3.2.2). The reason for this reduction in length is that for HHT, the spectrum amplitude and instantaneous frequency can be calculated for each point of time while the time-frequency distribution for DFT and MTM are calculated by using a sliding window. The frequency and amplitude are assigned to the center of the sliding window; hence, the time-frequency distributions of first half of the first and last half of the last sliding window are not available.

If we consider the 512-point data for example, with a sliding window length of 100-points, results in the 0∼50 points and 462 ($= 512 - 50$) ∼512 points are not available. There is a tradeoff in choosing the length of sliding window. Larger window captures the frequency better, but more information is lost in the beginning and ending data points. The other disadvantage of the sliding window is that the short time series produce coarse resolution in frequency and make the time-frequency representation of the DFT/MTM to be not as sharp as they are in HHT (Figure 3.2.3), which is calculated from the entire length of data.

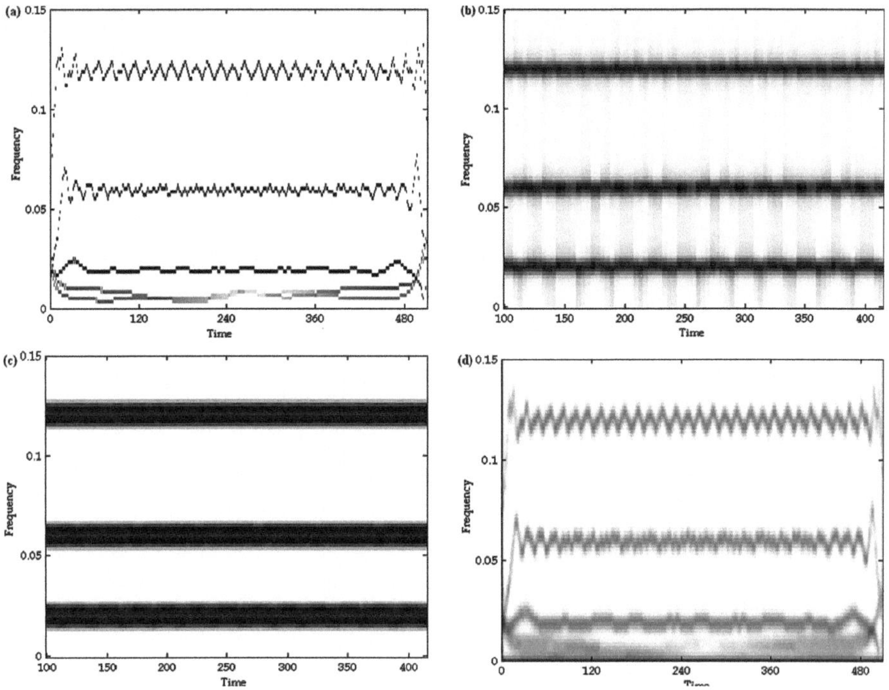

Figure 3.2.2. The time-frequency distribution of the signal in case 1 obtained by (a)HHT, (b)DFT and (c)MTM (d) a 5 × 5 Gaussian weighted filtered Hilbert spectrum

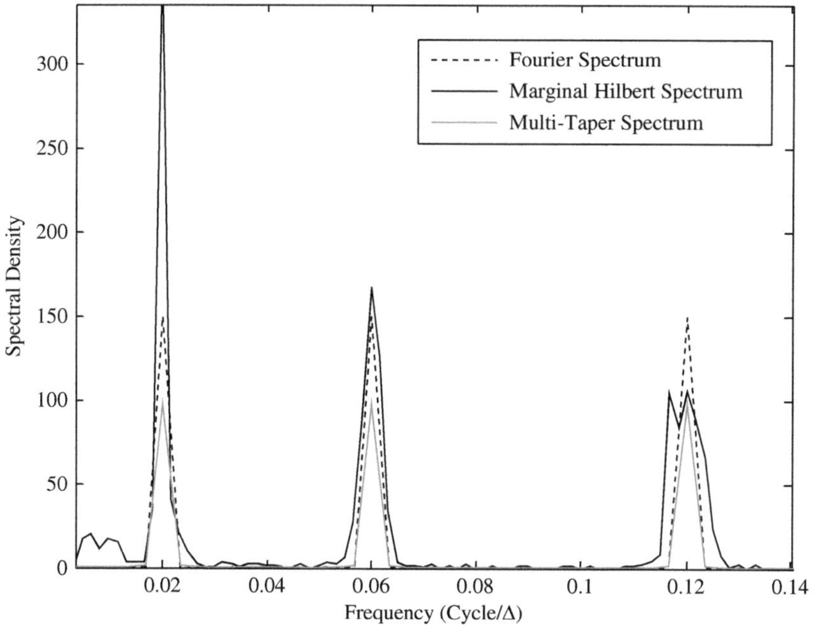

Figure 3.2.3. Comparison of the HHT marginal spectrum, Fourier spectrum and Multi-Taper spectrum of the signal in case 1

3.2.3 Decaying Signal

A 512-point decaying signal with decay rate 0.01 and a fundamental frequency 0.12 cycle/second Eq. (3.2.2) is examined. It is an amplitude modulated signal. The empirical mode decomposition is performed for this signal and five IMF components (c_1, c_2, c_3, c_4 and c_5) and one residual series (called r or c_6) are identified as shown in Figure 3.2.4.

$$x(t) = \cos(2\pi f t)\exp(-0.01t) \tag{3.2.2}$$

Since this case only has one fundamental frequency, the IMF 1 (c_1) in Figure 3.2.4 extracts almost the same oscillations as the original data. IMF 2~ IMF 5 are produced from the sifting process, but their signal amplitudes are relatively small. The HHT, DFT and MTM spectra are shown in Figure 3.2.5. For the spectrogram of DFT and MTM, a segment length of 150 is used, hence the total length of DFT and MTM are only available up to 362. With a decay rate, the peak frequency is not altered. Only the energy of spectrum decays with time. From Figure 3.2.5 we can see that the amplitude modulation generates the intrawave frequency modulation, but as we saw in case 1, the range is narrower than that in DFT. MTM also provides a more precise frequency location. DFT and MTM are like the smeared average of Hilbert spectrum and lack the details beyond the smoothed mean. The energy

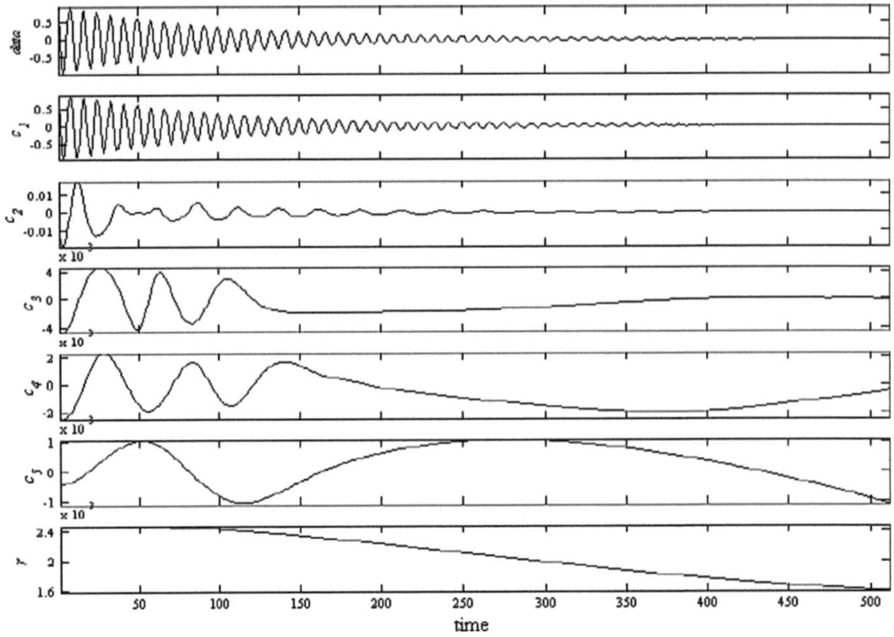

Figure 3.2.4. The exponentially decay signal (case 2) with its intrinsic mode functions ($c_1 \sim c_5$) and residue, r

spread-out situation in DFT and MTM can also be observed in Figure 3.2.6. The marginal Hilbert spectrum has a very sharp representation at the frequency 0.12, which is expected. Fourier-based analysis spreads energy to the higher frequency range for this nonlinear model.

3.2.4 A Signal with Three Close Frequencies

A simple harmonic wave (Eq. (3.2.3)) with three close fundamental frequencies: 0.04, 0.045 and 0.05 cycles/second is considered here. There are five IMF components (c_1, c_2, c_3, c_4 and c_5) and one residual series (called as r or c_6) as shown in Figure 3.2.7.

$$x(t) = \sin(2\pi \cdot 0.04t) + \sin(2\pi \cdot 0.045t) + \sin(2\pi \cdot 0.05t) \qquad (3.2.3)$$

The IMF represents simple oscillation modes embedded in the signal with zero mean. However, with this example, a problem is discovered. From Figure 3.2.7, we can see the first mode, c_1, actually involves these frequencies of oscillations together and cannot be successfully separated. The reason for this situation is because an IMF is not restricted to a narrow band signal; it can be both frequency and amplitude modulated as in c_1. To solve this problem, an intuitive

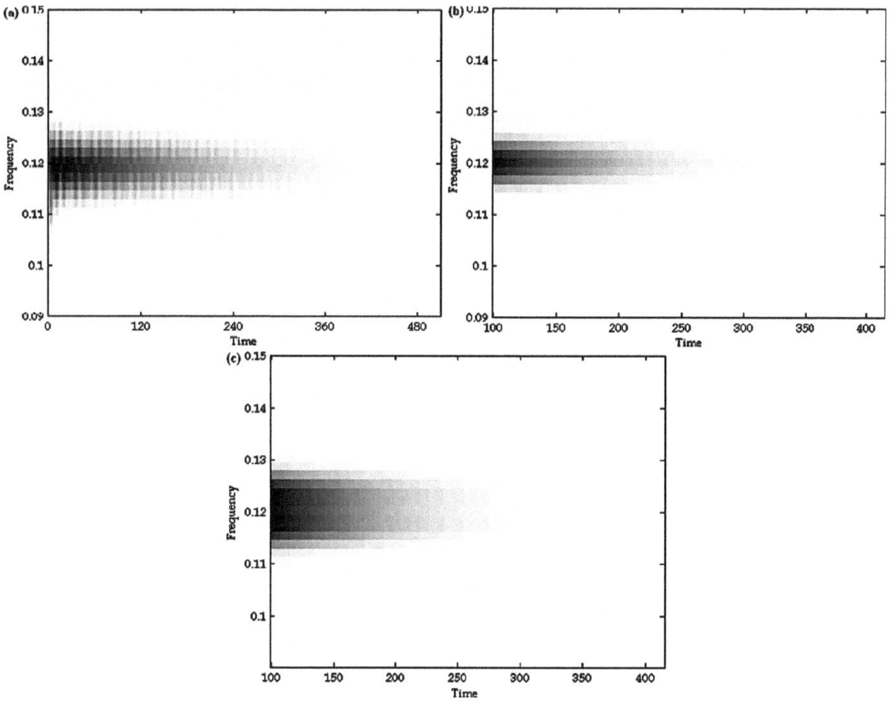

Figure 3.2.5. The time-frequency distribution of the signal in case 2 obtained by (a) HHT, (b) DFT and (c) MTM

idea is to decompose the signal to some narrow band signals first and then use EMD operation on each of these separated signals. Its time-frequency representation is shown in Figure 3.2.8. The HHT spectrum is sharp in defining the mean frequency, but it fails to separate the three frequencies. Although Fourier-based methods also fail to separate the frequencies, they give a range of the frequency distribution.

Furthermore, a non-stationary version of the signal was examined by adding 1 for the first half of the time series, and subtracting 1 for the second half. The data are shown in Figure 3.2.9 with the IMF components (left) and the check of completeness (right, solid line represents the reconstructed signal and the dotted line is the original signal). The residual does show the nonstationarity which is created by adding 1 for the first half of the time series and subtracting 1 for the second half.

Besides studying the residual, we investigated the relationship between the higher order IMF (c_5 in Figure 3.2.9) and the stepwise trends of the data. The time series $x(t)$ is divided into several segments based on the maxima and minima of c_5. The linear regressed slopes for each segment of the data are shown in Figure 3.2.10. These slopes yield quite consistent results as the empirical mode decomposition.

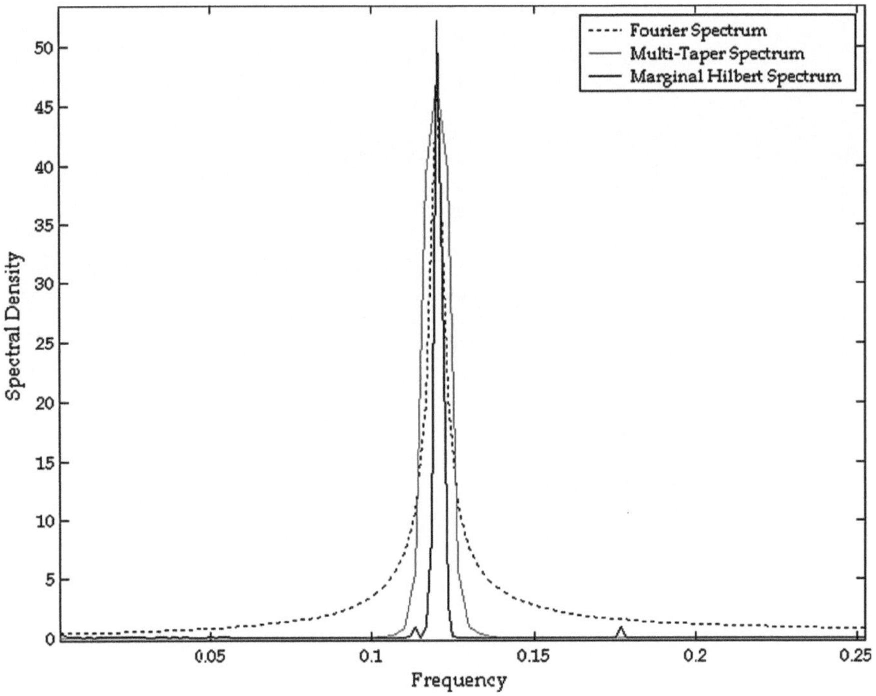

Figure 3.2.6. Comparison of Fourier, multitaper and marginal Hilbert spectra for case 2

The time-frequency representation (a, b and c) and spectrum (d) are shown in Figure 3.2.11.

Three methods all have the peak frequency at 0.04 but none of them can separate three close frequencies. The broad peak covering frequency from 0.001 to 0.002 represents the period of the full data length. Whether this can or cannot be treated as a genuine oscillation component is debatable. But the data certainly have such a period and HHT correctly identifies it. DFT and MTM poorly resolve the non-stationary information. They have very broad range from frequency 0 to 0.02 and also only indicate the transition section vary vaguely.

3.2.5 Autoregressive Model

Among all existing parametric models, Autoregressive (AR) model is widely used in parametric spectral analysis. The model assumes the signal under study $x(t)$ to be a linear combination of its past samples, $x(t-1), x(t-2)\ldots$, plus the noise $e(t)$.

$$x(t) = \sum_{k=1}^{p} a_k x(t-k) + e(t) \tag{3.2.4}$$

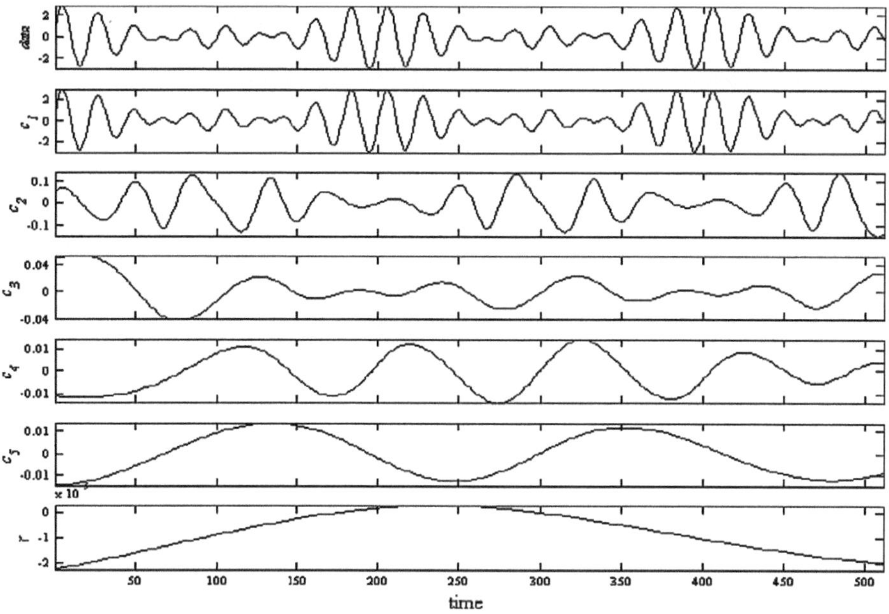

Figure 3.2.7. The signal with closed-frequency signal (case 3) with its IMFs

The coefficients a_k are the autoregressive parameters, p is the model order and $e(t)$ is the white noise. Once the AR parameters are estimated, the spectral estimator of the signal $x(t)$ is given by Eq. (3.2.5),

$$S(f) = \frac{\sigma_e^2}{\left[1 + \sum_{k=1}^{p} a_k e^{-i2\pi kf}\right]^2} \tag{3.2.5}$$

As an example, we consider a 2nd order autoregressive model signal defined with $a_1 = 0.75, a_2 = -0.5$ and the input white noise signal with variance 0.2. 8192 points are generated from the AR(2) model. Its intrinsic mode functions are shown in Figure 3.2.12.

The theoretical power spectrum of the autoregressive model is calculated by using Eq. (3.2.5) and is shown in Figure 3.2.13a. The high-pass HHT marginal spectrum, Fourier spectrum, and Multi-taper spectrum are plotted and are overlaid with the smoothed curves in Figure 3.2.13b–d, respectively. A peak occurs at frequency 0.15 in the theoretical spectrum. Although there is considerable leakage in FFT and MTM spectra, a 0.15 frequency still can be seen. The HHT spectrum in Figure 3.2.13b does not include all the IMF components, and it is estimated by using only the first two IMF components.

In this case, while calculating the marginal HHT spectrum following the regular procedure and integrating the spectrum over the time span for all IMD components,

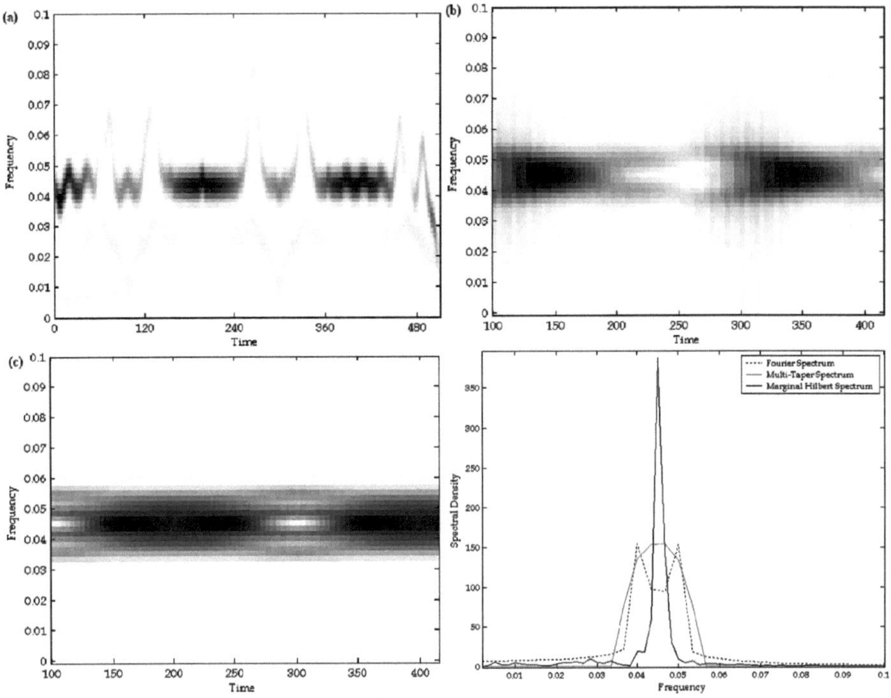

Figure 3.2.8. The time-frequency distribution of the signal in case 3 obtained by (a) a 5 × 5 Gaussian weighted filtered Hilbert spectrum, (b) DFT and (c) MTM; (d) the marginal Hilbert, DFT and MTM spectra

the spectrum is not estimated properly. The amplitude in the low-frequency is too strong and it is too strong even in high frequency band. This can be observed in Figure 3.2.14 that shows the peak spectrum moves from high frequency to low frequency with increasing order of IMFs. The higher-order IMF components with very small amplitudes do not affect the Hilbert spectrum, but the problem occurs when integrating over time with very narrow frequency range to get the marginal spectrum.

Therefore, the error in overestimating the spectra in low-frequency can be minimized if the high-order, low-frequency IMF components with very small amplitudes is not used in the marginal spectral calculation (Zhang, et al., 2004). The time-frequency representation of the spectrum obtained from the three methods is shown in Figure 3.2.15. Figure 3.2.15d is the Hilbert spectrum calculated from all the IMF components; the spurious energy in low frequency is overwhelming the spectrum and the previous discussion on the decomposition process reveals the reason. Figure 3.2.15a is calculated from the first two IMF components, and it has the same peak frequency around 0.15 as DFT and MTM spectrograms. From Figure 3.2.13, multi-taper method seems to have a better estimate and approaches the theoretical spectrum. MTM has less leakage in high frequency and more concentrated energy

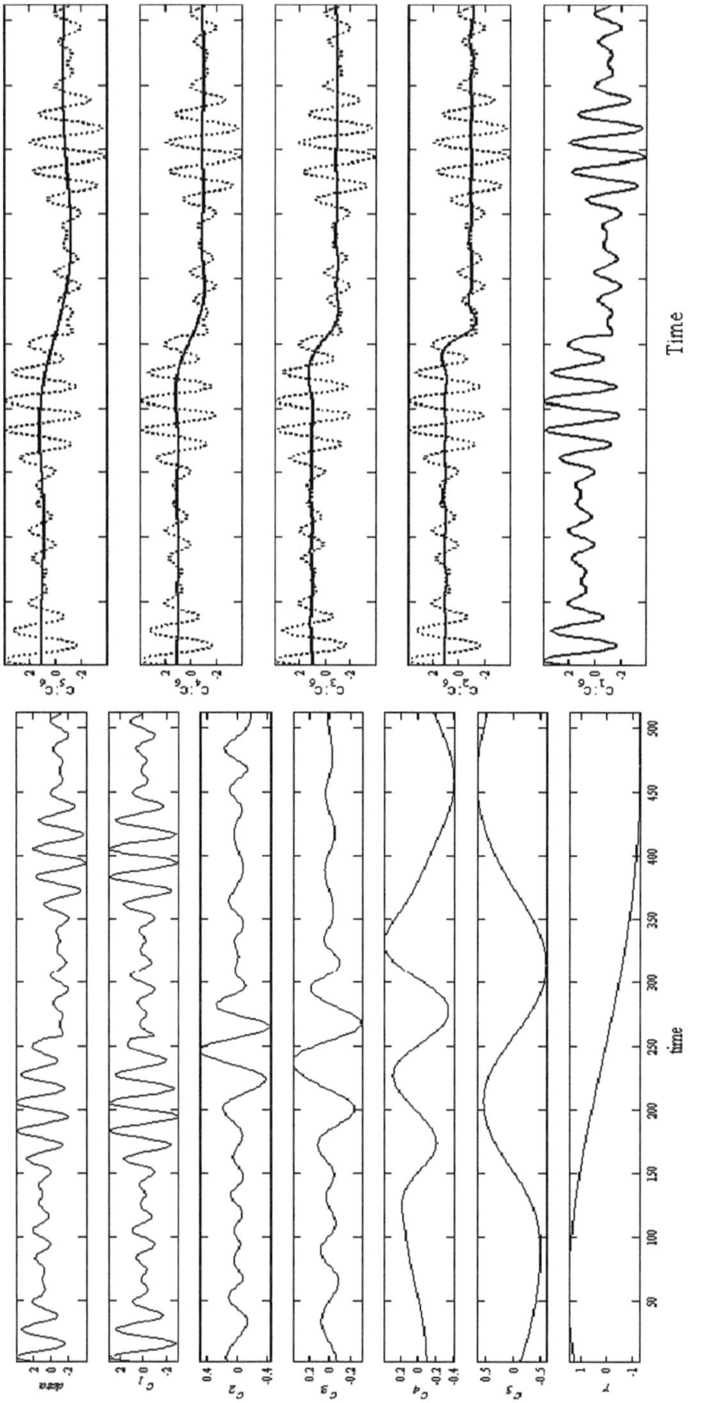

Figure 3.2.9. IMFs of case 3b (left) and the signal reconstruction (right)

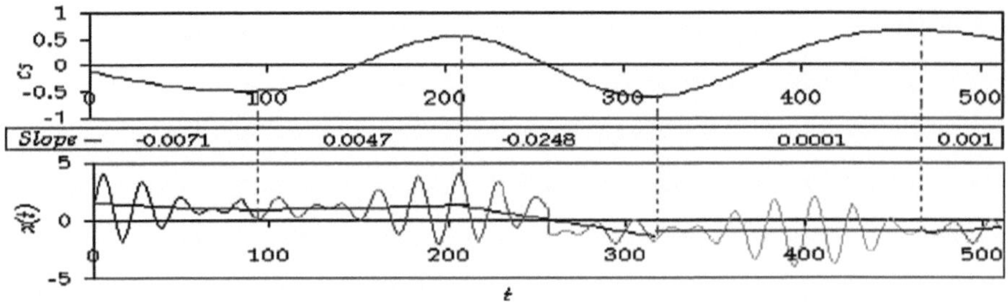

Figure 3.2.10. Higher order IMF (c_5 in Figure 3.2.9) and the stepwise trends of the data

in 0.15 cycle/time than the Fourier spectrum. It picks up the frequency 0.15 better than the others in Figure 3.2.13 as well. In conclusion, automatically including all the IMF components may not always ensure accurate estimates of spectra.

3.3. SIMULATION OF NONSTATIONARY RANDOM PROCESSES

3.3.1 Introduction

The empirical mode decomposition technique also makes it possible to simulate nonstationary processes. For most natural time series, there is only one realization available. Therefore it makes it impossible to calculate the ensemble statistical properties. However, the success of the simulation technique discussed below allows us to reproduce the nonstationary processes for further analysis. There are three major methods for simulation of nonstationary data (Wen and Gu, 2004). In the first method the random phase values, and in the second one the random phase and random amplitude are used. The third method, which is an improved Wen-Yeh method, separates noise from signal and generates more random elements than the other two methods. In this chapter, three methods are investigated by using different types of data and the properties of each method are examined.

3.3.2 Simulation with Random Phases

The Hilbert spectrum representation model suggests a method for simulation of nonstationary process. The Hilbert spectral representation of a signal,

$$x(t) = \text{Re}\left\{\sum_{j=1}^{n} a_j(t)\exp\{i\theta_j(t)\}\right\} + r_n(t) \tag{3.3.1}$$

which has a Hilbert spectrum characterized by $a_j^2(t)$ with instantaneous frequency $d\theta_j(t)/dt$, for $j = 1$ to n, suggests that the underlying random process can be represented by introducing random elements as follows:

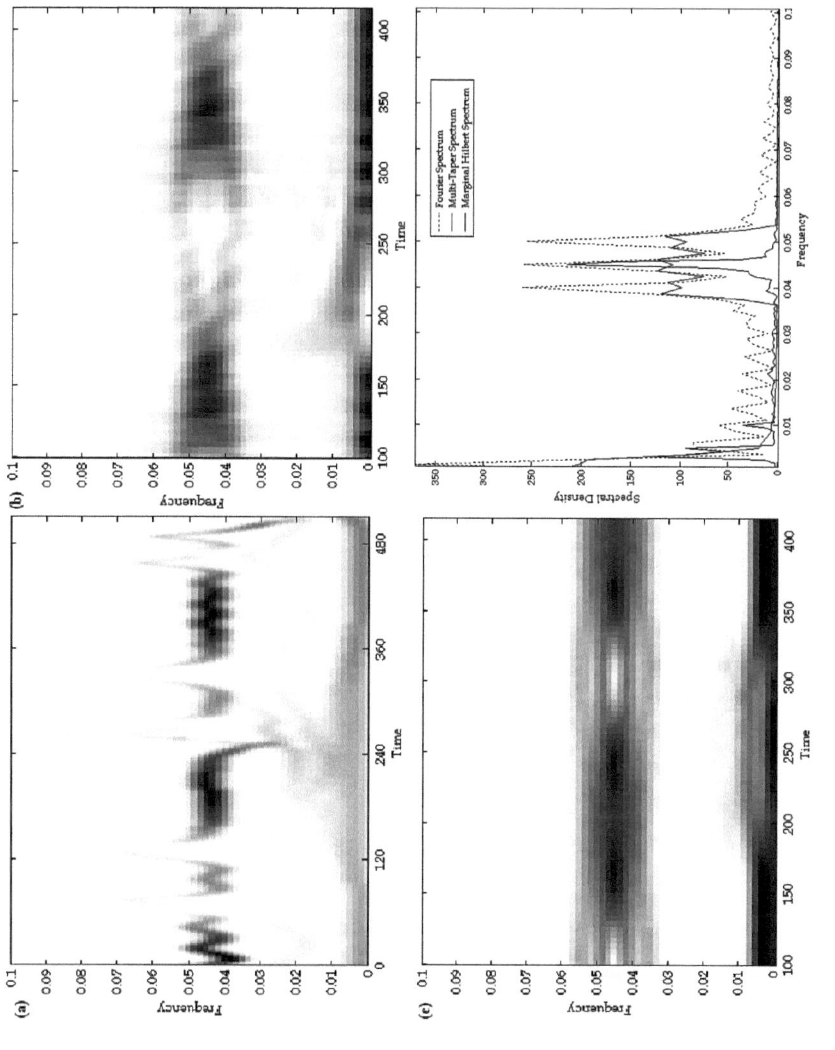

Figure 3.2.11. The time-frequency distribution of the trend-added signal in case 3 obtained by (a) a 5 × 5 Gaussian weighted filtered Hilbert spectrum, (b) DFT and (c)MTM; (d) the marginal Hilbert, DFT and MTM spectra

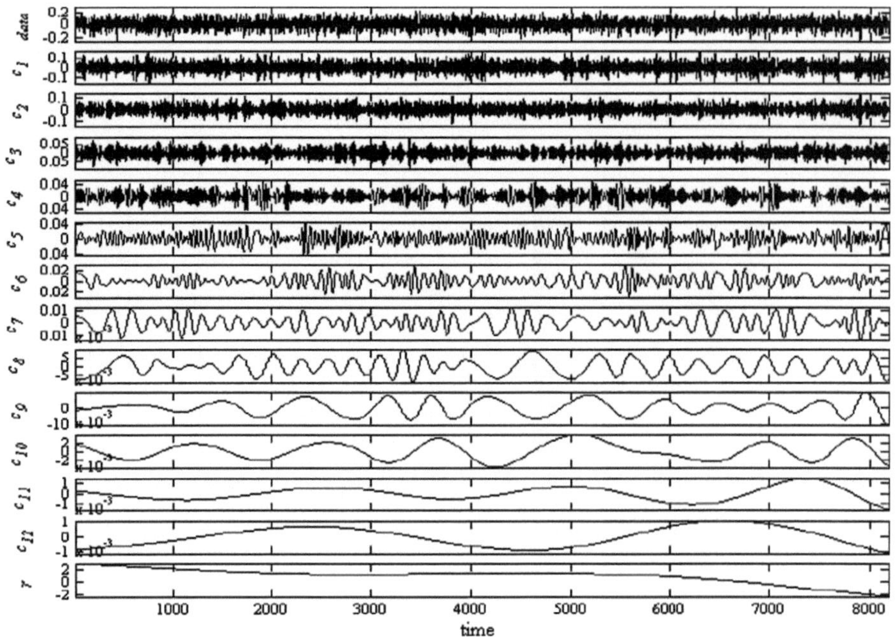

Figure 3.2.12. The synthetic AR(2) time series and its intrinsic mode functions

$$X(t) = \text{Re}\left\{\sum_{j=1}^{n} a_j(t) \exp\{i[\theta_j(t) + \phi_j]\}\right\} + r_n(t) \tag{3.3.2}$$

In Eq. (3.3.2) ϕ_j is an independent random phase angle uniformly distributed between 0 and 2π (Wen and Gu, 2005). $X(t)$ is a random process. One can generate the random phase angles and recombine the IMFs. Due to the central limit theorem, $X(t)$ approaches a Gaussian process for large n. This method is easy to implement. $a_j(t)$, $\theta_j(t)$ and $r_n(t)$ are obtained by the empirical mode decomposition, so ϕ_j is the only generated variable and it does not vary with time.

The process has the following mean, variance and covariance functions:

$$\mu_X(t) = \text{Re}\left\{\sum_{j=1}^{n} a_j(t) e^{i\theta_j(t)} E(e^{i\phi_j})\right\} + r_n(t) = r_n(t) \tag{3.3.3}$$

$$K_{XX}(t_1, t_2) = \frac{1}{2} \sum_{j=1}^{n} a_j(t_1) a_j(t_2) \cos[\theta_j(t_1) - \theta_j(t_2)] \tag{3.3.4}$$

$$\sigma_X^2(t) = E\{[x_s(t) - \mu_{x_s}(t)]^2\} = \frac{1}{2} \sum_{j=1}^{n} a_j^2(t) \tag{3.3.5}$$

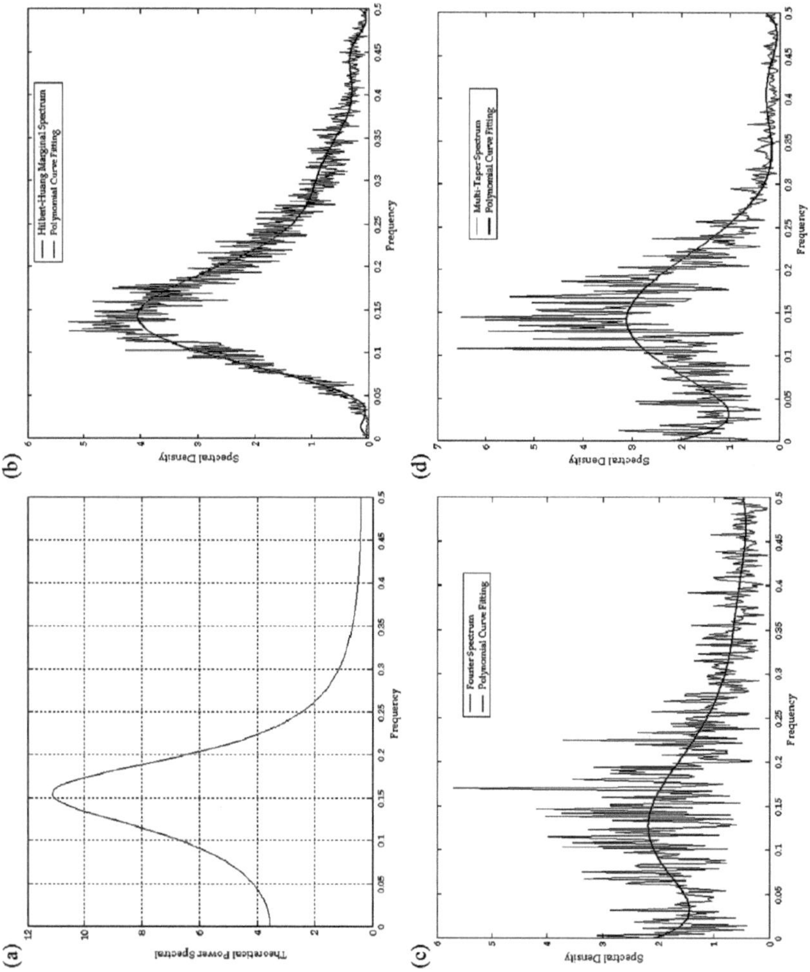

Figure 3.2.13. (a) The theoretical power spectrum of AR(2) model; (b) the HHT marginal spectrum obtained from the first two IMF components (c_1 and c_2); (c) the Fourier spectrum and (d) the Multi-Taper method spectrum. They all show a peak around frequency 0.15

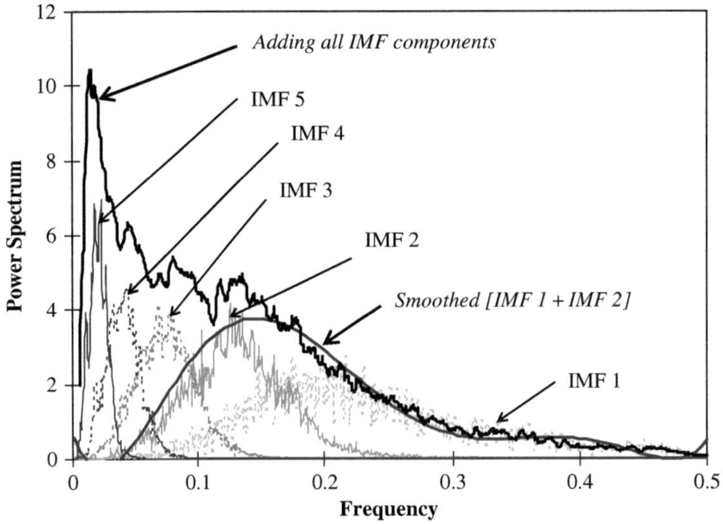

Figure 3.2.14. Marginal spectrum calculated for each IMF component. It shows that the peak of the spectrum moves from high frequency to low frequency with increasing order of IMFs

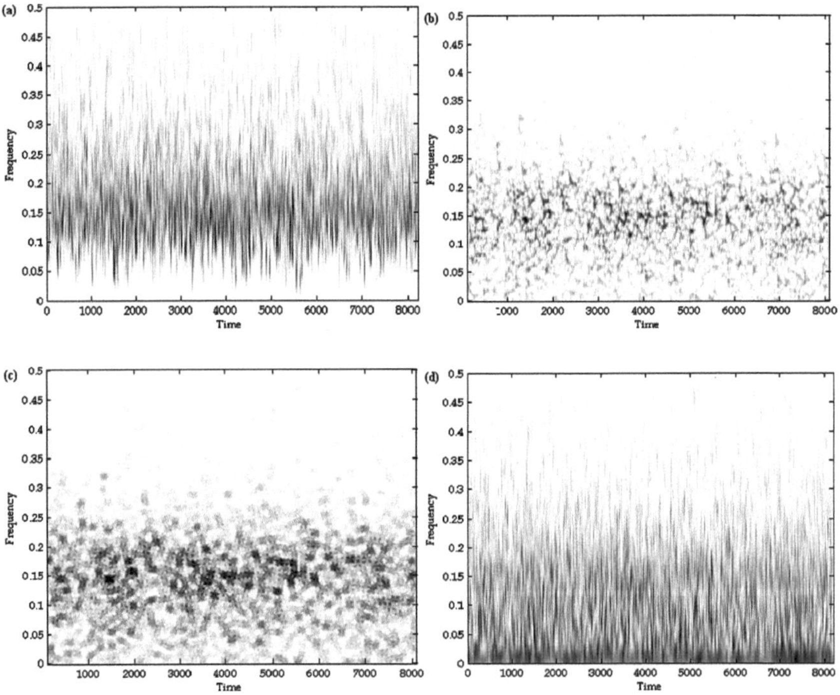

Figure 3.2.15. The time-frequency distribution of the signal obtained by (a) a 19 × 19 Gaussian filtered Hilbert spectrum by considering the first two IMFs, (b) DFT, (c) MTM and (d) a 19 × 19 Gaussian filtered Hilbert spectrum by considering all IMF components

The relationships can be extended to a vector process $X(t)$ of m components: $[X_1(t), X_2(t), \ldots, X_m(t)]$. The kth component has a Hilbert spectral representation given by Eq. (3.3.6),

$$X_k(t) = \text{Re}\left\{\sum_{j=1}^{n} a_{jk}(t)\exp\{i[\theta_{jk}(t) + \phi_{jk}]\}\right\} + r_{nk}(t) \tag{3.3.6}$$

The cross covariance between the pth and qth components is described by Eq. (3.3.7),

$$K_{X_pX_q}(t_1, t_2) = \frac{1}{2}\sum_{j=1}^{n}\sum_{k=1}^{n} a_{jp}(t_1)a_{kq}(t_2)E\{\cos[\theta_{jp}(t_1) + \phi_{jp}] \times \cos[\theta_{kq}(t_2) + \phi_{kq}]\} \tag{3.3.7}$$

The statistical properties of each simulation are examined by the histograms, autocorrelograms and spectral densities. The spectral density is an effective way to investigate the response of the simulated signals. For comparison, Fourier, Multi-taper and marginal Hilbert spectra are computed. Histograms of the data in time or spatial domain help us to examine the distribution of the simulated signal compared to the original signal. Autocorrelation function is used to compare the persistence of simulated and the original data. The autocorrelation functions (Box and Jenkins, 1976) are used to detect non-randomness in data. For given measurements, $y_1, y_2, \ldots, y_n$ at $t = 1, 2, \ldots, n$, the lag k autocorrelation function is defined in Eq. (3.3.8).

$$r_k = \frac{\sum_{i=1}^{n-k} (y_i - \bar{y})(y_{i+k} - \bar{y})}{\sum_{i=1}^{n} (y_i - \bar{y})^2} \tag{3.3.8}$$

In this section, method one, which is simulated only with random phase, is examined by using several sets of data. For different types of data, the results from one or two samples are used for demonstration. The data used for simulation are listed in Table 3.3.1. They are data of monthly rainfall, streamflow, temperature, daily peak wind speed and lake temperature versus water depth. Five series of data are generated for each series and they are shown in Figure 3.3.1. For each figure, the first row is the original data and the other five are the simulated data based on the IMFs obtained from the observed data.

Table 3.3.1. Data used for simulation

Data type	Gauging location	Time of sampling
Monthly rainfall	HCN120177, Indiana, USA	1895–2002
Monthly streamflow	USGS03276500, Indiana, USA	1915–2004
Monthly temperature	HCN120177, Indiana, USA	1895–2002
	Cadiz, Spain	1786–2000
Daily peak wind speed	Indianapolis, Indiana, USA	1888–2002
Lake temperature	Square Lake, Minnesota, USA	10/20/2004

The characteristics of the original and simulated data are summarized in Figure 3.3.2–3.3.7. The comparison of Fourier spectra, multitaper spectra, marginal Hilbert spectra, histogram, and autocorrelation are of interest. The areas under these spectra are computed as well. Fourier spectra are relatively flat compared to the other two spectra.

3.3.3 Simulation with Random Phases and Amplitudes

In the previous procedure, all realizations of underlying nonstationary random process have the same energy variation with time and frequency represented by the target Hilbert spectrum. To allow the variations from one to another, Wen and Gu (2004) introduced an additional element G_j to Eq. (3.3.2).

$$X(t) = \text{Re}\left\{\sum_{j=1}^{n} a_j(t)\cdot G_j \cdot \exp\{i[\theta_j(t)+\phi_j]\}\right\} + r_n(t) \tag{3.3.9}$$

Where G_j are assumed to be independent random variables with

$$E(G_j) = 1 \tag{3.3.10}$$

This assumption ensures that the ensemble average of the Hilbert spectra of the sample is equal to that of the target Hilbert spectrum. G_j is assumed to be 1 if only one record is available. If G_j are modeled with Rayleigh distribution, then $X(t)$ will be a Gaussian process. The Rayleigh distribution is a special case of the Weibull distribution. The Rayleigh probability density function is

$$y = f(x|b) = \frac{x}{b^2}\exp\left(-\frac{x^2}{2b^2}\right) \tag{3.3.11}$$

The estimate of Rayleigh parameter b is

$$b = \sqrt{\frac{1}{2n}\sum_{i=1}^{n} x_i^2} \tag{3.3.12}$$

The mean of Rayleigh distribution is $\sigma\sqrt{\pi/2}$ where $\sigma = \sqrt{\frac{1}{2n}\sum_{i=1}^{n} x_i^2}$, hence b is equal to $\sqrt{2/\pi}$ in this case.

Similar to the previous discussion, the data listed in Table 3.3.1 are used for simulation. Five series of data are generated for each series and they are shown in Figure 3.3.8. Since the overall amplitudes vary with each simulation as well as the phase, the simulated data have higher variability compared to those generated by using only random phase values.

From the results of method 2 presented in Fig. 3.3.8, the variation of the simulated data is much higher than the results from method 1. In Figure 3.3.8, there are

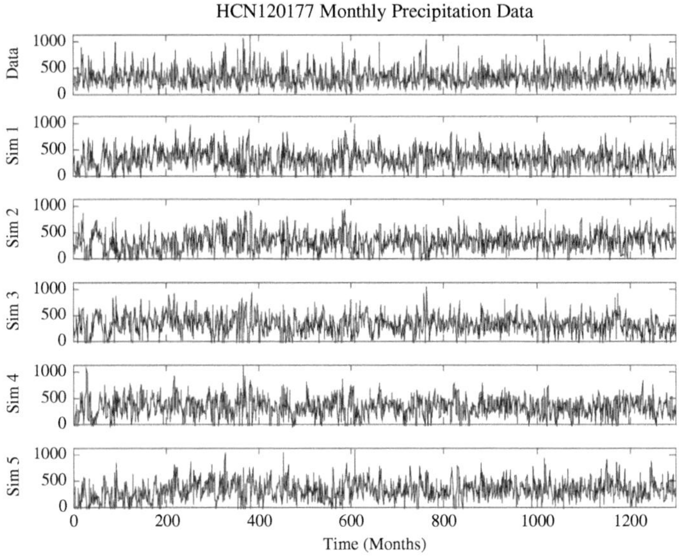

(a) HCN 120177 monthly rainfall data (unit: 0.01 inches)

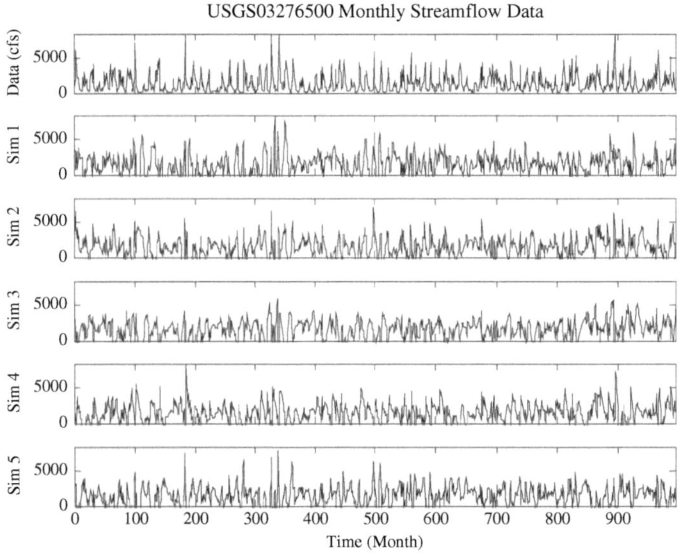

(b) USGS 03276500 monthly streamflow data

Figure 3.3.1. Five simulated series from method 1

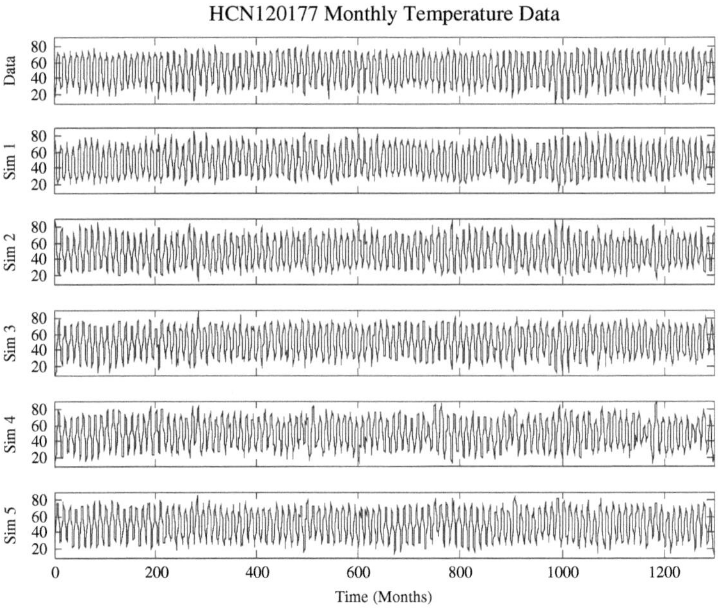

(c) HCN 120177 monthly temperature data (°F)

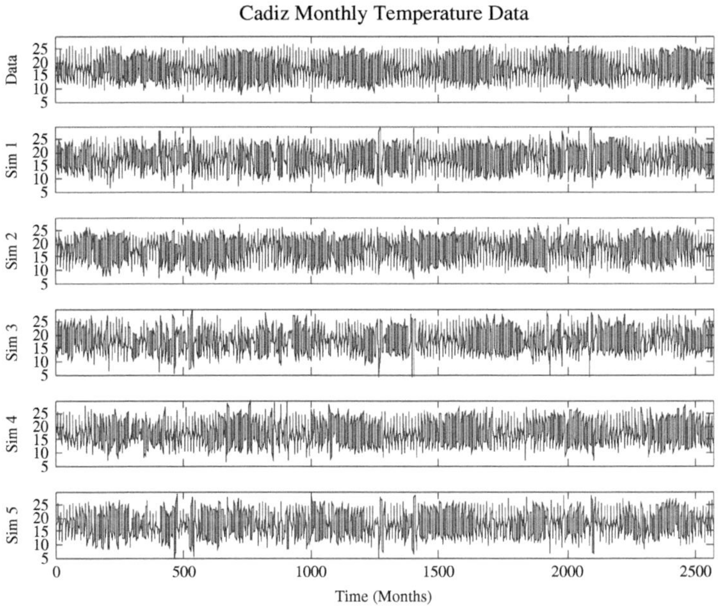

(d) Cadiz monthly temperature data (°C)

Figure 3.3.1. (*Continued*)

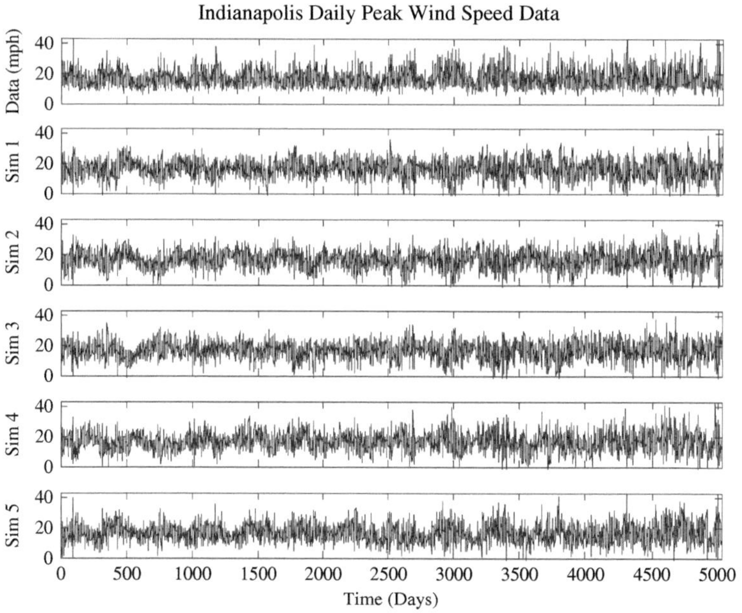

(e) Indianapolis daily peak wind speed data

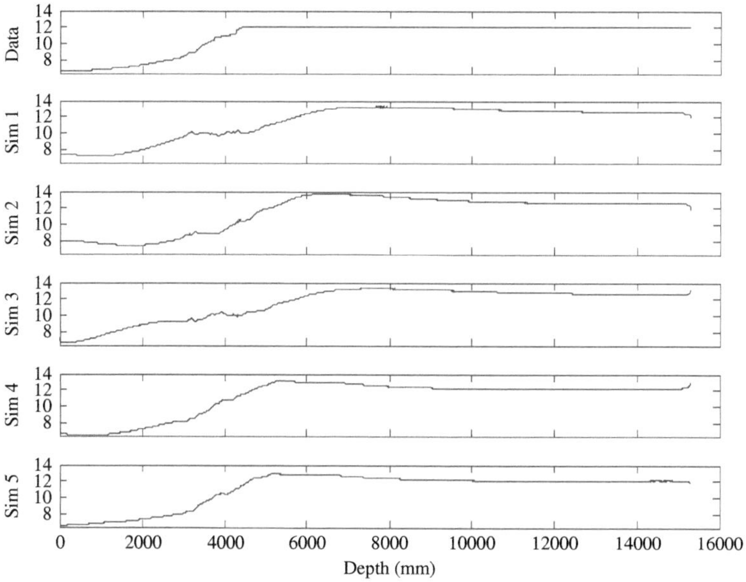

(f) Five simulated series for lake temperature data of date 10/20/04

Figure 3.3.1. (*Continued*)

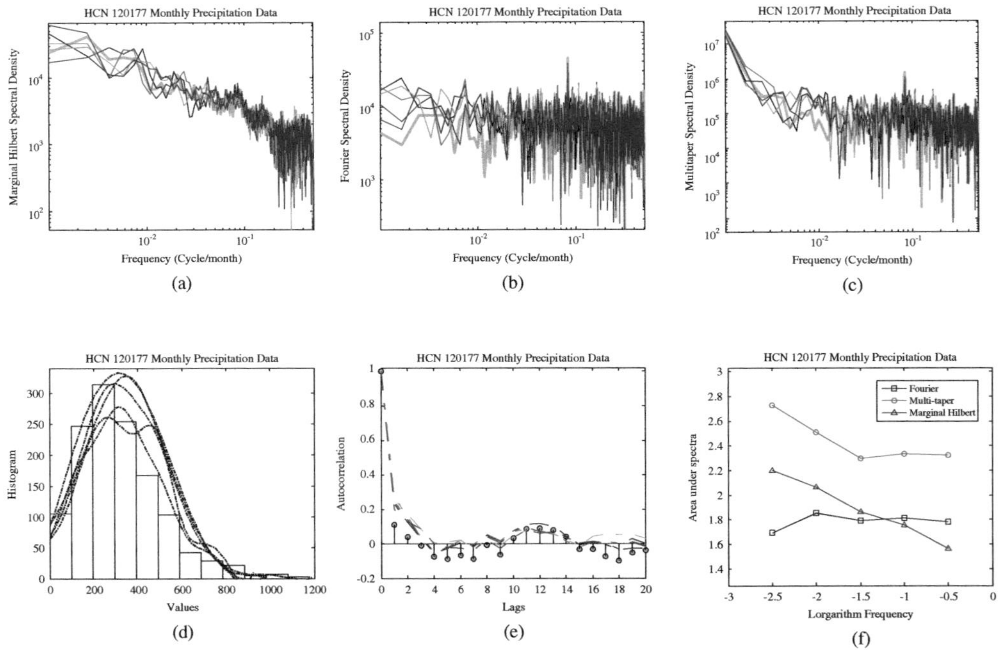

Figure 3.3.2. Comparison of characteristics of five simulated series for HCN 120177 monthly rainfall data by method 1. (a) HHT (b) Fourier (c) Multitaper (the gray bold solid line is calculated from original data while the others five thin lines are from simulated data) (d) histogram (e) Autocorrelogram (the background bars are calculated from original data) (f) Areas under spectra

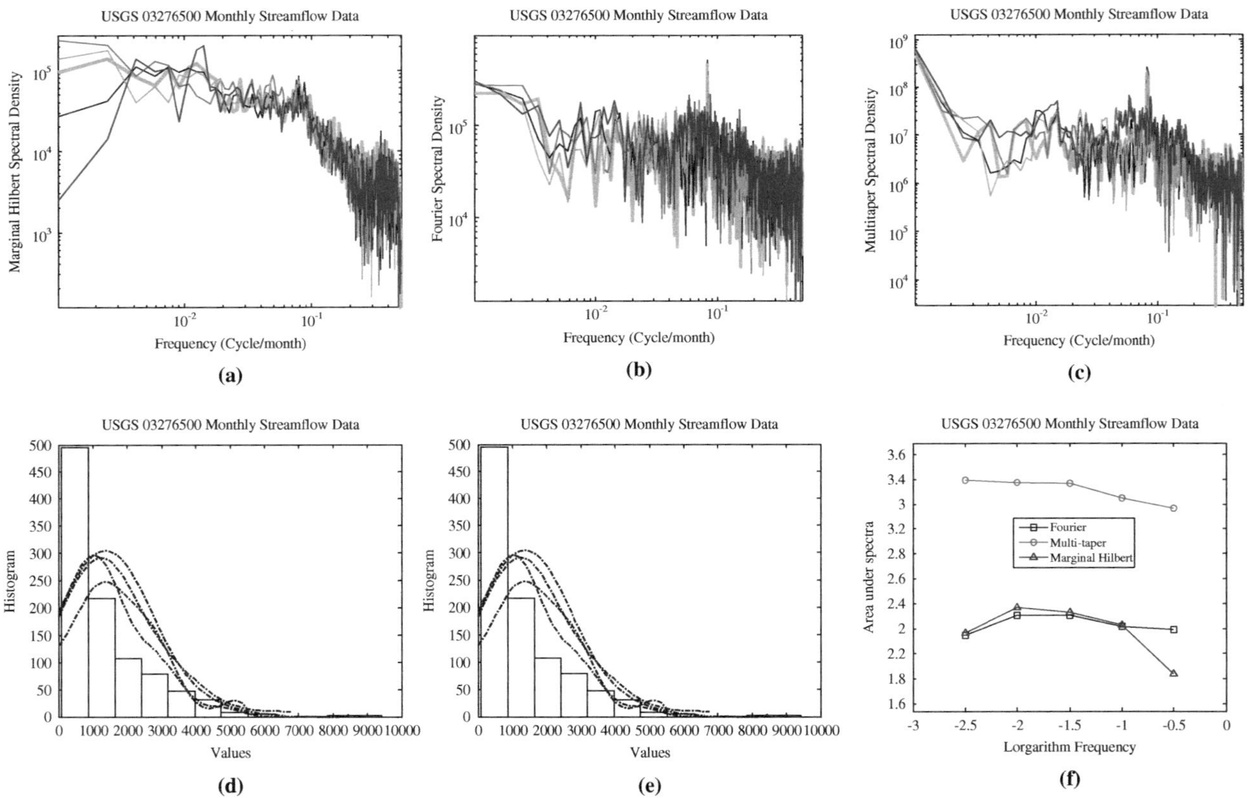

Figure 3.3.3. Comparison of characteristics of five simulated series for USGS03276500 monthly streamflow data by method 1 (a) HHT (b) Fourier (c) Multitaper (the gray bold solid line is calculated from original data while the others five thin lines are from simulated data) (d) histogram (e) Autocorrelogram (the background bars are calculated from original data) (f) Areas under spectra

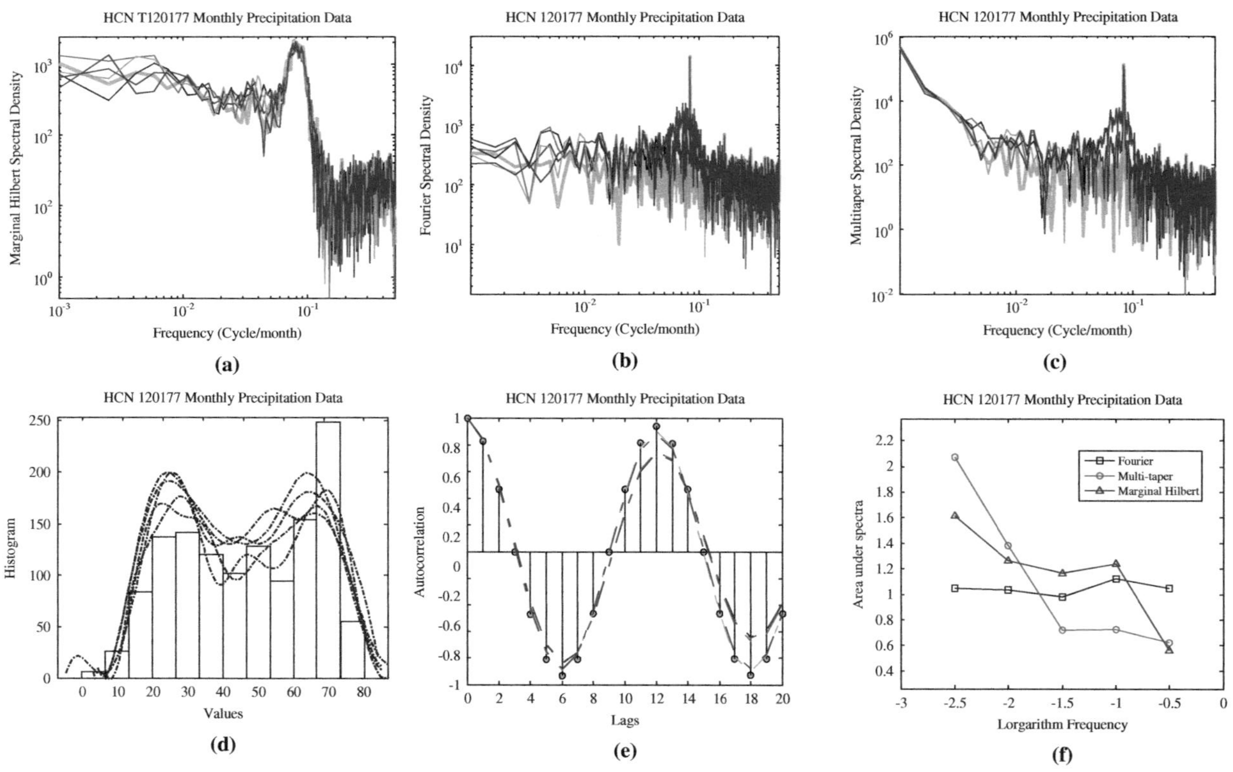

Figure 3.3.4. Comparison of characteristics of five simulated series for HCN 120177 monthly temperature data by method 1 (a) HHT (b) Fourier (c) Multitaper (the gray bold solid line is calculated from original data while the others five thin lines are from simulated data) (d) histogram (e) Autocorrelogram (the background bars are calculated from original data) (f) Areas under spectra

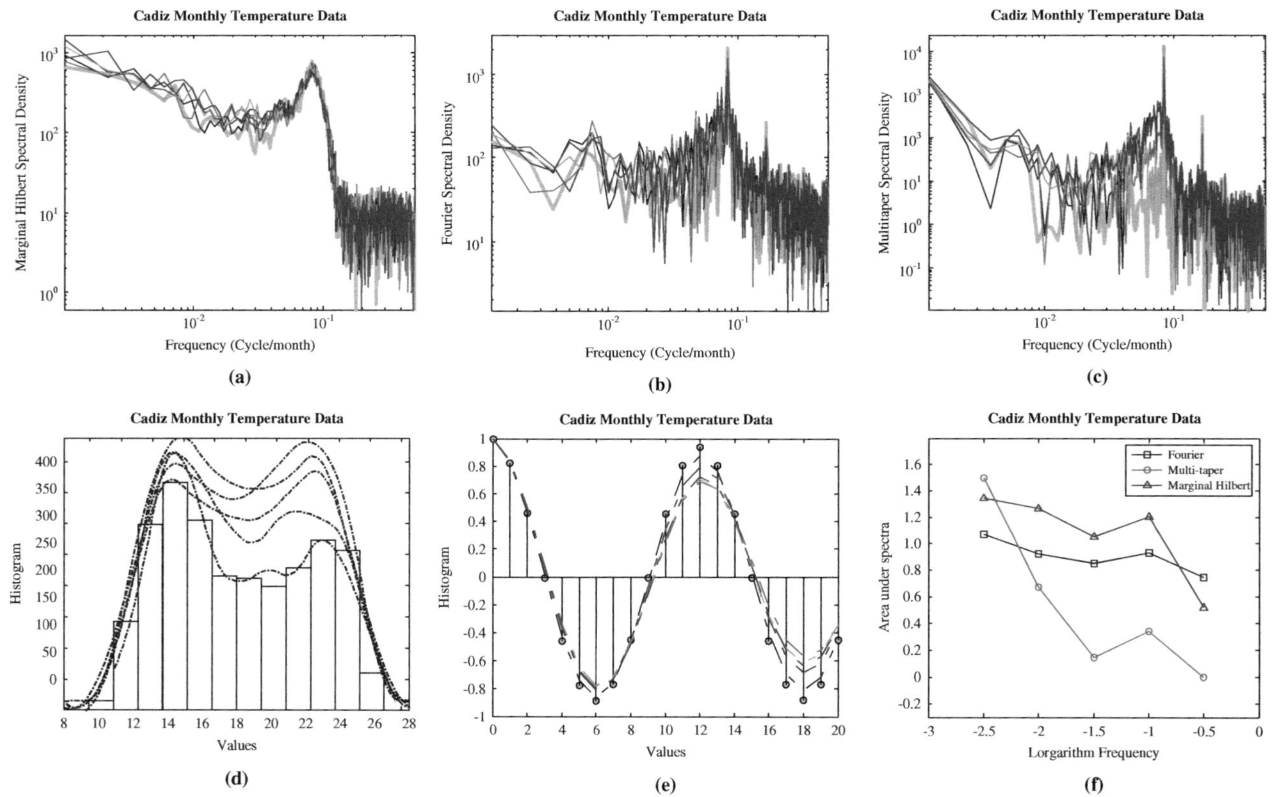

Figure 3.3.5. Comparison of characteristics of five simulated series for Cadiz, Spain monthly temperature data by method 1 (a) HHT (b) Fourier (c) Multitaper (the gray bold solid line is calculated from original data while the others five thin lines are from simulated data) (d) histogram (e) Autocorrelogram (the background bars are calculated from original data) (f) Areas under spectra

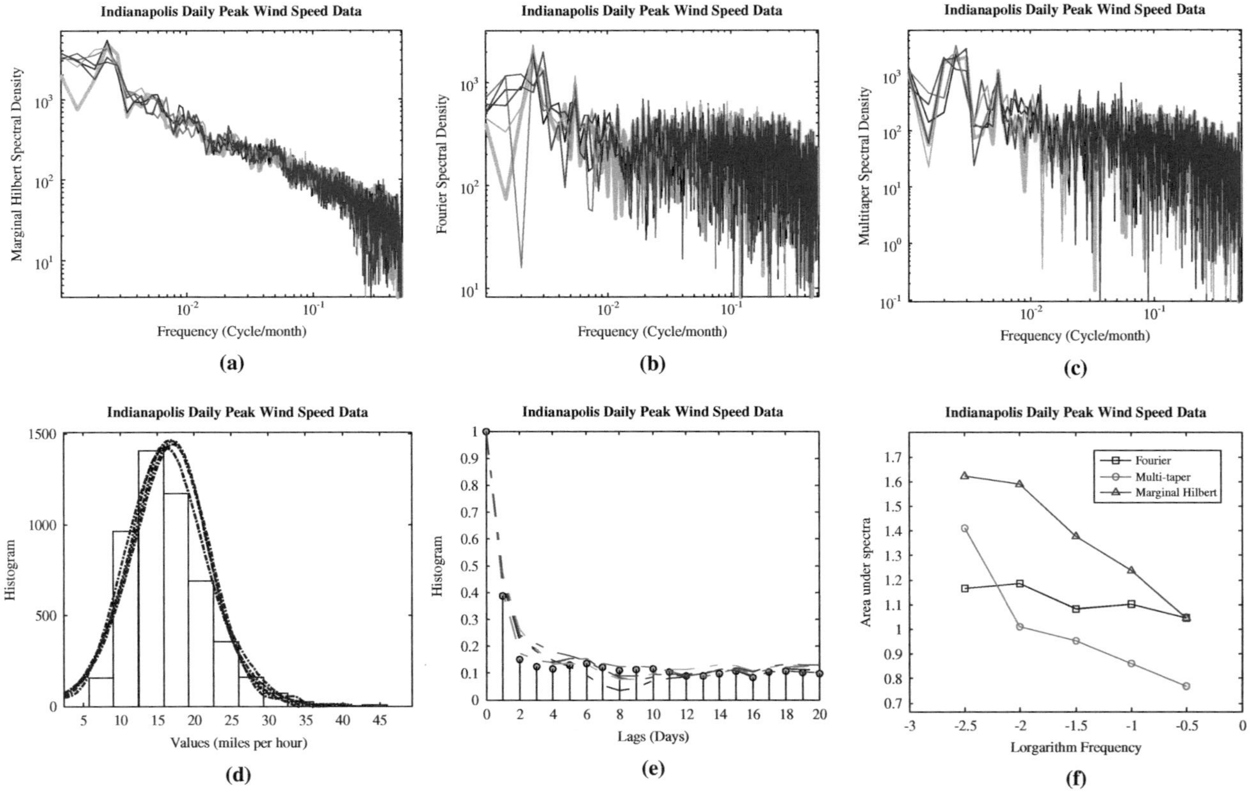

Figure 3.3.6. Comparison of characteristics of five simulated series for Indianapolis daily peak wind speed data by method 1 (a) HHT (b) Fourier (c) Multitaper (the gray bold solid line is calculated from original data while the others five thin lines are from simulated data) (d) histogram (e) Autocorrelogram (the background bars are calculated from original data) (f) Areas under spectra

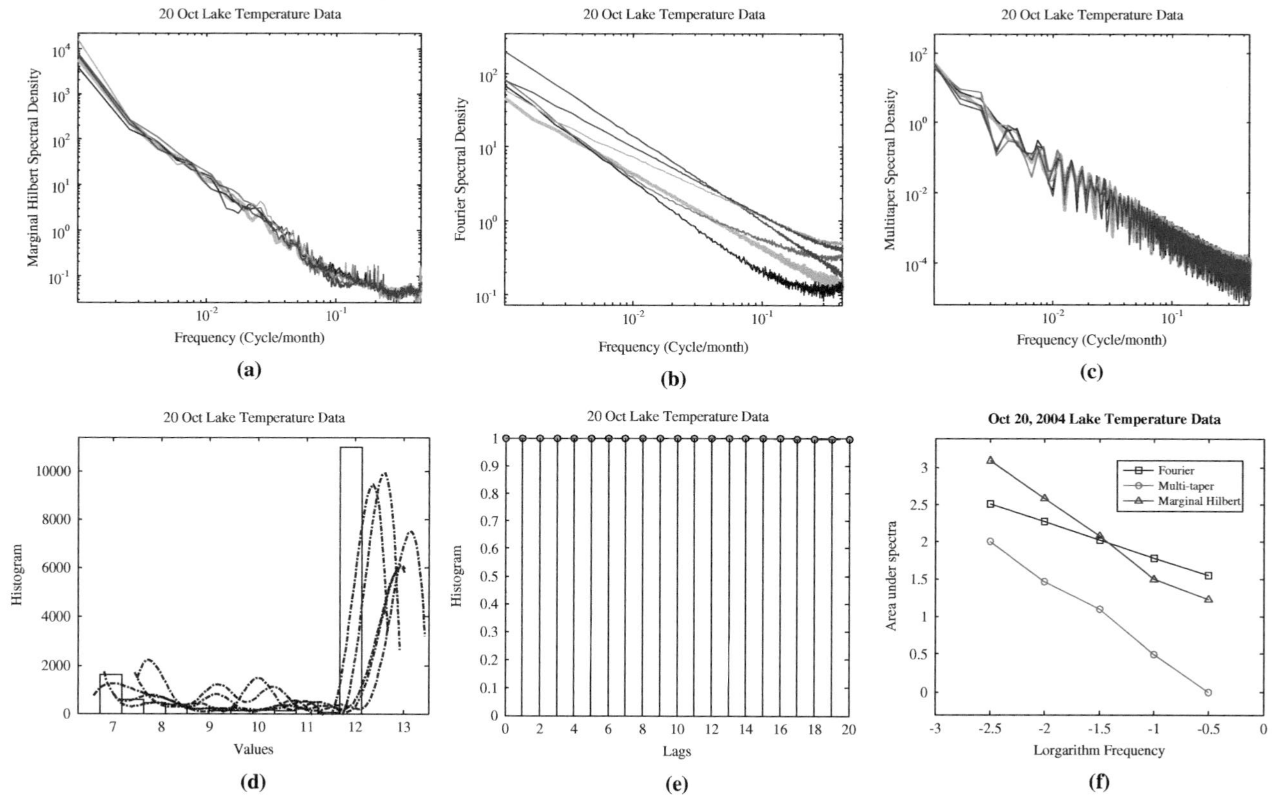

Figure 3.3.7. Comparison of characteristics of five simulated series for lake temperature data of 20 Oct by method 1 (a) HHT (b) Fourier (c) Multitaper (the gray bold solid line is calculated from original data while the others five thin lines are from simulated data) (d) histogram (e) Autocorrelogram (the background bars are calculated from original data) (f) Areas under spectra

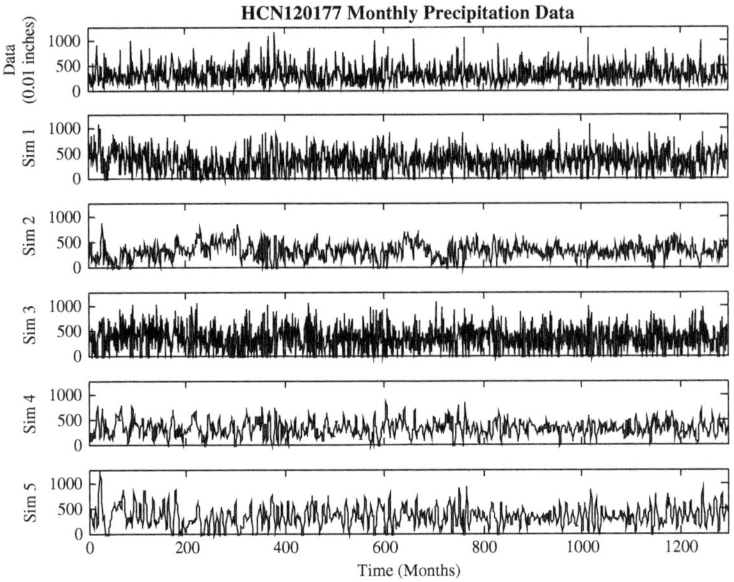

(a) HCN 120177 monthly rainfall data (unit: 0.01 inches)

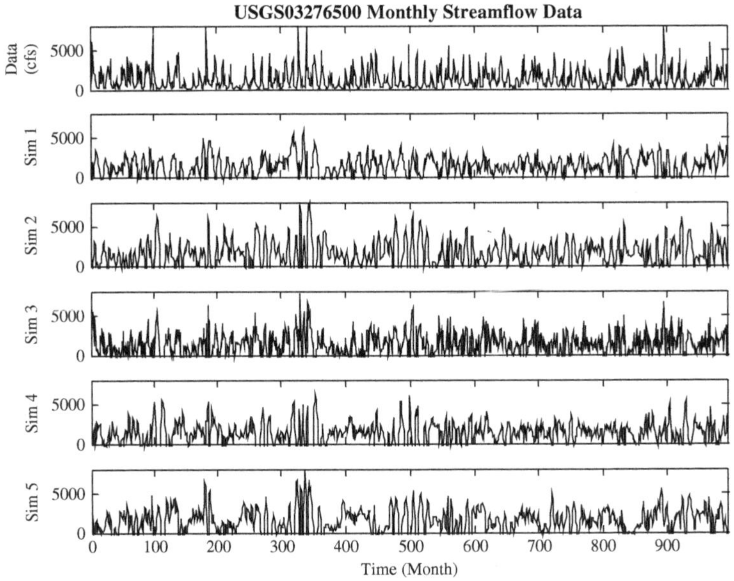

(b) USGS 03276500 monthly streamflow data

Figure 3.3.8. Five simulated series from method 2

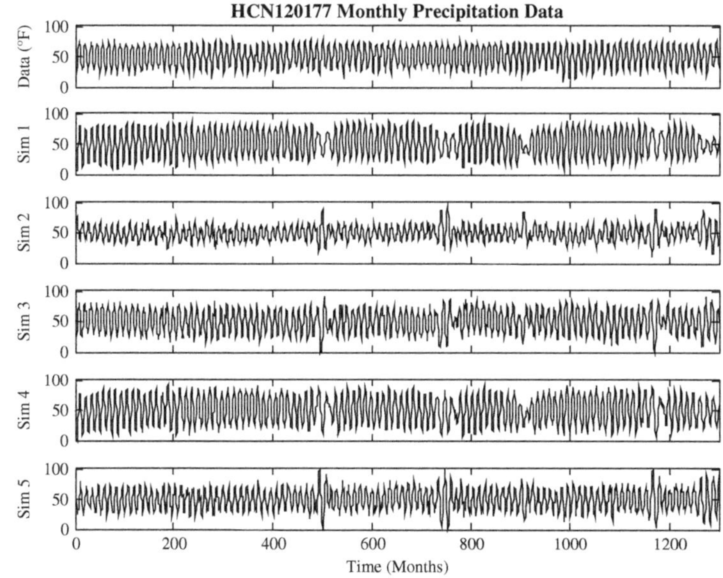

(c) HCN 120177 monthly temperature data (°F)

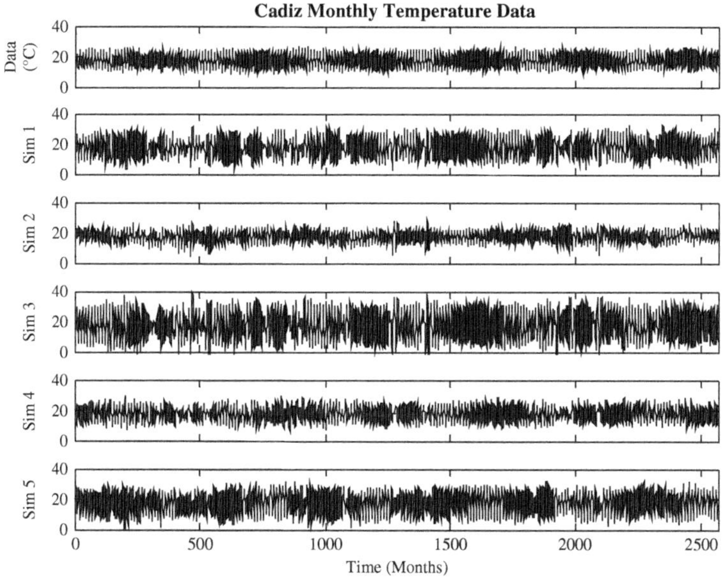

(d) Cadiz monthly temperature data (°C)

Figure 3.3.8. (*Continued*)

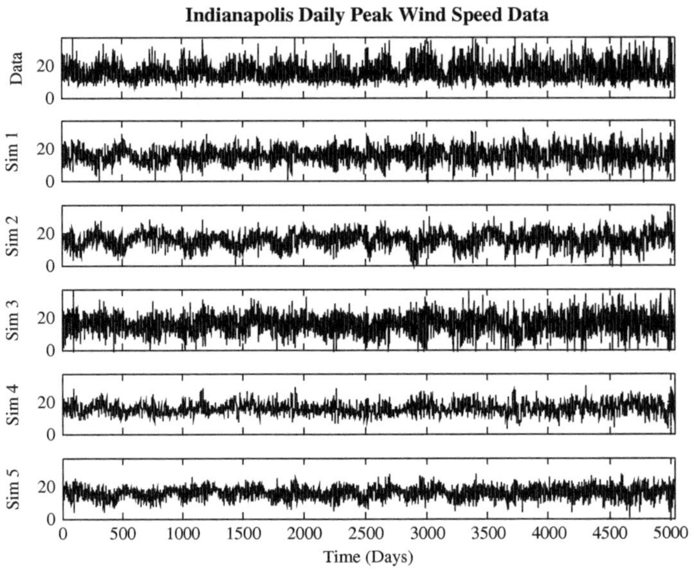

(e) Indianapolis daily peak wind speed data

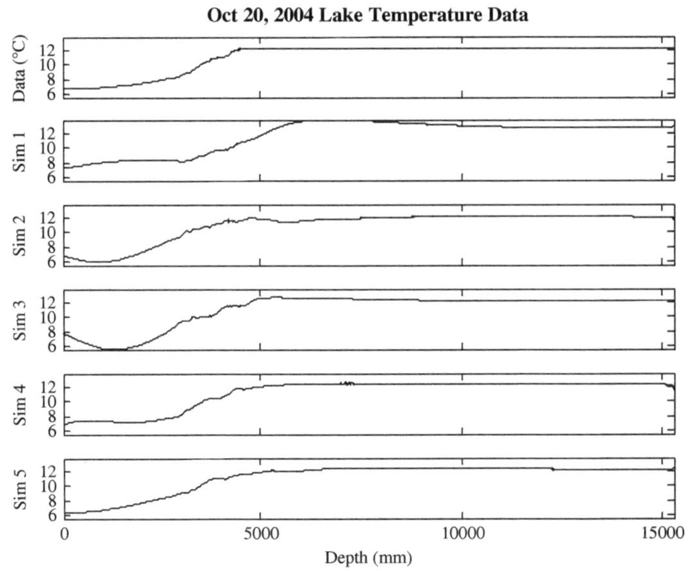

(f) Five simulated series for lake temperature data of date 10/20/04

Figure 3.3.8. (*Continued*)

considerable fluctuations in the simulated data. Another disadvantage of using method 2 is that there are more negative values than in method one. Since most data used in this study are positive time series, such as rainfall, runoff and temperature in Fahrenheit, these negative simulated values are not acceptable. The characteristics of the original and simulated data are presented in Figures 3.3.9 to 3.3.14. These results are similar to those in Figures 3.3.2–3.3.7.

3.3.4 Simulation by Wen-Yeh Method

In the previous two methods, only one (for phase) or two (for phase and amplitude) random components are generated for each IMF. Hence, only a small number of random elements are considered. Wen-Yeh method is developed to include more random components. It separates the noise from the underlying signal by smoothing the amplitudes and instantaneous frequencies. They propose the following procedure. (Gu and Wen, 2005).

(1) Decompose the signal into n intrinsic mode functions (IMFs), $C_j(t)$, $j = 1, 2, .., n$ and Hilbert transform of each IMF to determine $a_j(t)$ and $\omega_j(t)$.
(2) Smooth $a_j(t)$ and $\omega_j(t)$, and denote the smoothed function as $a_{js}(t)$ and $\omega_{js}(t)$, respectively. The reduced process $C_{jr}(t)$ is this obtained by removing the amplitude modulation $a_{js}(t)$ from $C_j(t)$ and changing the time scale by $\omega_{js}^{-1}(t)$. The way to change the time scale of a function $y(t)$ is to first integrate the smoothed instantaneous frequency function $\omega_{js}(t)$ to obtain the $\theta(t)$ function. The purpose of doing this is to make the signal a function of θ instead of t, i.e. $y(\theta)$. Since $y(\theta)$ is not evenly spaced, the signal has to be resampled to make it evenly spaced. The resulting series is the reduced process of $y(t)$, i.e., the process obtained after removing the frequency modulation.
(3) Simulate the reduced process $C_{jr}(t)$ as a stationary process and obtain samples of the process $S_{jr}(t)$.
(4) Restore the time scale by using the function $\omega_{js}(t)$ in $S_{jr}(t)$, and restore the amplitude modulation by using $a_{js}(t)$. The result is the simulated jth IMF $S_j(t)$. Restoring time scale is accomplished by first resampling the underlying stationary process of θ and then expressing the signal as a function of time t.
(5) Finally, add all $S_j(t)$, $j = 1, 2, .., n$, to construct the simulated signal.

To illustrate these procedures, the monthly rainfall data from station HCN 120177 is used. For the second step above, the smoothed instantaneous frequencies and amplitudes of the 2nd IMF component are shown in Figure 3.3.15. It is smoothed by a 24-point moving average window.

The reduced process $C_{jr}(t)$ is obtained by removing the amplitude modulation $a_{js}(t)$ from $C_j(t)$ and changing the time scale by $\omega_{js}^{-1}(t)$. The effect of change in time scale by removing frequency modulation for the amplitude-reduced second

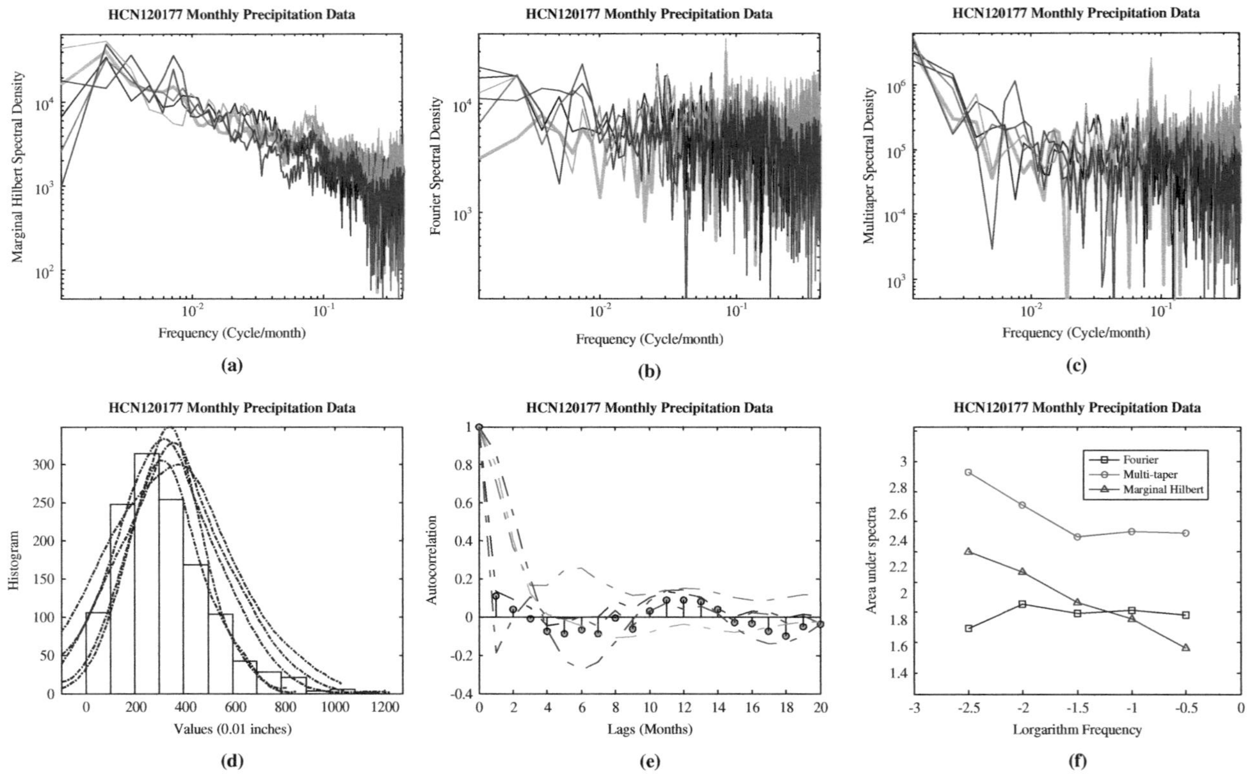

Figure 3.3.9. Comparison of characteristics of five simulated series for HCN 120177 monthly rainfall data by method 2. (a) HHT (b) Fourier (c) Multitaper (the gray bold solid line is calculated from original data while the others five thin lines are from simulated data) (d) histogram (e) Autocorrelogram (the background bars are calculated from original data) (f) Areas under spectra

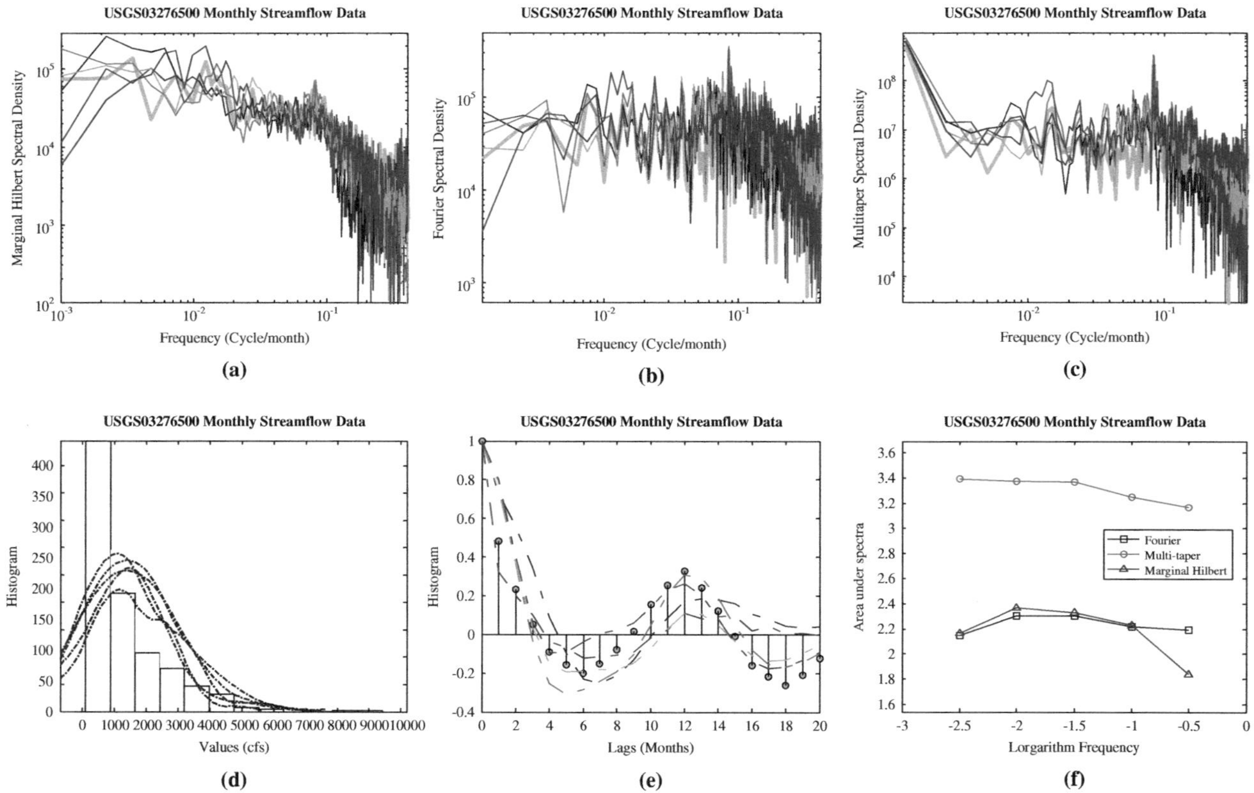

Figure 3.3.10. Comparison of characteristics of five simulated series for USGS 03276500 streamflow data by method 2 (a) HHT (b) Fourier (c) Multitaper (the gray bold solid line is calculated from original data while the others five thin lines are from simulated data) (d) histogram (e) Autocorrelogram (the background bars are calculated from original data) (f) Areas under spectra

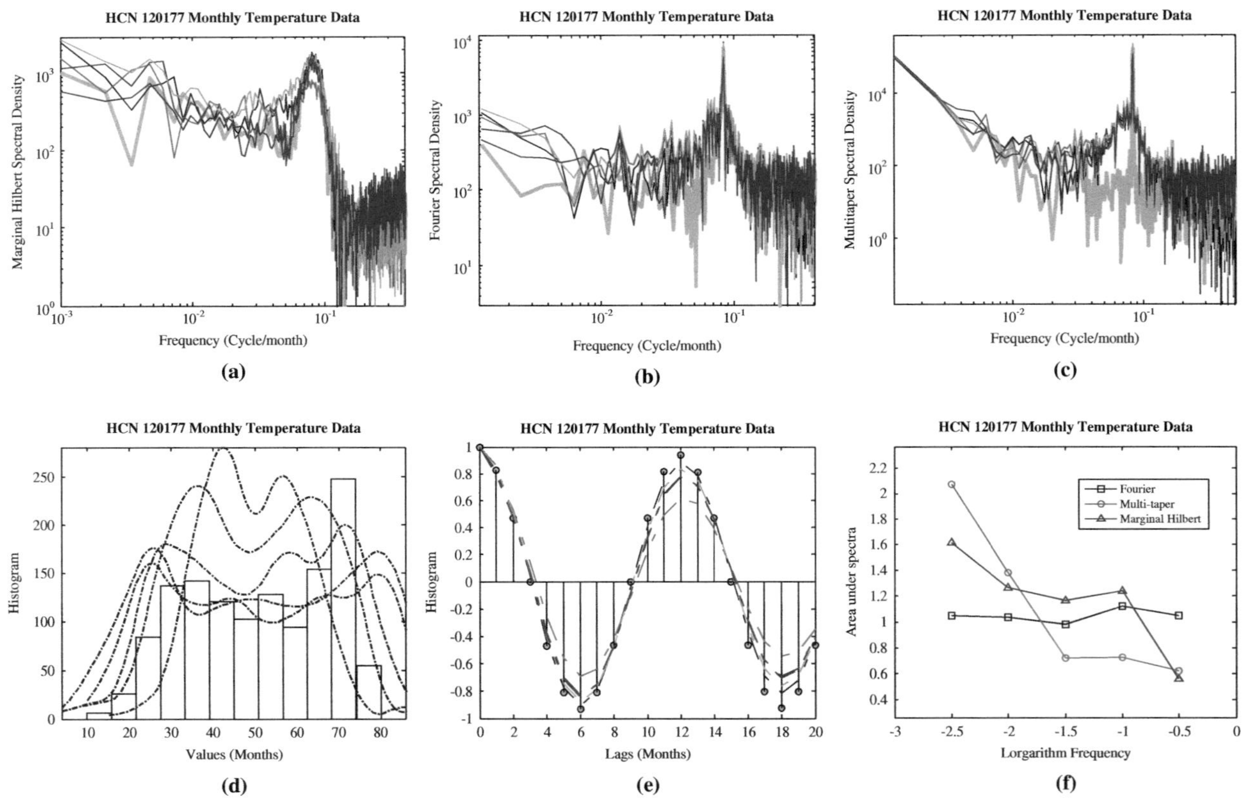

Figure 3.3.11. Comparison of characteristics of five simulated series for HCN 120177 monthly temperature data by method 2 (a) HHT (b) Fourier (c) Multitaper (the gray bold solid line is calculated from original data while the others five thin lines are from simulated data) (d) histogram (e) Autocorrelogram (the background bars are calculated from original data) (f) Areas under spectra

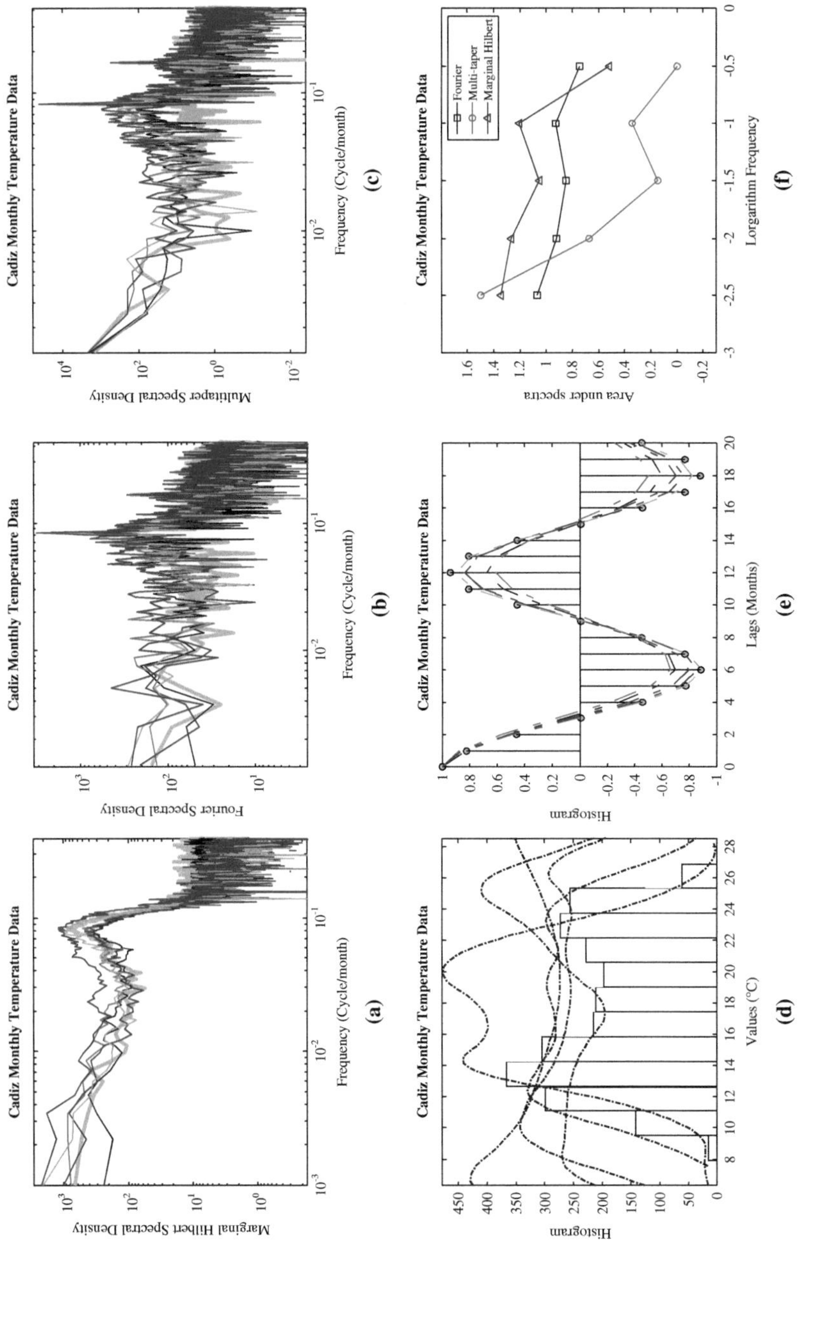

Figure 3.3.12. Comparison of characteristics of five simulated series for Cadiz monthly temperature data by method 2 (a) HHT (b) Fourier (c) Multitaper (the gray bold solid line is calculated from original data while the others five thin lines are from simulated data) (d) histogram (e) Autocorrelogram (the background bars are calculated from original data) (f) Areas under spectra

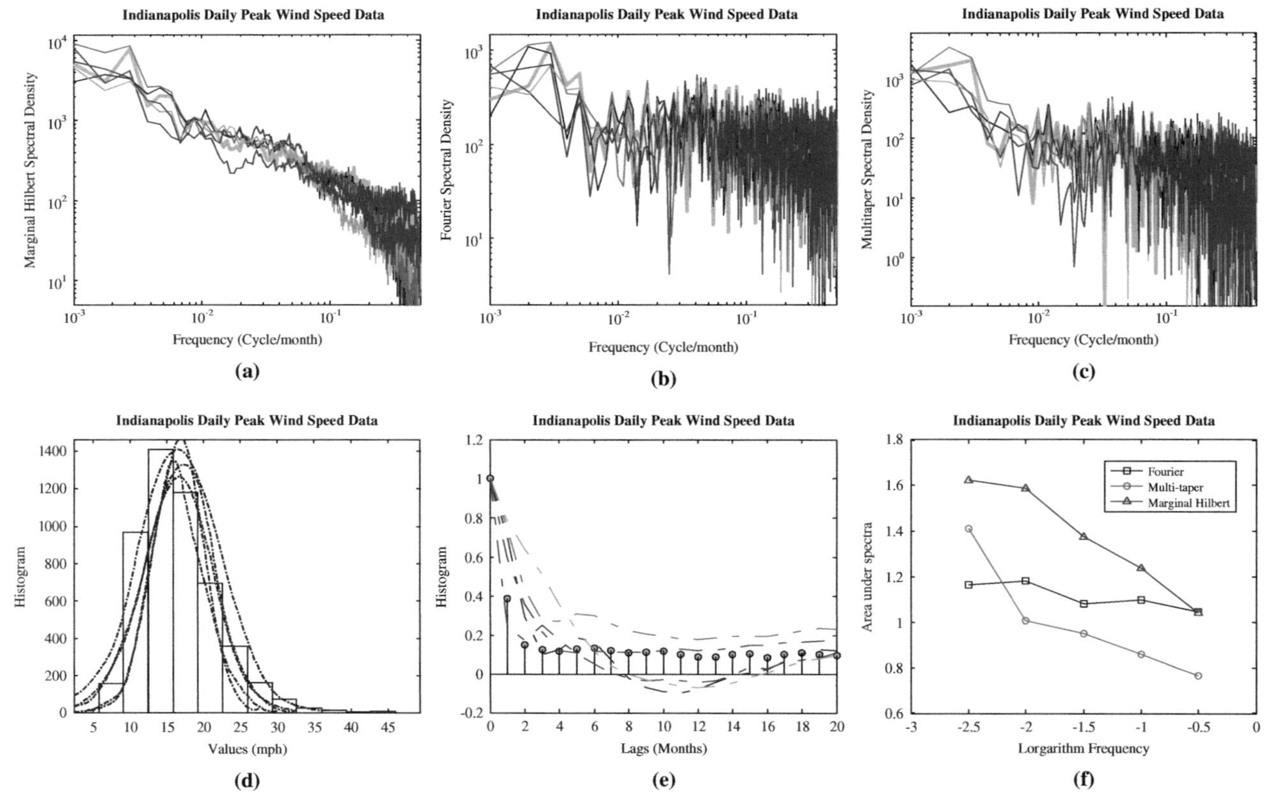

Figure 3.3.13. Comparison of characteristics of five simulated series for Indianapolis daily peak wind speed data by method 2 (a) HHT (b) Fourier (c) Multitaper (the gray bold solid line is calculated from original data while the others five thin lines are from simulated data) (d) histogram (e) Autocorrelogram (the background bars are calculated from original data) (f) Areas under spectra

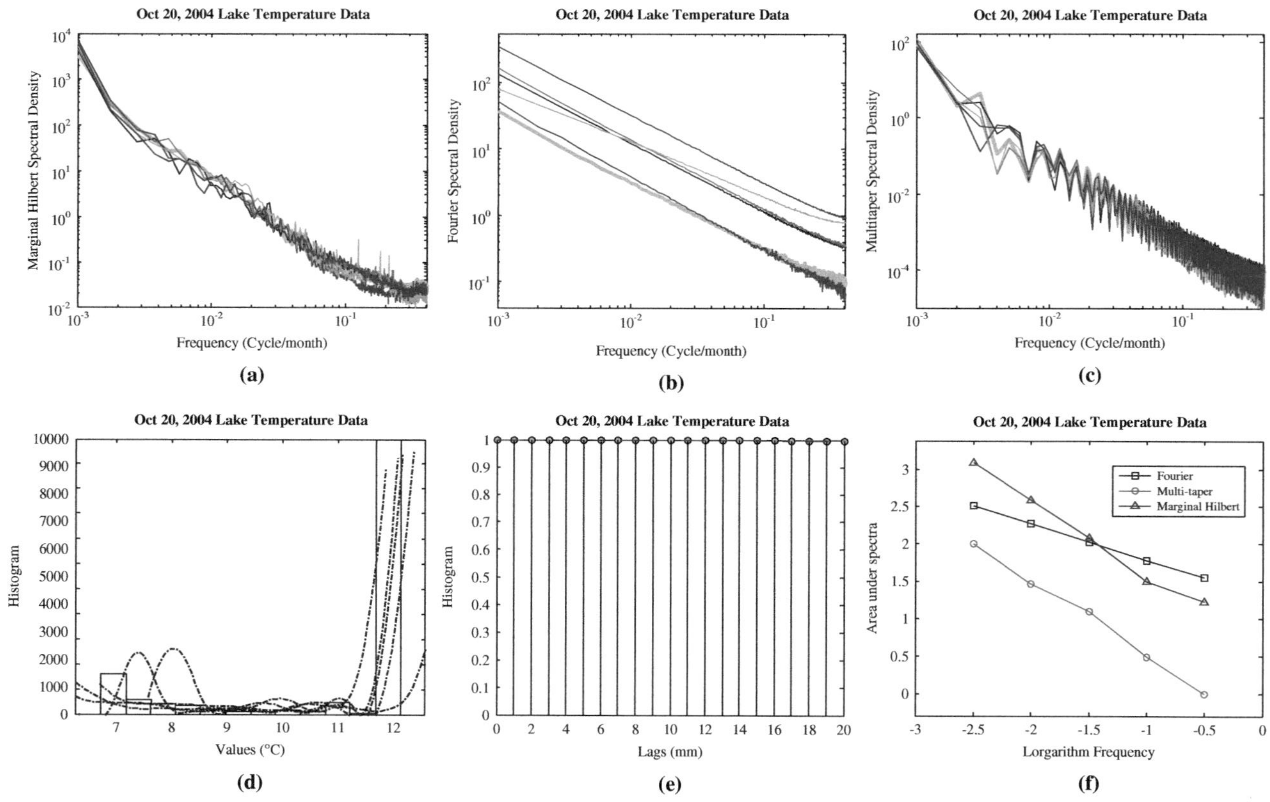

Figure 3.3.14. Comparison of characteristics of five simulated series for Oct 20, 2004 lake temperature data by method 2 (a) HHT (b) Fourier (c) Multitaper (the gray bold solid line is calculated from original data while the others five thin lines are from simulated data) (d) histogram (e) Autocorrelogram (the background bars are calculated from original data) (f) Areas under spectra

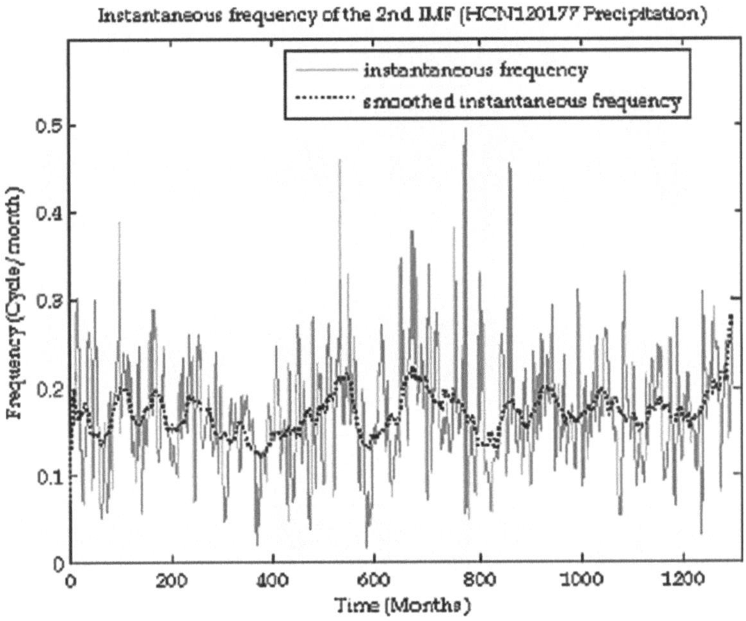

(a)

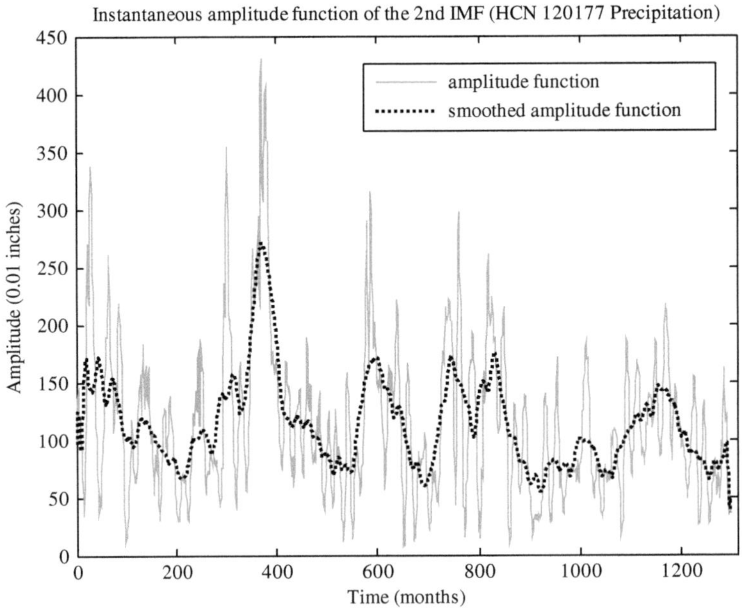

(b)

Figure 3.3.15. Smoothed instantaneous frequency and amplitude

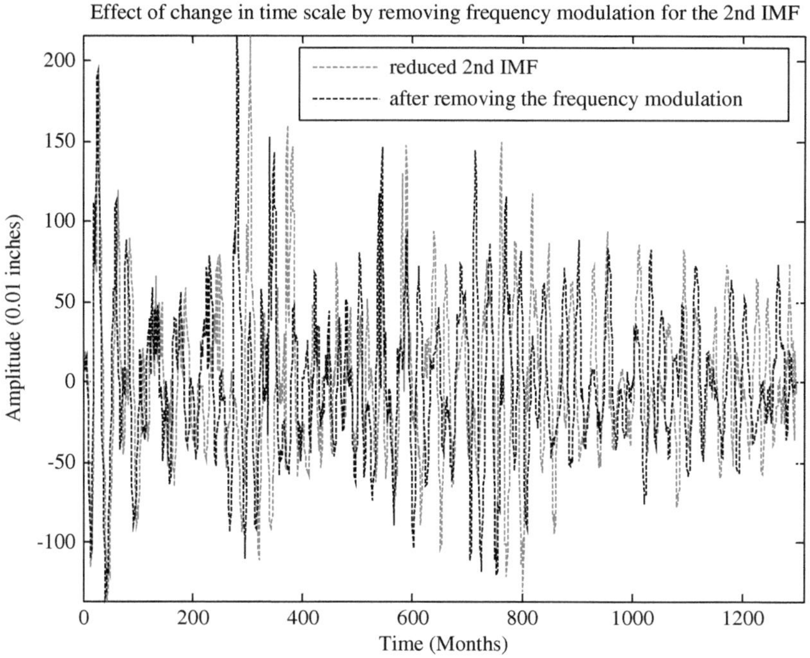

Figure 3.3.16. Rescaled reduced-2nd IMF component

IMF is shown in Figure 3.3.16. This step makes the signal a function of θ instead of t. Therefore there is some shifting and amplitude difference due to even spacing.

After removing these modulations, the 2nd IMF thus can be simulated as a stationary random process (the 3rd step). Then, the time scale is restored by using the function and the amplitude modulation is restored by using $a_{js}(t)$. The simulated 2nd IMF is obtained as shown in Figure 3.3.17. This is only a demonstration for a single IMF component. Similar operation has to be performed for each IMF. Addition of all the components yields the final simulated series.

For the six natural time and spatial series, there are five simulations conducted for each one of them. The results are shown in Figure 3.3.18. The characteristics of the original data and simulated data are compared in Figure 3.3.19~Figure 3.3.24. The comparison of Fourier spectra, multitaper spectra, marginal Hilbert spectra, histogram, and autocorrelation are of interest. The areas under these spectra are computed as well.

Of the three methods described previously to simulate the nonstationary processes, method 2, which is based on using the random phases and amplitudes, is not recommended. Based on the results shown in Figure 3.3.8, although it provides a wider range of amplitude variations and satisfies overall statistics, the characteristics of simulated data do not compare well with those of the original data. Some of

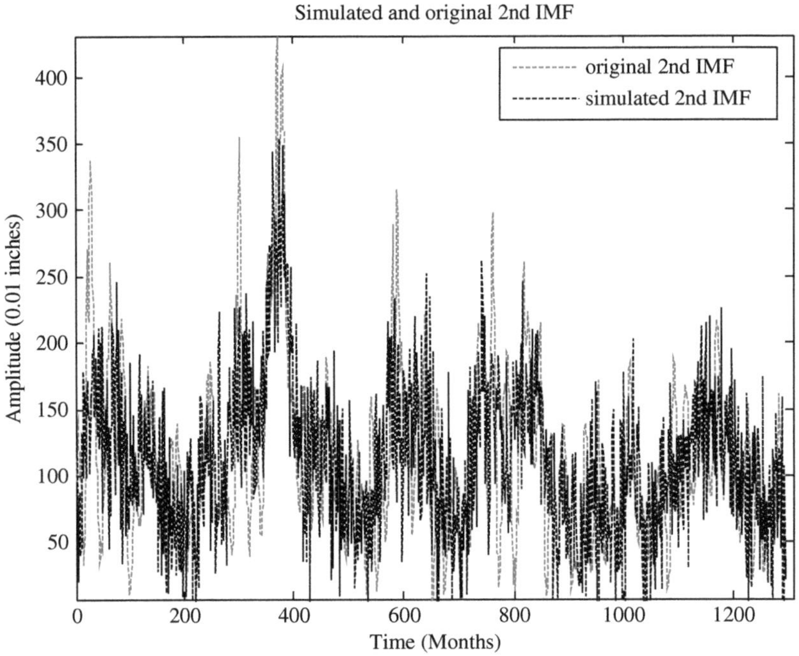

Figure 3.3.17. Simulated 2nd IMF compared to the original 2nd IMF

them even have differences up to the order of ten and that makes it very unreliable to compute the confidence limits.

The signal variability is well captured by methods 1 and 3. Based on the simulated signals in Figure 3.3.1 and Figure 3.3.18, it is clear that the results from method 1 have more variation than the results from method 3. It makes sense because the simulation in method 3 is based on the reduced signal. Although there are more random components used in method 3 for simulation, the results depend on the smoothing window length used for producing the reduced signal. While the amplitude and variation of the reduced signal are not large, there is no significant difference between the simulated and the original signals. Therefore, we may conclude that the variation of the simulated amplitude compared to the original signal is greatest in method 2 and least in method 3. Consequently, the range of confidence limits is greatest with method 2 and least with method 3. In other words, method 2 produces the widest confidence limits while method 3 yields the smallest confidence limits. Identifying significant periodicities is an important goal of spectral analysis. The results from method 1 are not as good as those from method 3, but the computations in method one are not as complex as in method 3. Hence, method 1 is used for computing the confidence limits. It is straightforward and provides reasonable simulated signals.

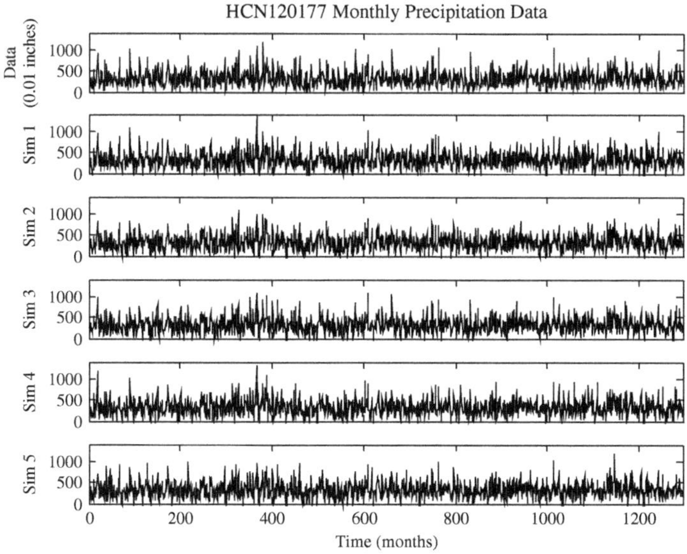

(a) HCN 120177 monthly rainfall data (unit: 0.01 inches)

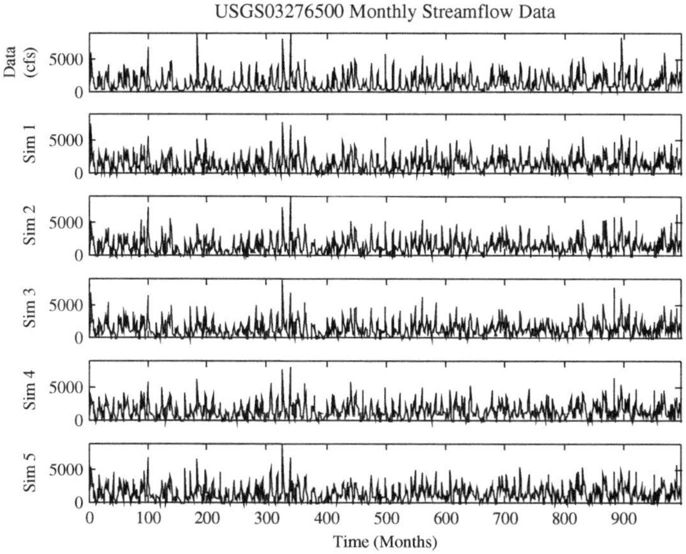

(b) USGS 03276500 monthly streamflow data

Figure 3.3.18. Five simulated series from method 3

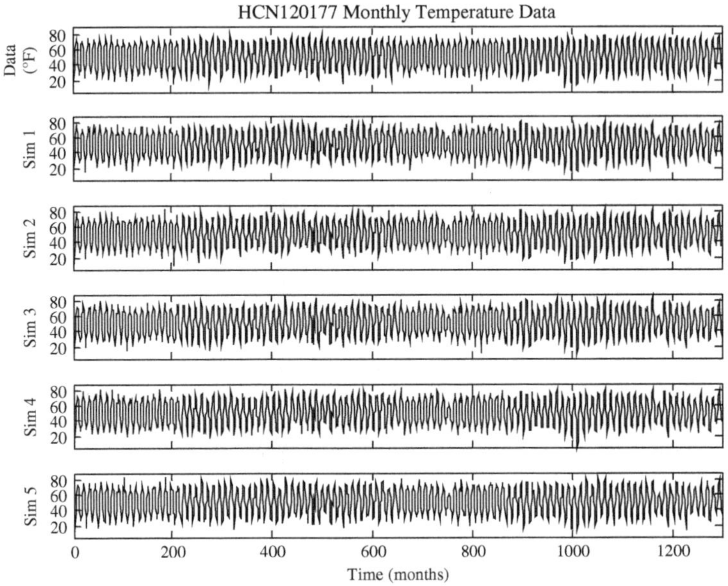

(c) HCN 120177 monthly temperature data (°F)

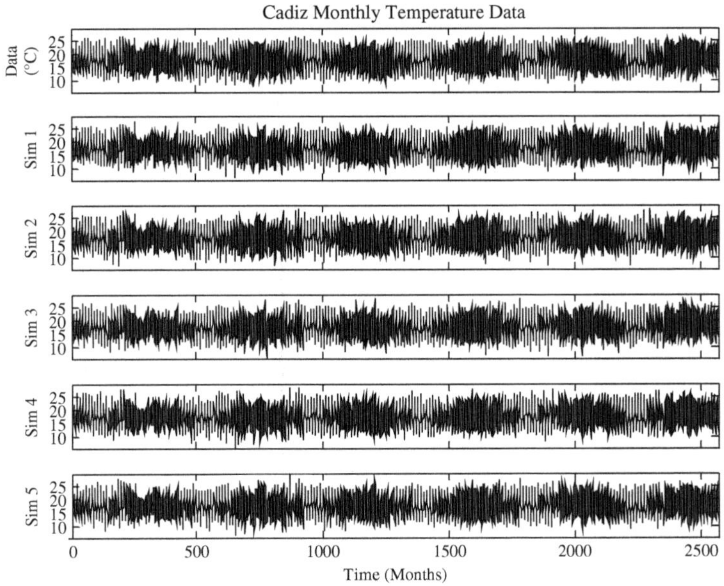

(d) Cadiz monthly temperature data (°C)

Figure 3.3.18. (*Continued*)

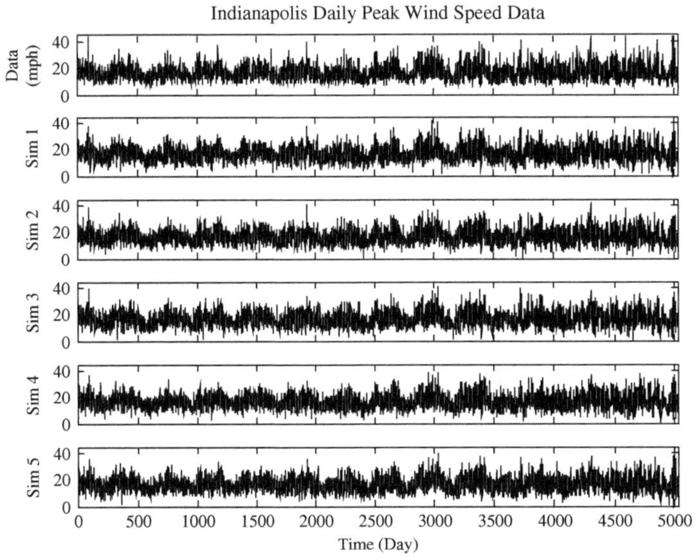

(e) Indianapolis daily peak wind speed data

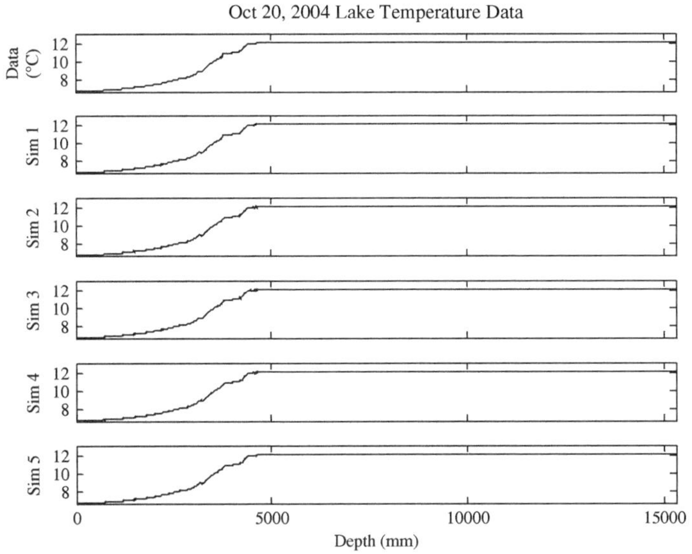

(f) Five simulated series for lake temperature data of date 10/20/04

Figure 3.3.18. (*Continued*)

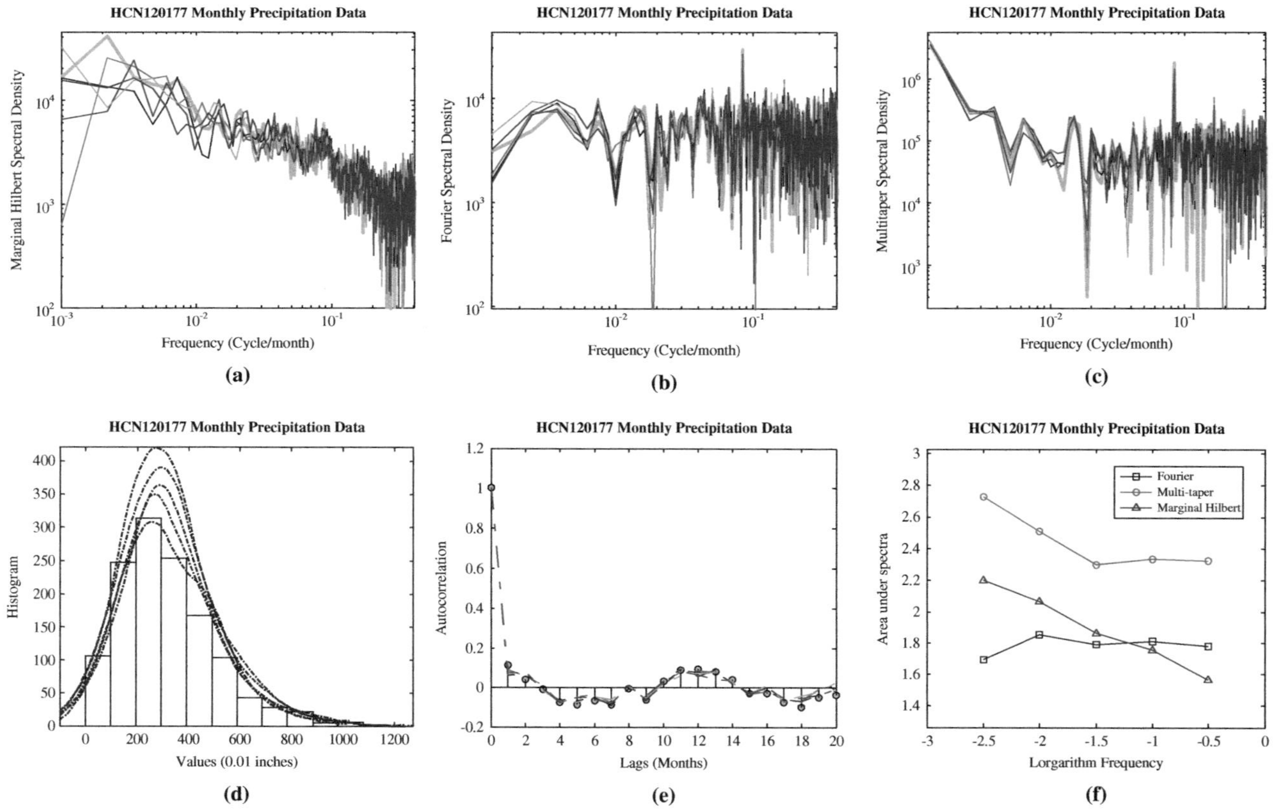

Figure 3.3.19. Comparison of characteristics of five simulated series for HCN120177 monthly rainfall data by method 3 (a) HHT (b) Fourier (c) Multitaper (the gray bold solid line is calculated from original data while the others five thin lines are from simulated data) (d) histogram (e) Autocorrelogram (the background bars are calculated from original data) (f) Areas under spectra

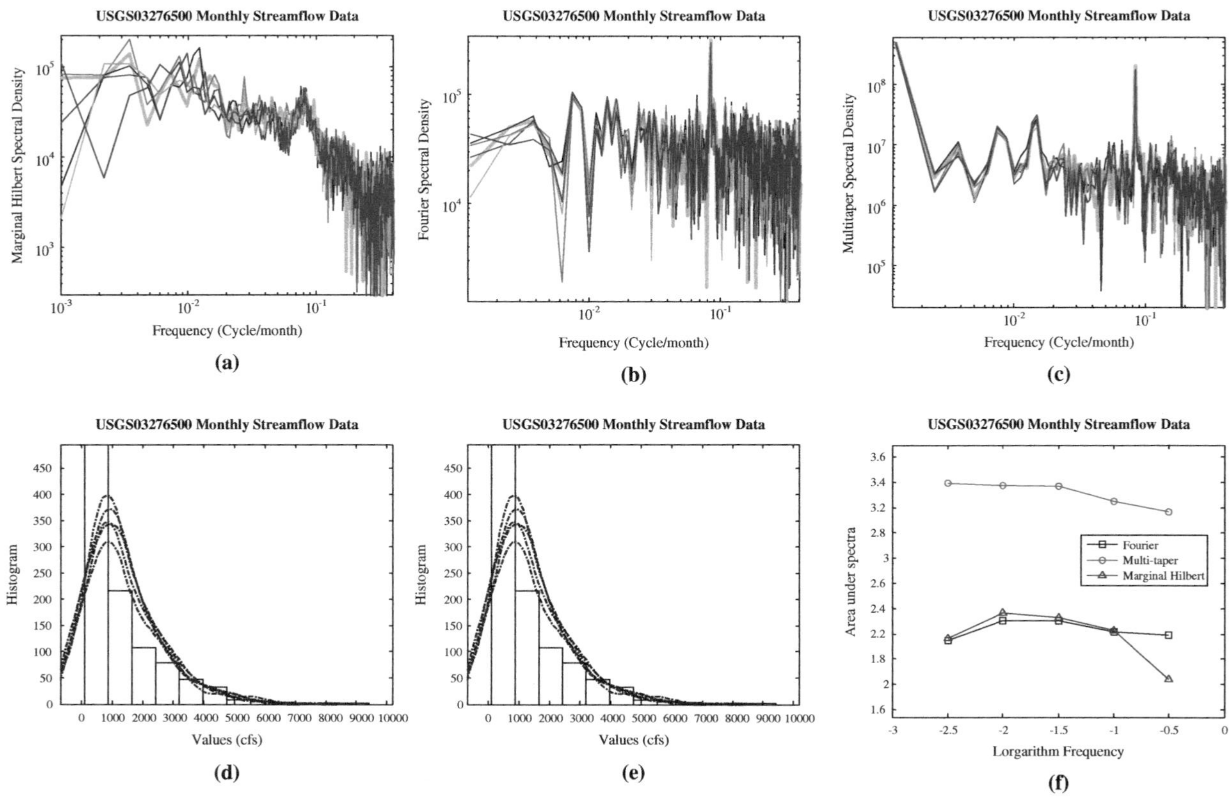

Figure 3.3.20. Comparison of characteristics of five simulated series for USGS 032276500 streamflow data by method 3 (a) HHT (b) Fourier (c) Multitaper (the gray bold solid line is calculated from original data while the others five thin lines are from simulated data) (d) histogram (e) Autocorrelogram (the background bars are calculated from original data) (f) Areas under spectra

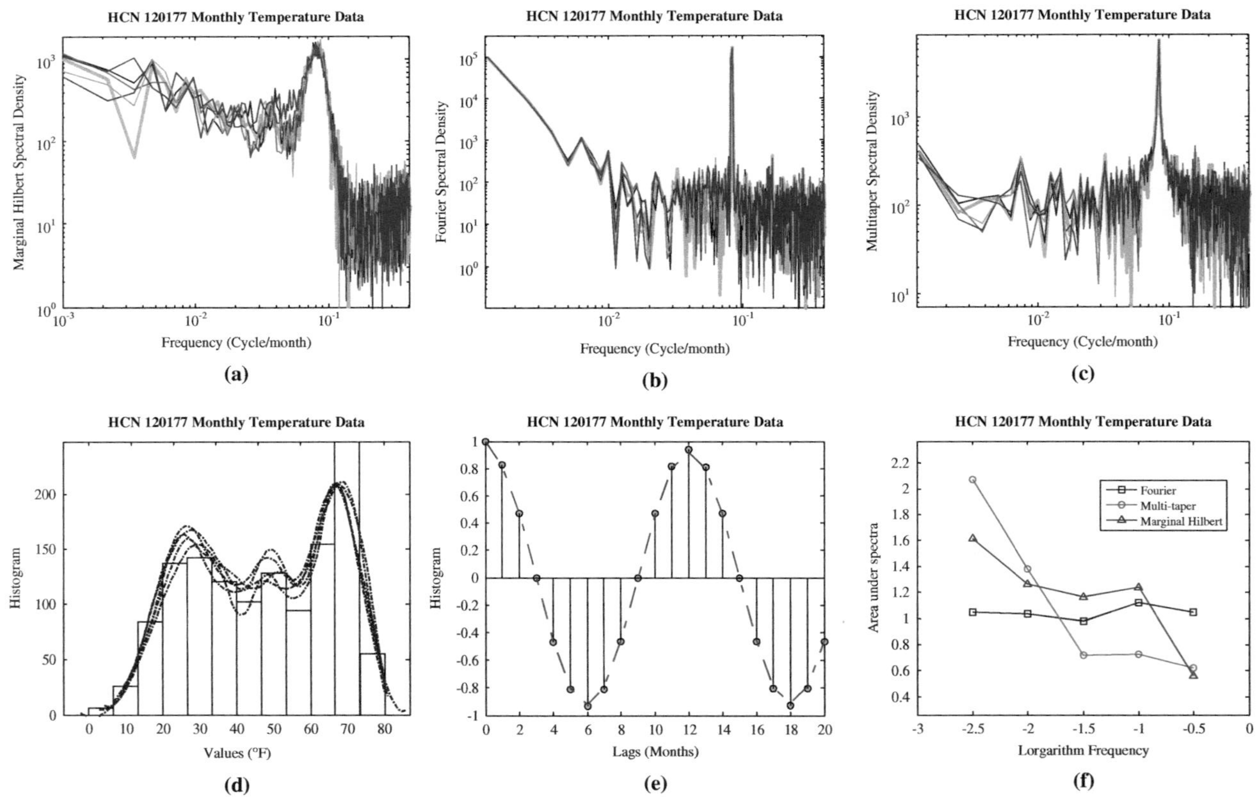

Figure 3.3.21. Comparison of characteristics of five simulated series for HCN120177 monthly temperature data by method 3 (a) HHT (b) Fourier (c) Multitaper (the gray bold solid line is calculated from original data while the others five thin lines are from simulated data) (d) histogram (e) Autocorrelogram (the background bars are calculated from original data) (f) Areas under spectra

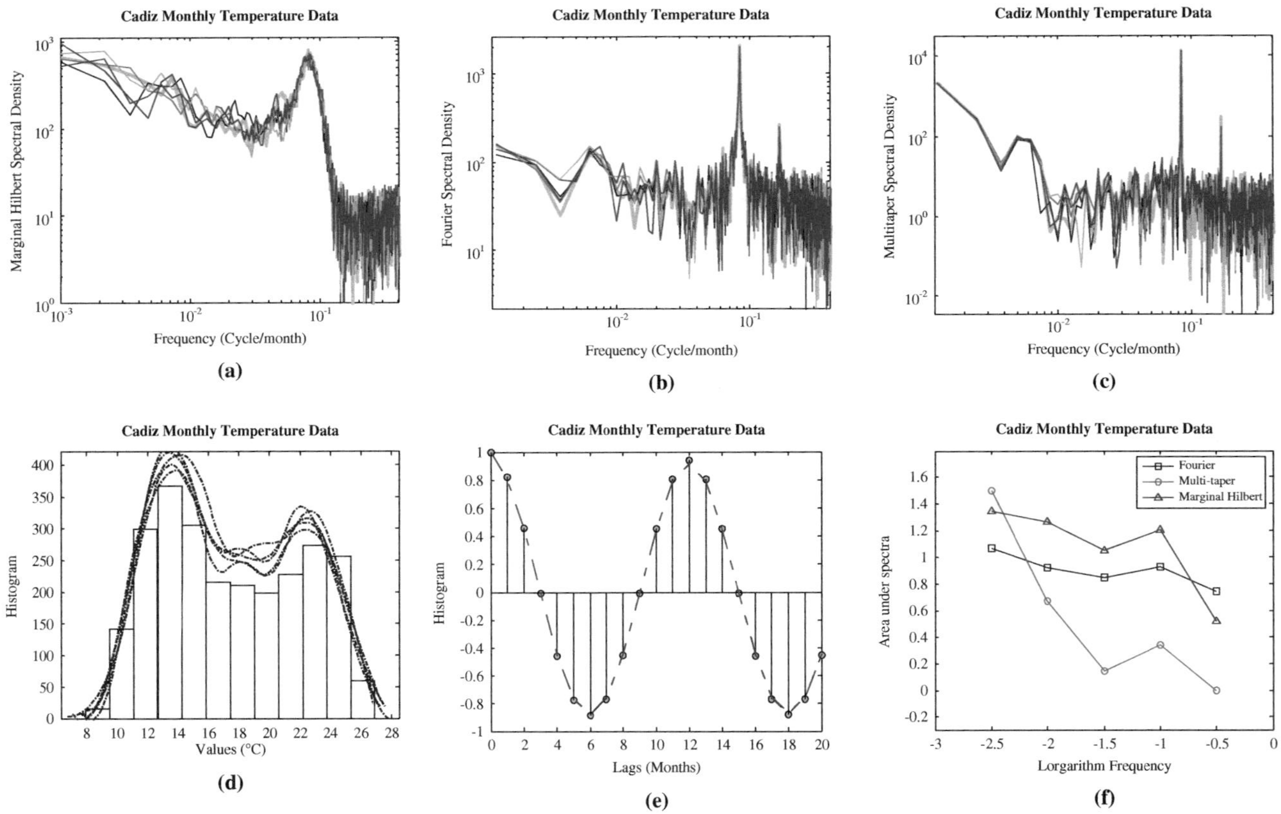

Figure 3.3.22. Comparison of characteristics of five simulated series for Cadiz monthly temperature data by method 3 (a) HHT (b) Fourier (c) Multitaper (the gray bold solid line is calculated from original data while the others five thin lines are from simulated data) (d) histogram (e) Autocorrelogram (the background bars are calculated from original data) (f) Areas under spectra

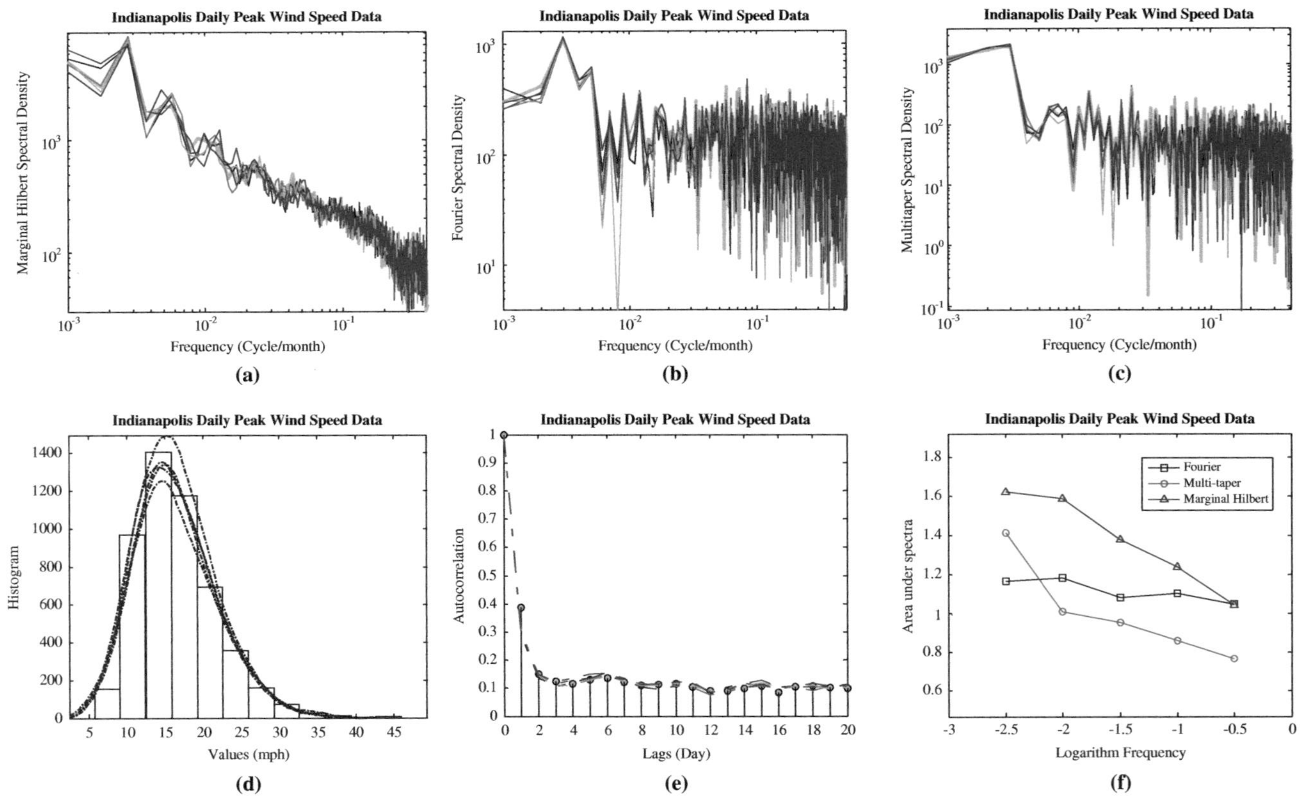

Figure 3.3.23. Comparison of characteristics of five simulated series for Indianapolis daily peak wind speed data by method 3 (a) HHT (b) Fourier (c) Multitaper (the gray bold solid line is calculated from original data while the others five thin lines are from simulated data) (d) histogram (e) Autocorrelogram (the background bars are calculated from original data) (f) Areas under spectra

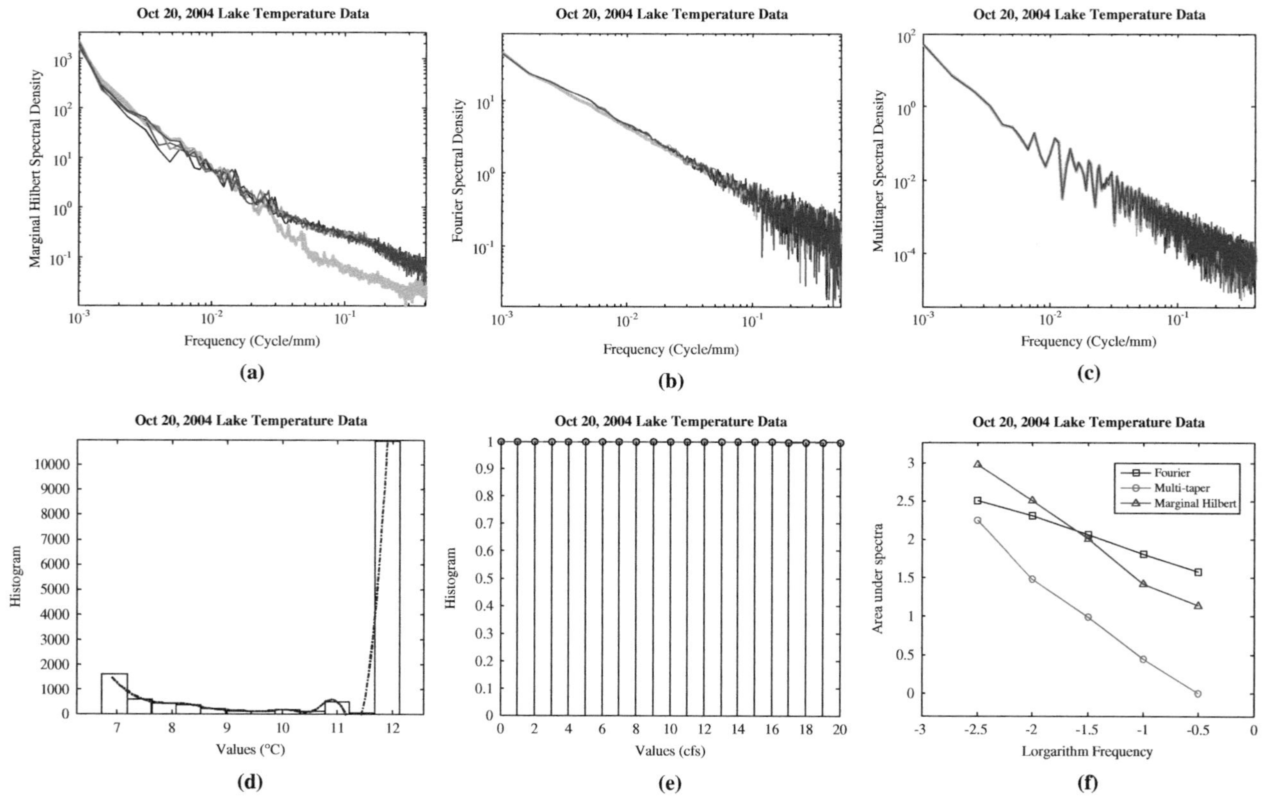

Figure 3.3.24. Comparison of characteristics of five simulated series for Oct 20, 2004 Lake temperature data by method 3 (a) HHT (b) Fourier (c) Multitaper (the gray bold solid line is calculated from original data while the others five thin lines are from simulated data) (d) histogram (e) Autocorrelogram (the background bars are calculated from original data) (f) Areas under spectra

3.4. CONFIDENCE INTERVALS FOR MARGINAL HILBERT SPECTRUM

The success in simulation of observed data by using IMFs also suggests a useful method which may be used in spectral analysis. By using these simulated spectra, the confidence intervals can be constructed for marginal Hilbert spectra. Through the simulation, the statistical variation of spectra can be estimated. Once the confidence limits are estimated, the significant periodicities can be systematically identified.

For each time series discussed herein, 25 simulated series are generated for the calculation of confidence intervals by using method 1. This appears to be sufficient to compare the statistical variability of the spectrum. The confidence intervals are obtained by the following procedure. Twenty-five series of data are simulated by using method one, which is by random phase generation. The marginal Hilbert spectra are then calculated for each time series. Once the mean and standard deviation are estimated for the 25 marginal Hilbert spectra, the confidence intervals are defined by assuming the student's t-distribution with 95% confidence limits.

For a population with unknown mean and unknown standard deviation, a confidence interval for the population mean, based on a simple random sample of size n, is as shown in Eq. (3.4.1).

$$CL = \bar{x} \pm t_{\alpha/2,n-2} \frac{s}{\sqrt{n}} \tag{3.4.1}$$

Where $\bar{x}$ is the sample mean, s is the standard deviation and $t_{\alpha/2,n-2}$ is the value of the student's t-distribution for a 100(1–α) percent of confidence interval with n–2 degrees of freedom. As the sample size n increases, the t distribution becomes closer to the normal distribution, since the standard error approaches the true standard deviation for large n.

The results include the marginal Hilbert spectrum of the original data, and the mean and the 95% confidence intervals of the simulated signals. In addition to these, a straight line is fit to the spectra. It is the best fit line computed from the resampled spectrum. The reason the spectrum is resampled is because there are more points in the high frequency range. A graphical illustration is shown in Figure 3.4.1. The spectrum is divided into three bands in the order of ten. There are 800 points in the high frequency band while only 20 points in the low frequency. If the regression is done by using the original or full spectrum, the points in high frequency dominate the fitting. The impact from the low frequency region is very small. Hence, the spectrum is uniformly resampled in the log-scale. For example, ten points are sampled from each band in this case. The regression is done by using the resampled spectrum. It carries equal weights for each frequency band. It is observed in Figure 3.4.1 that the best fit line for the resampled spectrum does shift down in the low frequency band. The best fit line indicates the variation in power distribution and aids in further comparison analysis.

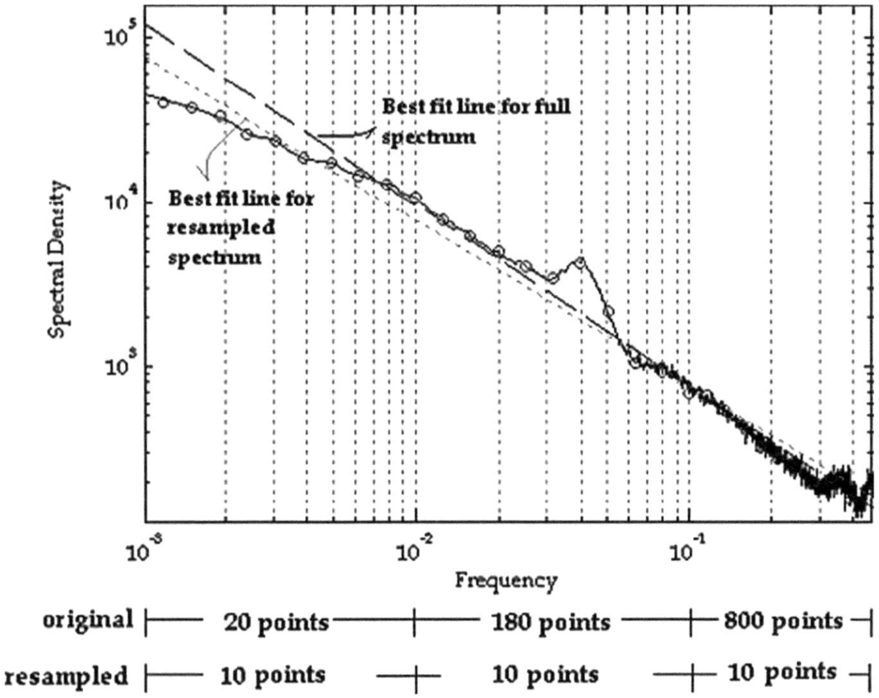

Figure 3.4.1. Illustration of the best fit line

To be consistent, in this study 20 points are taken for each frequency band in the resampled spectrum.

The results for the six examples are shown in Figure 3.4.2. Some points in marginal Hilbert spectra exceed the confidence limits in Figure 3.4.2. They are identified as the significant periodicities in these signals. These significant periodicities are summarized in Table 3.4.1. As for the lake temperature data, simulation of the nonstationary process rather than the identification of significant periodicities is of greater interest.

Table 3.4.1. Periodicities detected in rainfall series

Data	Detected Periodicities
HCN 120177 monthly rainfall	1.0, 1.9, 4.3, 12.3, 47.6 (years)
USGS 03276500 monthly streamflow	1.1, 1.8, 3.0, 4.7, 7.1, 30.2 (years)
HCN 120177 monthly temperature	1.0, 7.8, 17.5 (years)
Cadiz monthly temperature	1.0, 1.6, 10.8 (years)
Indianapolis daily peak wind speed	6.5, 23, 44, 174, 363 (days)

HCN120177 Monthly Precipitation Data

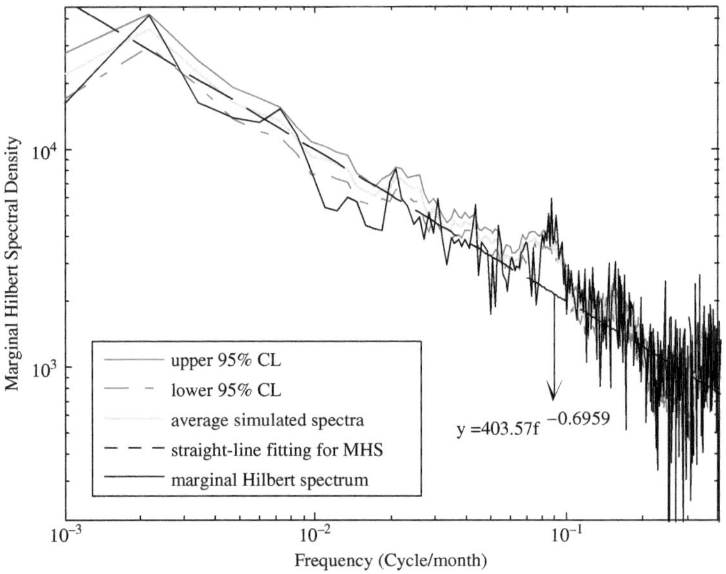

(a) HCN 120177 monthly rainfall data

USGS03276500 Monthly Streamflow Data

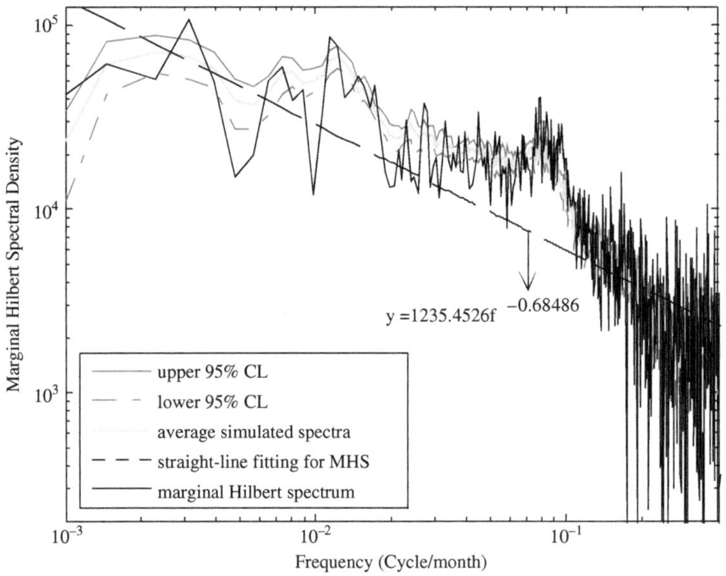

(b) USGS 03276500 monthly streamflow data

Figure 3.4.2. 95% confidence limits for marginal Hilbert spectra

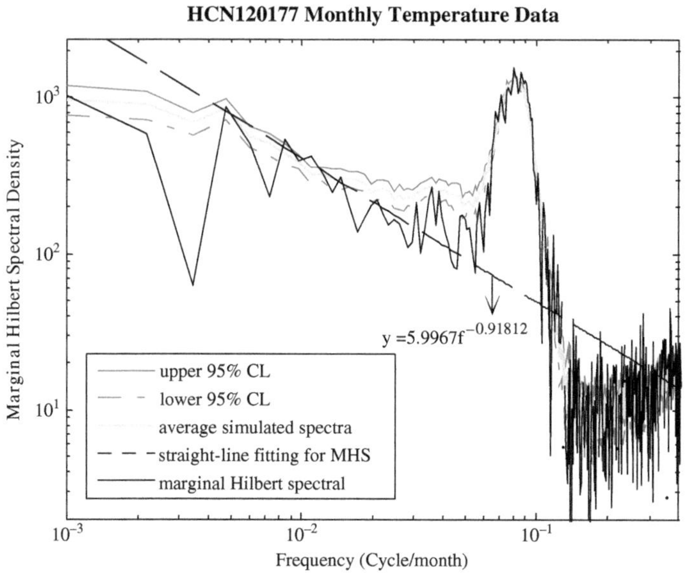

(c) HCN 120177 monthly temperature data

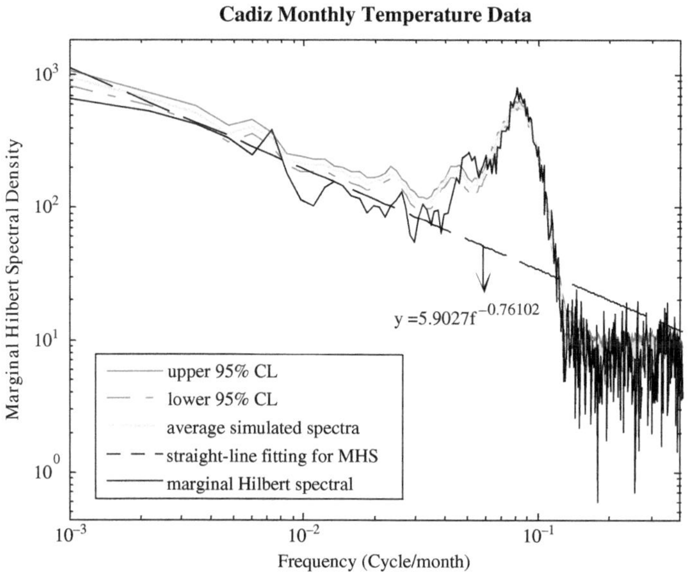

(d) Cadiz monthly temperature data

Figure 3.4.2. (*Continued*)

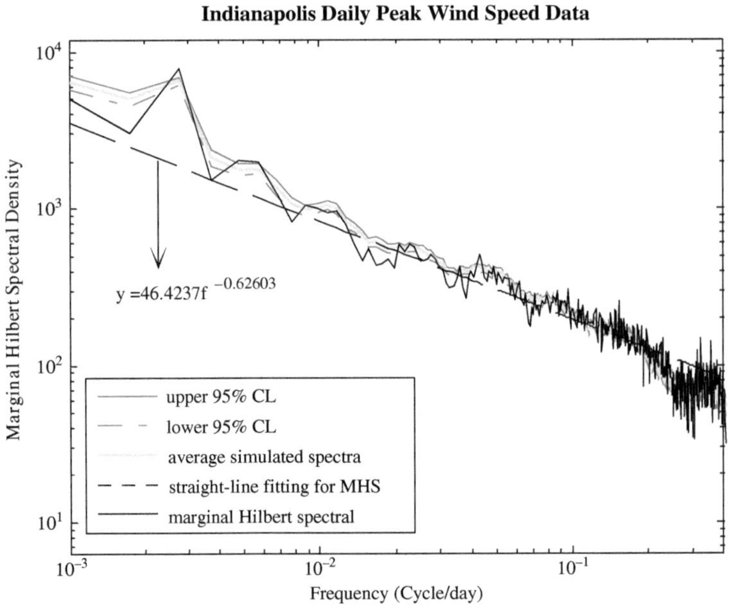

(e) Indianapolis daily peak wind speed data

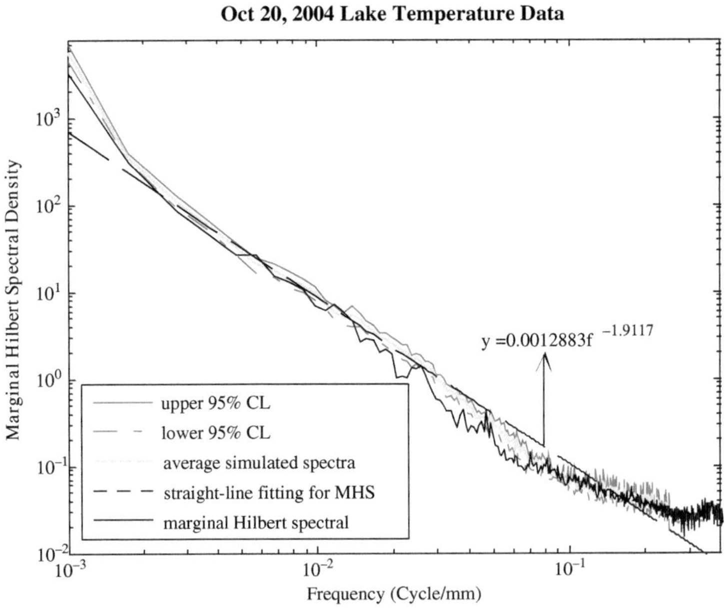

(f) Five simulated series for lake temperature data of date 10/20/04

Figure 3.4.2. (*Continued*)

3.5. CONCLUDING COMMENTS

Several synthetic time series are used to examine the performance of Hilbert-Huang Transform method. For the series with three well-separated frequencies, DFT, MTM and HHT methods are able to identify these frequencies. However, when the frequencies are close, HHT method fails to distinguish them and yields an average frequency instead. This is due to the fluctuation in empirical mode decomposition. The mixing of frequencies makes them not easily distinguishable. For the case of decaying signal, the HHT spectra show their advantage in identification of the fundamental frequency despite the interference of the decaying energy. The synthetic nonstationary signal is another case in which the advantage of HHT technique is obvious. The conventional spectral analysis methods are very sensitive to nonstationary data and they produce spurious harmonic waves in low frequency. Therefore in nonstationary data, the long-term oscillations cannot be accurately interpreted.

The simulation of the autoregressive model raises an important issue in the computation of marginal Hilbert spectra. Following the usual procedure, the marginal Hilbert spectrum is computed by integrating the "entire" time span for "all" intrinsic mode functions. This synthetic case indicates the problem of considering all IMF components. Some of these components may be redundant for estimating the marginal spectra. The issue of selecting IMF components for computing the marginal Hilbert spectra needs further investigation.

Three simulation methods are studied. The method 2, in which the signal is simulated with random phases and random amplitudes, is the least preferred method since the generated signals vary quite a lot from one to another. Methods 1 and 3 yield similar results. However, method 1 is preferred since it has fewer assumptions than method 3 and is easy to implement. The assumption made in method 1 is the uniform distribution of the random phase. This is the basis for computing the confidence intervals. The method of estimation of confidence intervals merits further investigation.

CHAPTER 4

RAINFALL DATA ANALYSIS

4.1. INTRODUCTION AND DATA USED

The nonstationarity and periodicity in the rainfall data in the State of Indiana, U.S.A. are discussed in this chapter. There are two major sources of Indiana rainfall data. One is the data from Historical Climatology Network (HCN) and the other is the data from National Climate Data Center (NCDC). The monthly data are used in this study. Data from nine stations for HCN data and data from nine NCDC subdivisions in Indiana are investigated.

4.1.1 U.S. Historical Climatology Network (U.S. HCN)

The U.S. Historical Climatology Network (U.S. HCN) was compiled in response to the need for accurate, unbiased, modern climate record for climate change research. Department of Energy and the National Climatic Data Center (NCDC) of the National Oceanic and Atmospheric Administration (NOAA) established a network of 1219 stations in the contiguous United States for the specific purpose of compiling a data set suitable for detecting and monitoring climate change over the past two centuries. This network, known as the U.S. Historical Climatology Network (U.S. HCN), and the resulting data set were initially documented by Quinlan et al. (1987) and made available free of charge through the Carbon Dioxide Information Analysis Center (CDIAC), which includes the World Data Center for Atmospheric Trace Gases and is the primary global-change data and information analysis center of the U.S. Department of Energy. The USHCN database contains monthly maximum, minimum, and mean temperature data (degrees F) and rainfall (inches). The monthly rainfall for stations in the state of Indiana is investigated here. Data from these stations are used in the study by Hamed and Rao (1998). The locations of these stations are shown in Figure 4.1.1 and they are listed in Table 4.1.1. The annual average time series of HCN rainfall are plotted in Figure 4.1.2.

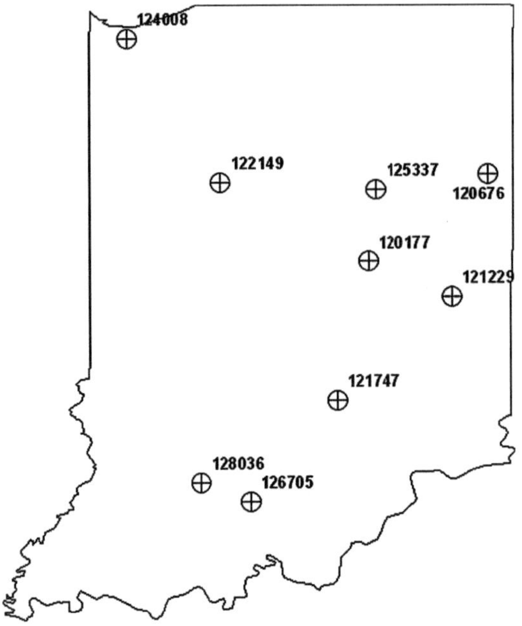

Figure 4.1.1. Location of the HCN stations in the state of Indiana, U.S.A.

Table 4.1.1. HCN stations for Indiana rainfall time series

No.	State	Station No.	Station Name	Record Year	Length
1	IN	120177–05	Anderson Sewage Plant	1895–2002	108
2		120676–03	Berne	1916-2002	87
3		121229–06	Cambridge City	1892–2002	111
4		121747–05	Columbus	1884–2002	119
5		122149–02	Delphi 3NNE	1885–2002	118
6		124008–01	Hobart 2WNW	1919–2002	84
7		125337–05	Marion 2N	1886–2002	117
8		126705–08	Paoli	1898–2002	105
9		128036–07	Shoals Highway 50 Bridge	1911–2002	92

4.1.2 NCDC Average Divisional Rainfall Data

The statewide values are available for the 48 contiguous States and are computed from the divisional values weighted by area. Monthly averages within a climatic division have been calculated by giving equal weight to stations reporting both temperature and rainfall within a division. In the U.S., observers at cooperative stations often take one observation per day, and the ending time of the climatological day at any station can vary from station-to-station as well as year-to-year. Details about discussion of data adjustment and bias correction are found in

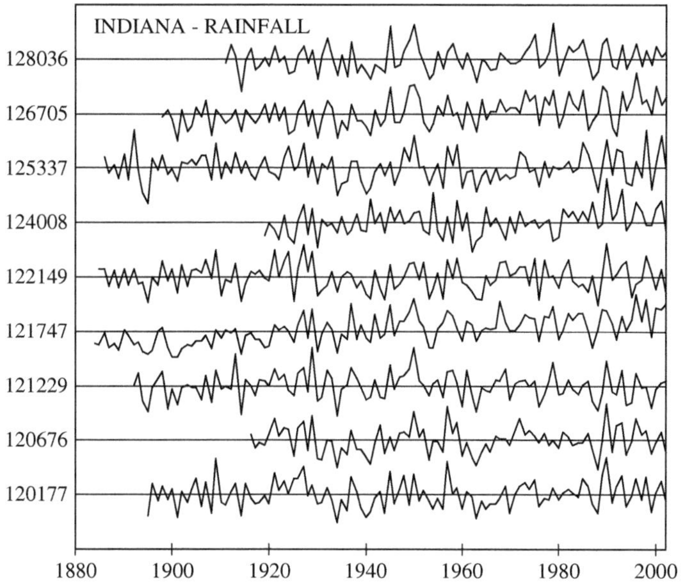

Figure 4.1.2. HCN monthly rainfall time series (annually averaged)

Karl et al. (1986). The data reported are sequential statewide, regional, and national monthly rainfall and monthly average temperature. The period of record is 1895 through the latest month available. The file is available online and is updated monthly. The data in this file are used for historical perspectives in the Climate Variation Bulletin (Historical Climatology Series 4–7). The borders of each subdivision in state of Indiana are illustrated in Figure 4.1.3.

Figure 4.1.3. Subdivisions for NCDC rainfall data in Indiana, U.S.A.

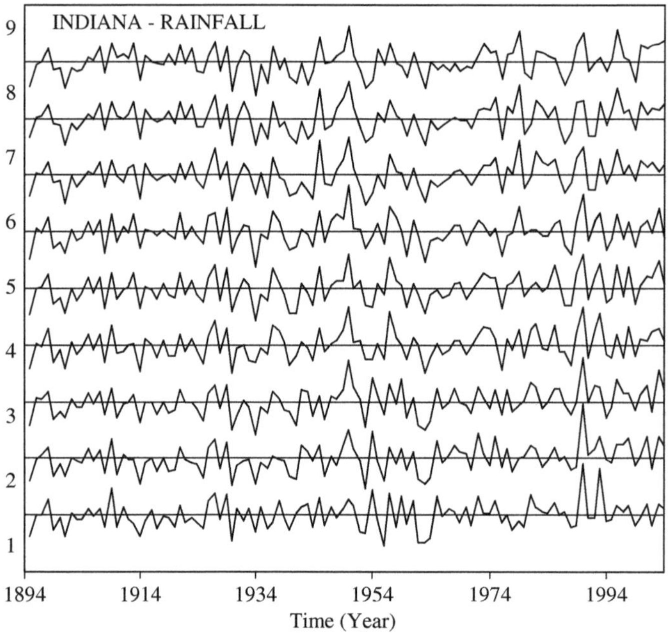

Figure 4.1.4. NCDC annual average rainfall time series

The annual average rainfall time series are plotted in Figure 4.1.4. The monthly data are used in this analysis. Since the NCDC data are obtained as regional averages, it has more stable features and smaller variation from region to region than HCN data. HCN data are site dependent and are influenced by local conditions. This effect can be seen by comparing Figure 4.1.2 and Figure 4.1.4.

4.2. HCN RAINFALL DATA

4.2.1 Long-Term Oscillations

The intrinsic mode functions of the HCN rainfall data are shown in Figure 4.2.1. Starting from the top of each figure, the original data, the IMF components from short to long period (c_1, c_2, ...) and the residual (r) are plotted. The highest order IMF for each station shows a very long term oscillation. In the time domain, the modified Mann-Kendall method is applied to the time series based on the segmentation of the last IMF component. The results of the trend test are summarized in Table 4.2.1. The trends are investigated by the modified Mann-Kendall tests. The segmentation of the original time series is based on the turning points in the last IMF component. The result from modified Mann-Kendall test is of most interest since it indicates whether the trend is upward or downward.

(a)

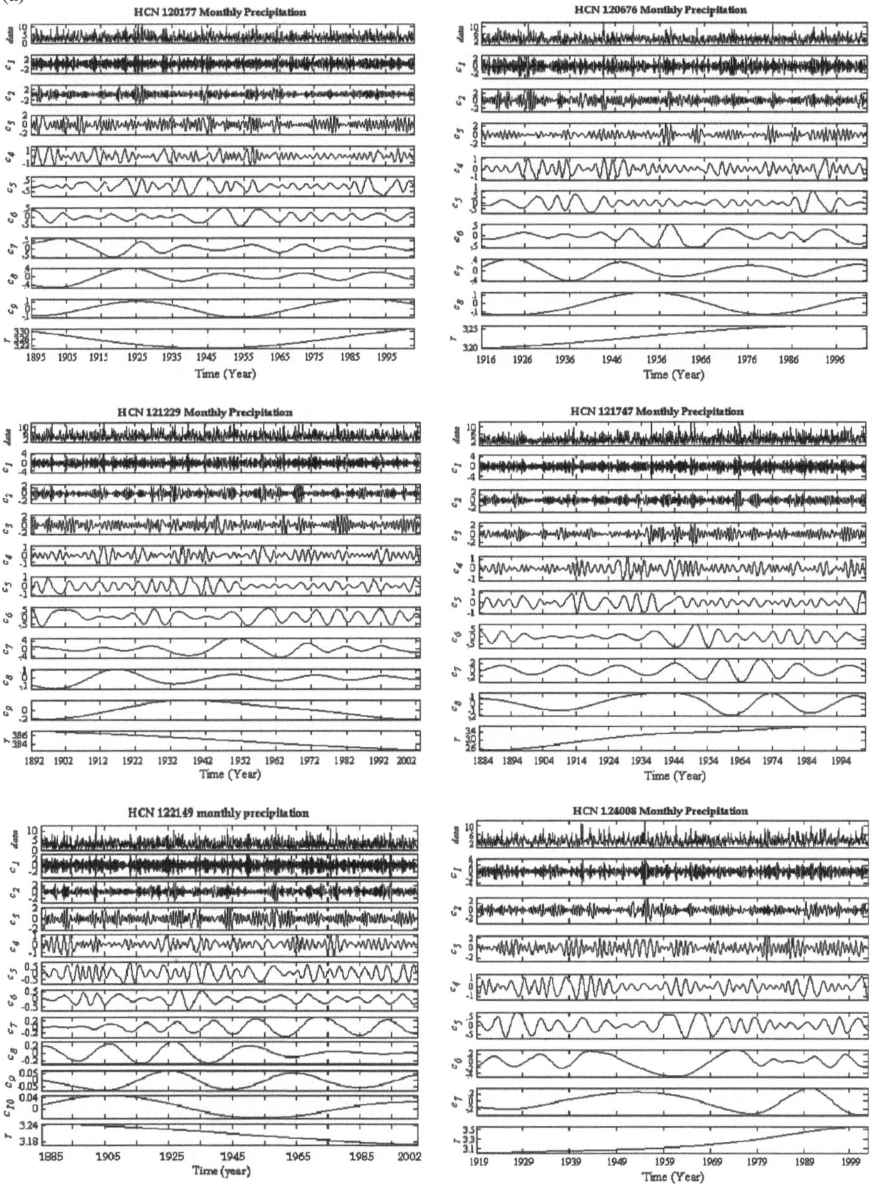

Figure 4.2.1. Intrinsic mode function components of HCN rainfall data

(b)

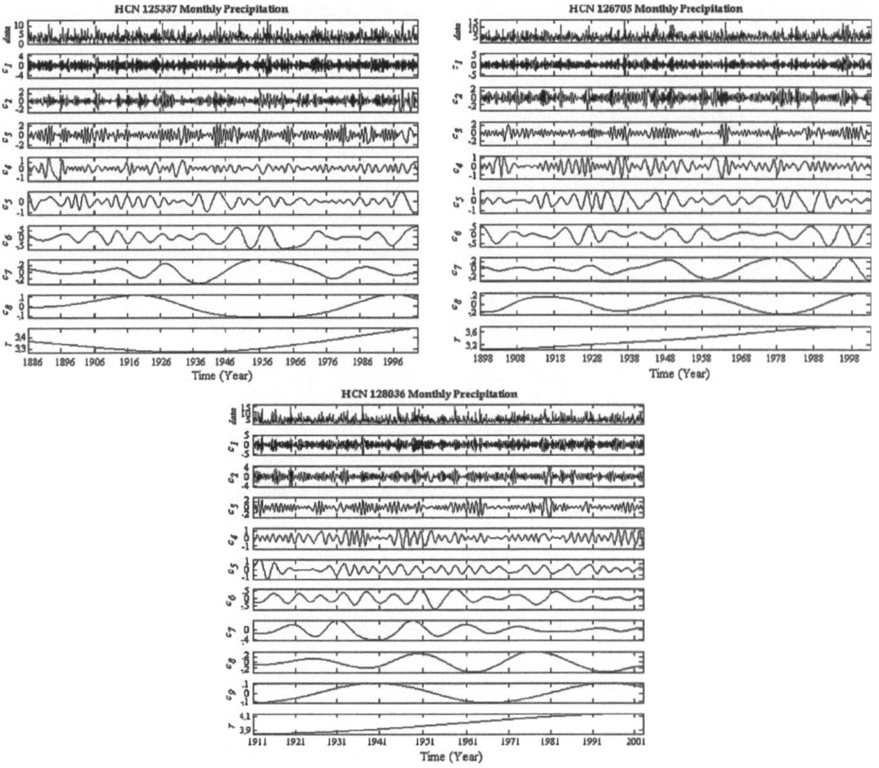

Figure 4.2.1. (*Continued*)

To compare the results from Mann-Kendall test and the actual slope of each segment, the segments of the last IMF components are plotted in Figure 4.2.2 along with the positive and negative signs from Mann-Kendall test (i.e. z value). The beginning and end of the last segments are not as stable as others especially when the length of segment is not long. Hence, they could be ignored. This is because of end effects in empirical mode decomposition. Investigating the trend and the segment slope, the ratio of consistency to inconsistency is 31:9. Consequently there is about 78% consistency in these results.

For the HCN monthly rainfall data, the empirical mode decomposition takes from seven to ten IMF components to completely reconstruct the signal. The first two IMF components are likely to capture much of the monthly variations while the annual and biannual periods are seen in the third and fourth IMFs. The higher order IMFs contain information about oscillations which vary from years to decades. The residual is obtained by subtracting all these zero-mean IMFs from original data. It represents the trend in the entire time series.

Table 4.2.1. Trend test results for HCN monthly rainfall time series

Station	Segment	Modified Mann-Kendall
120177	1	Stationary (z = 0.406)
	2	Stationary (z = −0.011)
	3	Stationary (z = 0.344)
	4	Stationary (z = 1.409)
120676	1	Stationary (z = 1.621)
	2	Stationary (z = 1.058)
	3	Stationary (z = 1.076)
	4	Stationary (z = 0.458)
121229	1	Stationary (z = −1.589)
	2	Stationary (z = 1.400)
	3	Stationary (z = −1.604)
121747	1	Stationary (z = −1.012)
	2	Stationary (z = 0.431)
	3	Stationary (z = −0.362)
	4	Upward trend (z = 2.017)
	5	Stationary (z = 0.083)
	6	Upward trend (z = 2.058)
	7	Upward trend (z = 2.157)
122149	1	Stationary (z = 0.552)
	2	Stationary (z = −1.408)
	3	Upward trend (z = 1.834)
124008	1	Stationary (z = −1.314)
	2	Stationary (z = 1.086)
	3	Stationary (z = 0.968)
	4	Stationary (z = 1.468)
	5	Stationary (z = −0.237)
125337	1	Stationary (z = 0.380)
	2	Stationary (z = −0.408)
	3	Upward trend (z = 1.834)
	4	Stationary (z = 0.549)
126705	1	Stationary (z = 0.763)
	2	Stationary (z = 0.434)
	3	Stationary (z = −0.904)
	4	Upward trend (z = 1.875)
	5	Stationary (z = 1.629)
	6	Stationary (z = 1.484)
128036	1	Stationary (z = 1.188)
	2	Stationary (z = 0.133)
	3	Stationary (z = 0.588)
	4	Stationary (z = 0.736)

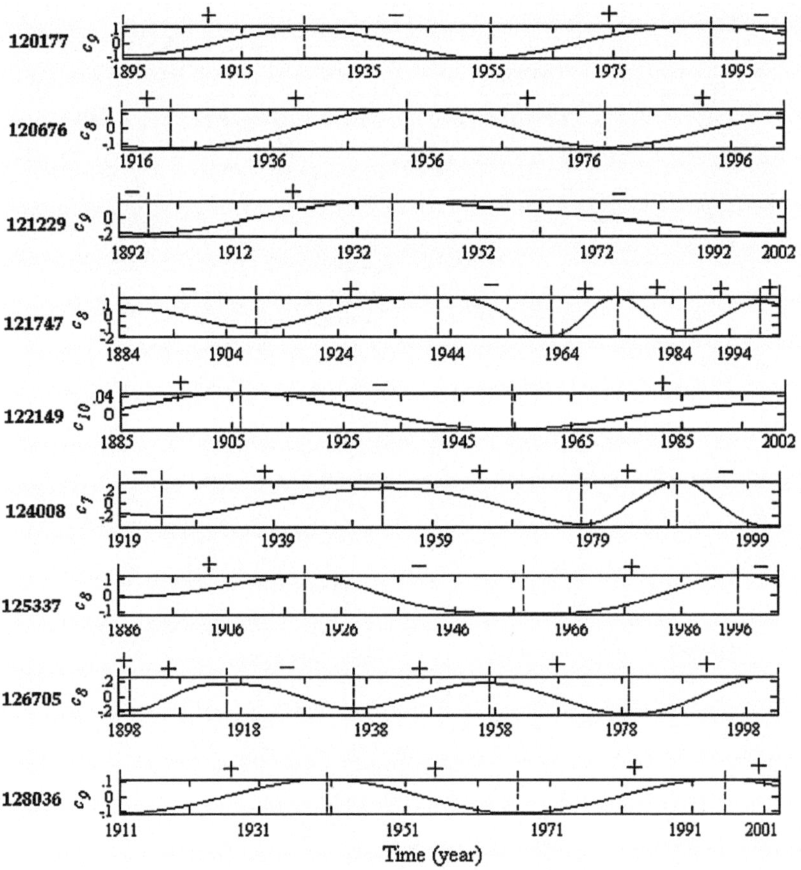

Figure 4.2.2. The last IMF component of HCN monthly rainfall and the results of Mann-Kendall test for each segment

To study the overall trend, a straight line is fitted to the entire data. The slopes obtained for these series are 0.0001 (No.120177), 0.0001 (No.120676), –0.00005 (No.121229), 0.0009 (No.121747), –0.00003 (No.122149), 0.0005 (No.124008), 0.00004 (No.125337), 0.0007 (No.126705) and 0.0004 (No.128036). The results are exactly consistent with those in the residuals in Figure 4.2.1. The residuals show a negative trend in station No.121229 and No.122149 and positive trends in others. All residuals are monotonic except for No.120177. Although the residual in No.120177 changes its value from positive to zero and to being positive at the end, it has an overall positive trend.

4.2.2 Time-Frequency Distribution

Once the intrinsic mode functions are obtained, the Hilbert transform is applied to all the components. Thus, the Hilbert spectrum which provides the information of time-frequency distribution is obtained. The Hilbert spectra of HCN monthly rainfall data are shown in Figure 4.2.3. Along the time axis, apart from the Hilbert spectrum, the original data, volatility and instantaneous energy (IE) are plotted. In the instantaneous energy plot, bold gray line indicates the average energy at the corresponding time while the dashed black line refers to the standard deviation of the instantaneous energy. It is of interest to observe the times with high standard deviation lines to investigate the relationships among the results in these plots (data, volatility, IE and Hilbert spectrum). The dark shaded areas of Hilbert spectrum indicate regions where the signal is quite strong. Along the frequency axis, the degree of stationarity (DS) corresponding to each frequency is shown. The smaller degree of stationarity indicates that it is more stationary at that frequency. The relatively low degree of stationarity is observed in the low frequency part and there is a dip-down in the frequency of 0.07–0.09 cycle/month, which corresponds to the annual cycle.

In Figure 4.2.3, there are transparently shaded boxes (⬚) added in each plot. They indicate some strong peaks in the time series. In order to investigate the details, these boxes are zoomed in and the results are as shown in Figure 4.2.4. On the top of these segments, they are marked as ①②③④, which corresponds to the segments in Figure 4.2.3 from left to right. For these highlighted time segments, the corresponding volatility, instantaneous energy, standard deviation of instantaneous energy, and Hilbert spectrum can be easily compared. For example, in station 12-0177, there are higher rainfall values around the year 1925. The volatility as well as the standard deviation of instantaneous energy is also fairly high around the year 1925. However, the instantaneous energy does not vary much in time.

From these segments in Figure 4.2.4, it is seen that the peaks or dips of the volatility do not exactly correspond to the peaks or dips in the data. There are lags between the volatility and data. Volatility does not have strong variation while there is a strong variation in the time series, such as the segment ② in Station 120676. The reason for this situation may be that the volatility is computed by selected IMF components. The IMF containing that peak may have dropped. Such results make interpretation of volatility results problematic. However, it still provides a picture of where these unusual variations are.

The instantaneous energy (bold gray line) is too flat to indicate any variation, but the standard deviation of instantaneous energy (dashed black line) indicates the variations well. It appears to indicate the time of variation better than volatility. Volatility is obtained from intrinsic mode functions while the instantaneous energy and degree of stationarity are computed from Hilbert spectrum. Hence, the Hilbert spectrum helps us to see the energy dispersion or concentration with respect to time and frequency. When the signal is larger, the amplitude of Hilbert spectrum is also larger (the shades in Figure 4.2.3 becomes darker). The stronger spectral amplitude concentrated in low frequency and weaker amplitude in high frequency increases the variance of instantaneous energy.

It is also of interest to study the correlation among these series from different locations in the state of Indiana. The volatility and the instantaneous energy are summarized in Figure 4.2.5a, b, respectively. There are several peaks observable

(a)

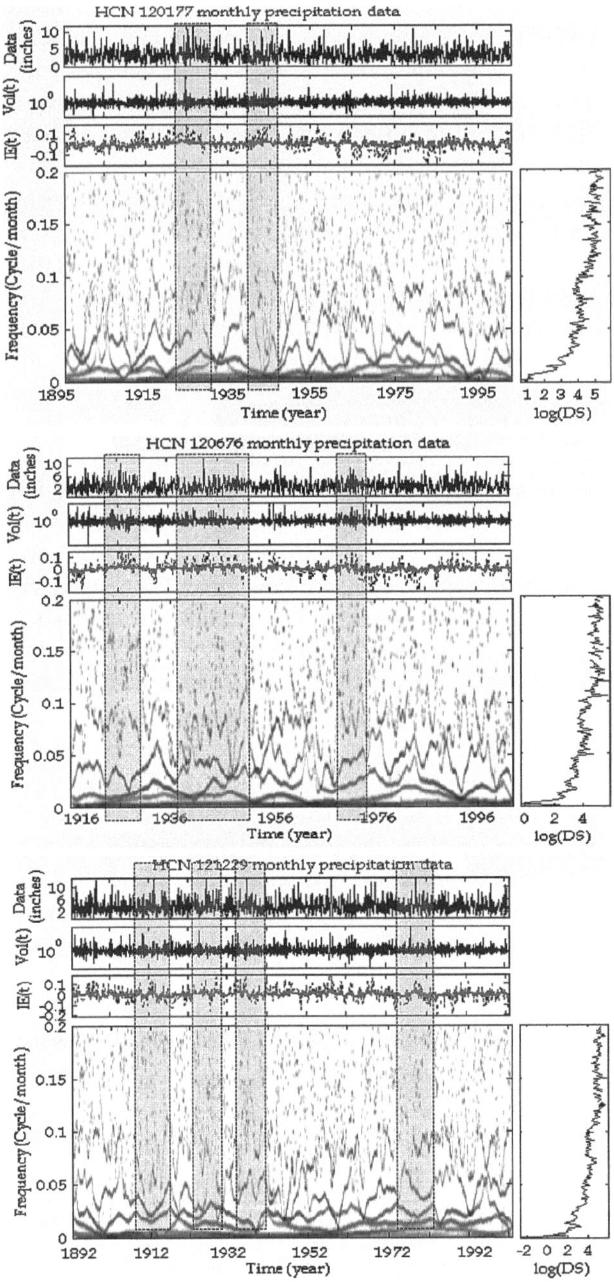

Figure 4.2.3. Time-frequency distributions of HCN rainfall data

(b)

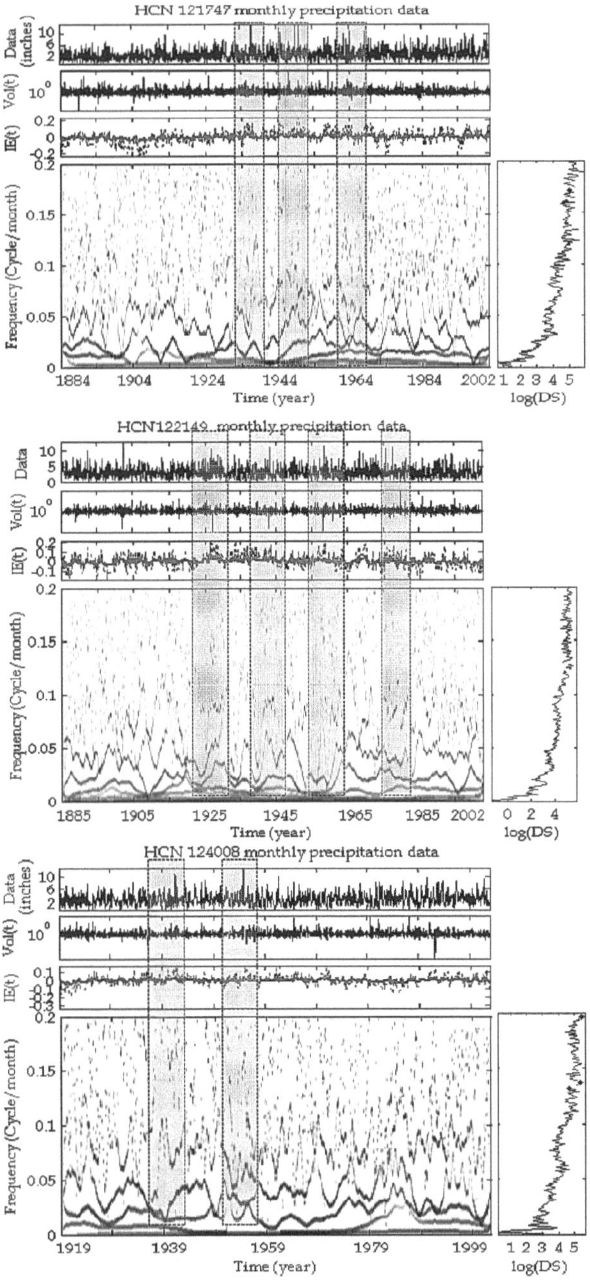

Figure 4.2.3. (*Continued*)

(c)

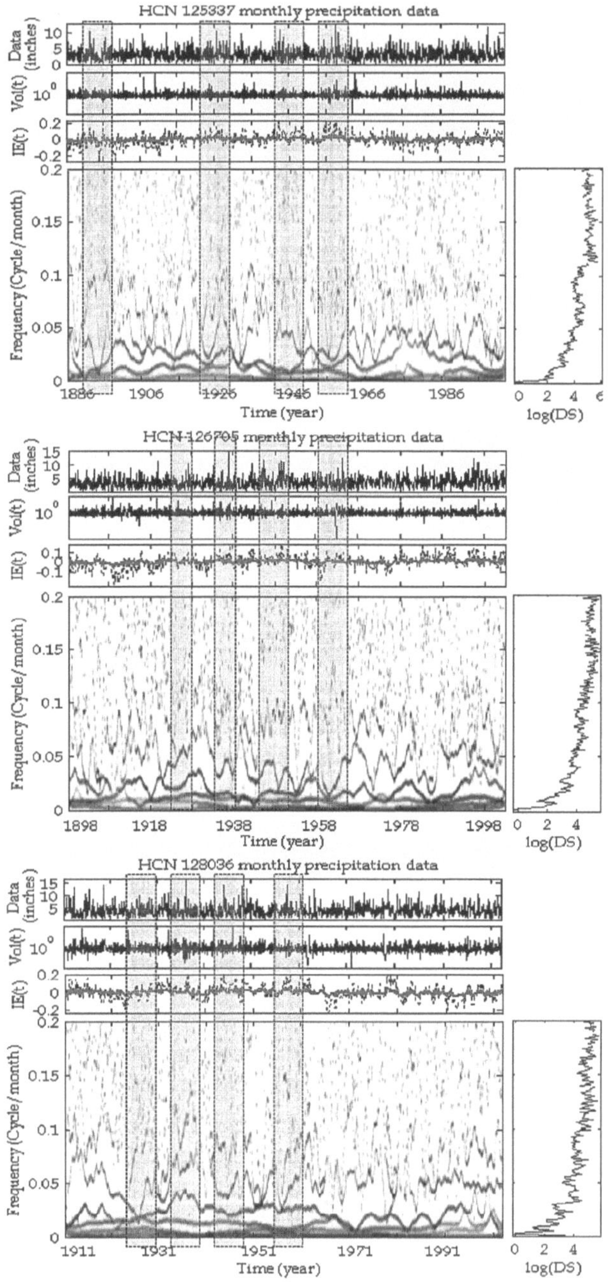

Figure 4.2.3. (*Continued*)

(a)

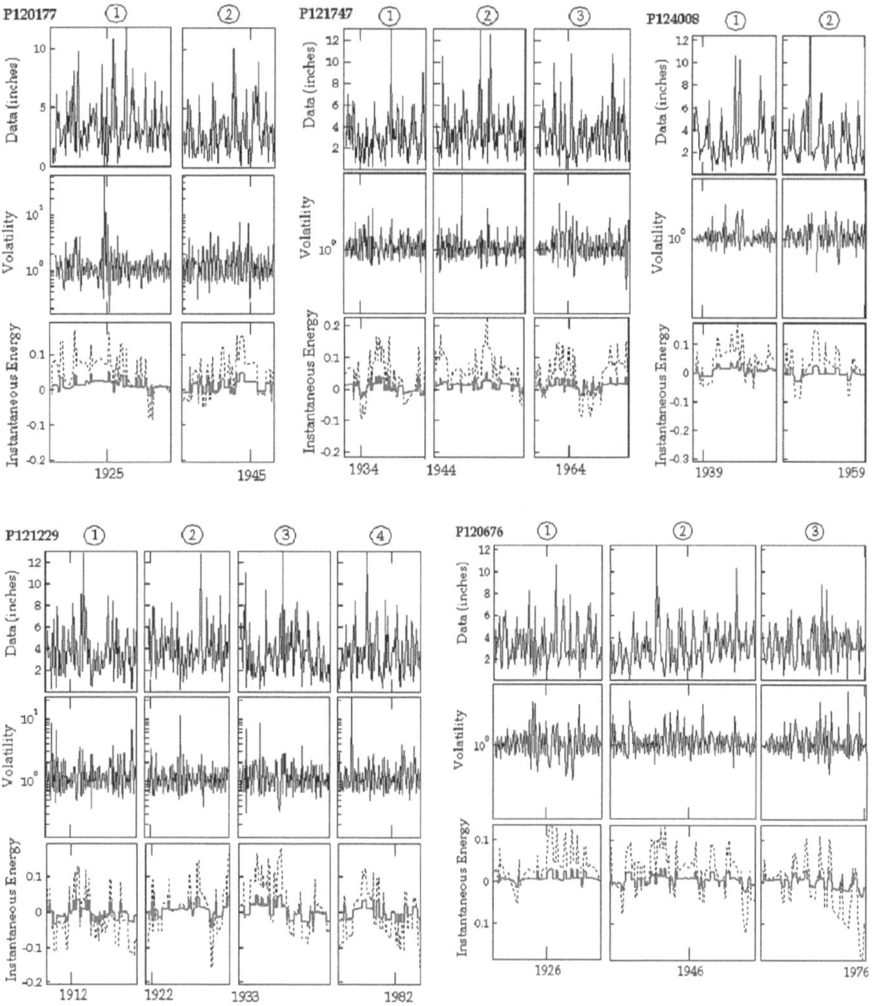

Figure 4.2.4. Time segments of HCN rainfall data

from the volatility and different stations yield similar peaks. Six peaks in year 1907, 1925, 1935, 1947, 1964 and 1980 are used as example. The results are shown in Table 4.2.2. If large variation of volatility is observed in a station, a circle mark ("O") is used to indicate it. Otherwise, a cross mark ("X") is used. If the data are not available in that year, a mark ("–") is used. For these 6 years, more than half of stations yield as high as variation of volatility as the others. This variability, however, is not easily seen in the variation in instantaneous energy (Figure 4.2.5b), which does not vary much. The standard deviation of IE offers a better indicator

(b)

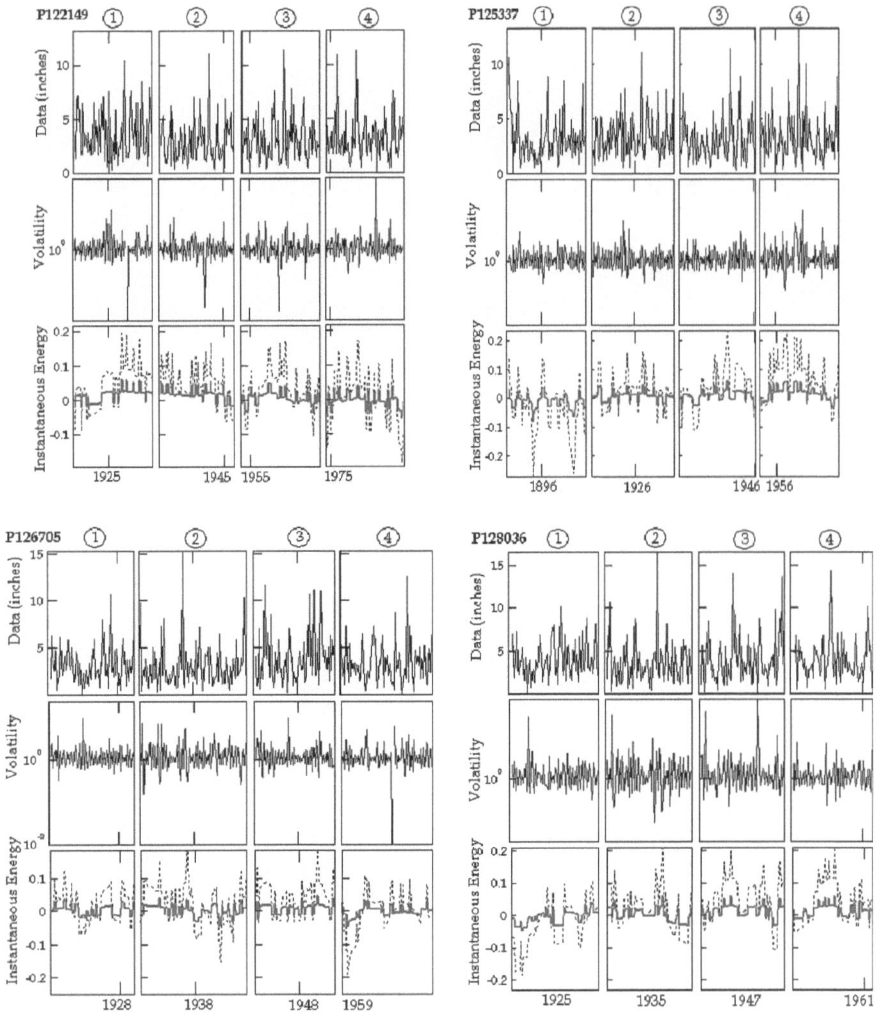

Figure 4.2.4. (*Continued*)

than the average IE. However, over the entire time span of the data, too many variations in IE make the results difficult to interpret.

For the purpose of further comparison to NCDC data, the HCN data are assigned to the NCDC subdivisions to which they belong. From subdivision 1 to 9, the corresponding HCN stations are 124008 (1), 125337 (2), 120676 (3), 122149 (4), 120177 (5), 121229 (6), 128036 (7), 126705 (8) and 121747 (9). To make it clear to associate with the spatial distribution, a shadow is marked if the peak in data is detected. This is

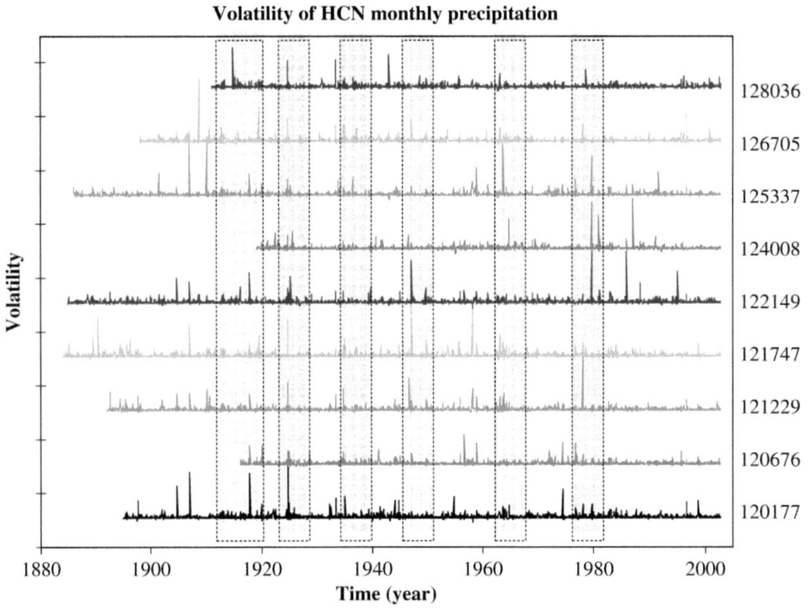

(a)

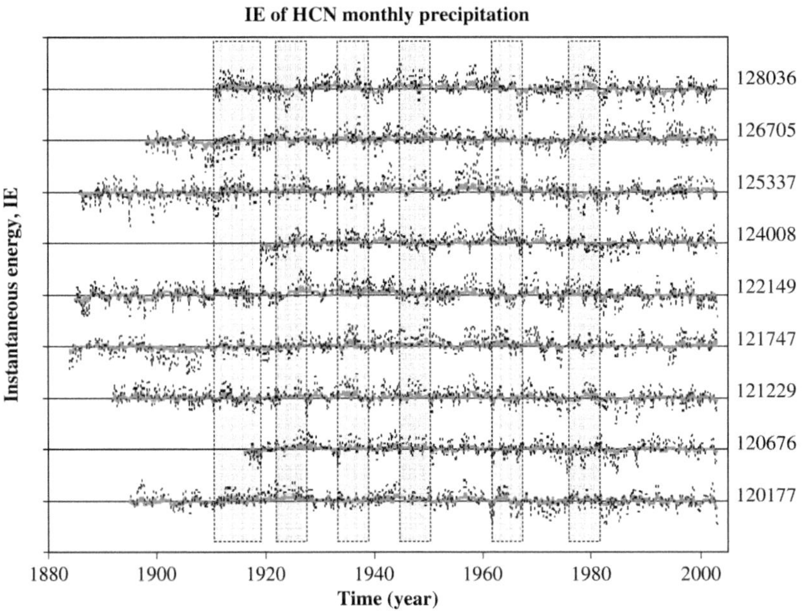

(b)

Figure 4.2.5. Volatility and instantaneous energy of HCN rainfall data (a) (b) (-: IE; —: standard deviation of Hilbert spectrum)

Table 4.2.2. Consistent volatilities in HCN monthly rainfall time series

year	120177	120676	121229	121747	122149	124008	125337	126705	128036
1980	O	O	O	X	O	O	O	X	O
1964	O	X	O	O	X	O	O	X	O
1947	O	X	O	O	O	X	X	O	O
1935	O	O	O	O	O	–	O	O	O
1925	O	O	O	O	O	O	O	O	O
1907	O	–	O	O	O	–	O	O	–

"O": a relatively high variation in that year; "X": absence of high variation in that year

shown in Figure 4.2.6. The NCDC subdivision is marked as shadow if the volatility of the HCN station in that region is significant and is left blank otherwise.

Although a HCN station is assigned to a NCDC subdivision as shown in Figure 4.2.6, it helps to investigate the spatial continuity of the volatility. For year 1925, the consistency is well observed in the study region. Years of 1907, 1935, 1947 and 1980 have spatial continuity but the result in 1964 is not continuous but scattered. Later the NCDC data are studied and the results from these years are investigated. The results in this figure are used for further comparison.

4.2.3 Frequency Domain Analysis

The marginal Hilbert spectra for the nine stations are shown in Figure 4.2.7. The 95% confidence intervals computed from the procedure mentioned in Chapter 3 are shown as dash-dot lines. The multi-taper spectrum is plotted for comparison. The best fit line for the resampled spectrum is shown in these plots. The equation and R-square value for the best fit line are also shown in the figure. The significant peak detected based on the 95% confidence intervals are pointed out by arrows and the corresponding periods. As the interest in long-term oscillation is higher, periodicities longer than the annual cycle are of interest. Annual cycle (around 12 months) is clearly detected in all stations. Other periods which occur are around 24, 50, 100 months and others. However, these are not consistent. These periods are summarized in Table 4.2.3 for comparison.

From Table 4.2.3, there is a 1 year, 2 year and an approximately 4 year and a 10–12 year period is present in these data. The annual cycle is not precisely detected

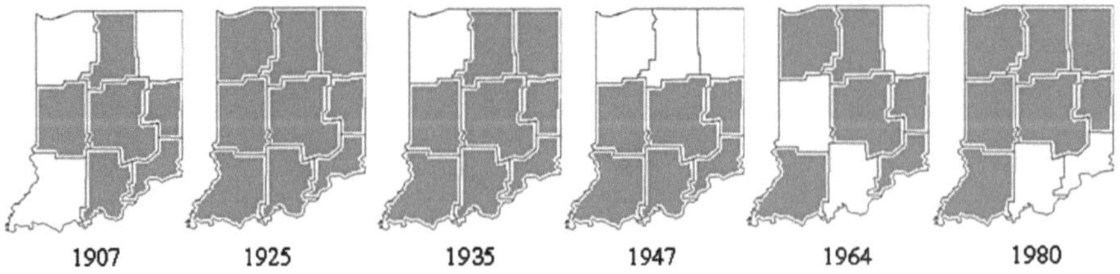

Figure 4.2.6. Significant volatilities detected for HCN rainfall data

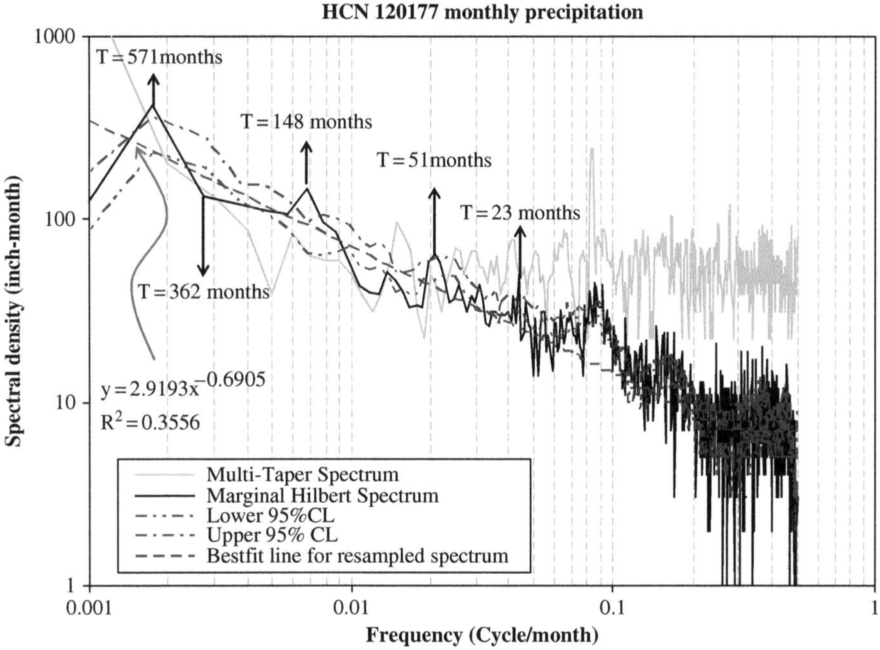

Figure 4.2.7. Marginal Hilbert spectrum and its confidence limits for HCN monthly rainfall data

Table 4.2.3. Detected periodicities in HCN monthly rainfall time series (Unit: years; the bracket after the station no. is the assigned NCDC subdivision)

No.\ Period	1 year		2 years		3~7 years			10 years	> 15 years
120177 (5)	0.9	1.0	1.9		4.3			12.3	47.6
120676 (3)	1.0	1.1	1.9	2.3					30.2
121229 (6)	0.9	1.0	1.8		3.1	3.7	6.5		
121747 (9)	1.0		1.7		3.3	5.3	6.5	12.3	
122149 (4)	1.1		1.8						
124008 (1)	1.0	1.1	1.9		3.7				
125337 (2)	0.9	1.1	2.3		3.3	3.7		8.5	
126705 (8)	0.8	1.1	2.5		3.1	5.0	6.5	9.5	
128036 (7)	0.9	1.0	2.1		3.1	3.8		10.8	

by the marginal Hilbert spectrum. It is well located by multi-taper method. It is also seen that multi-taper spectra have more energy in high frequency region than marginal Hilbert spectra. The 2 year period is detected in all HCN stations. For the 3~7 year oscillation, the 3, 4, 5, 6 and 7 year periods are detected but the values vary from site to site. The 10 year period is detected in data from 120177, 121747, 125337, 126705 and 128036. A 48 year period is detected in data from 120177 and a 30 year period is detected in data from 120676.

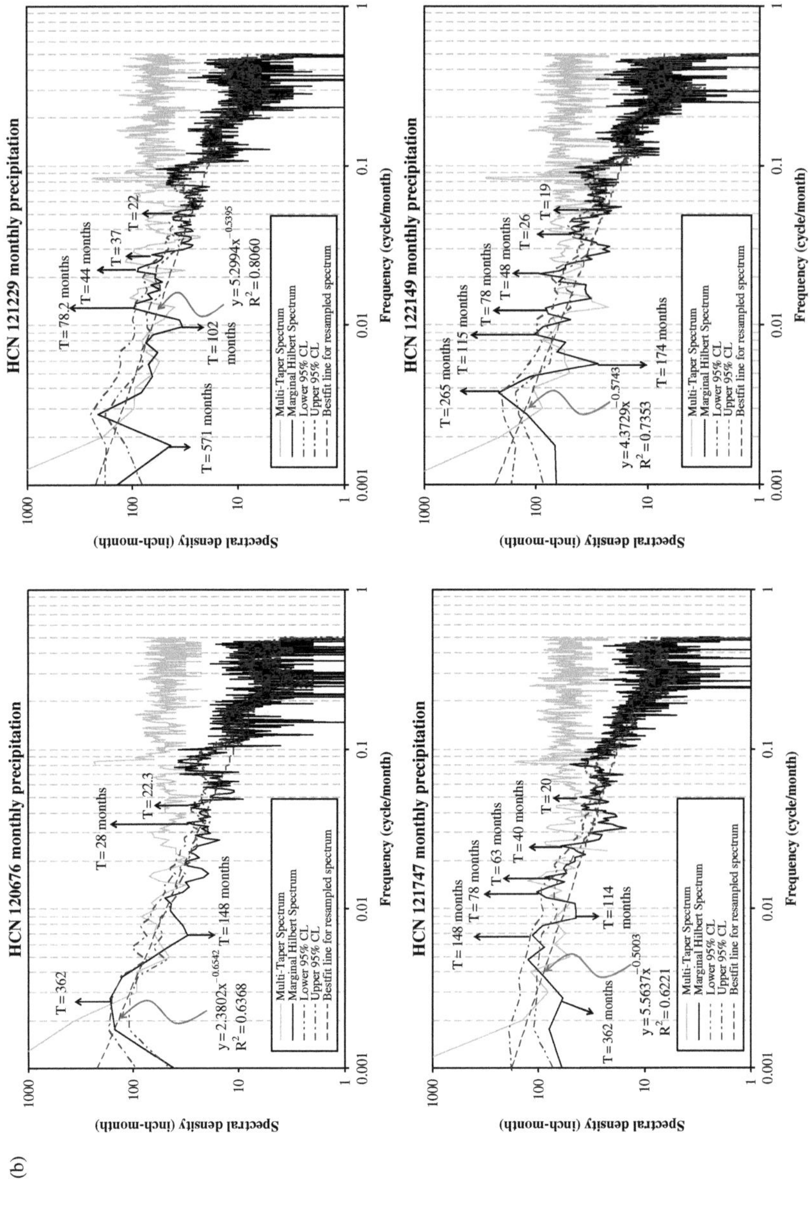

Figure 4.2.7. (Continued)

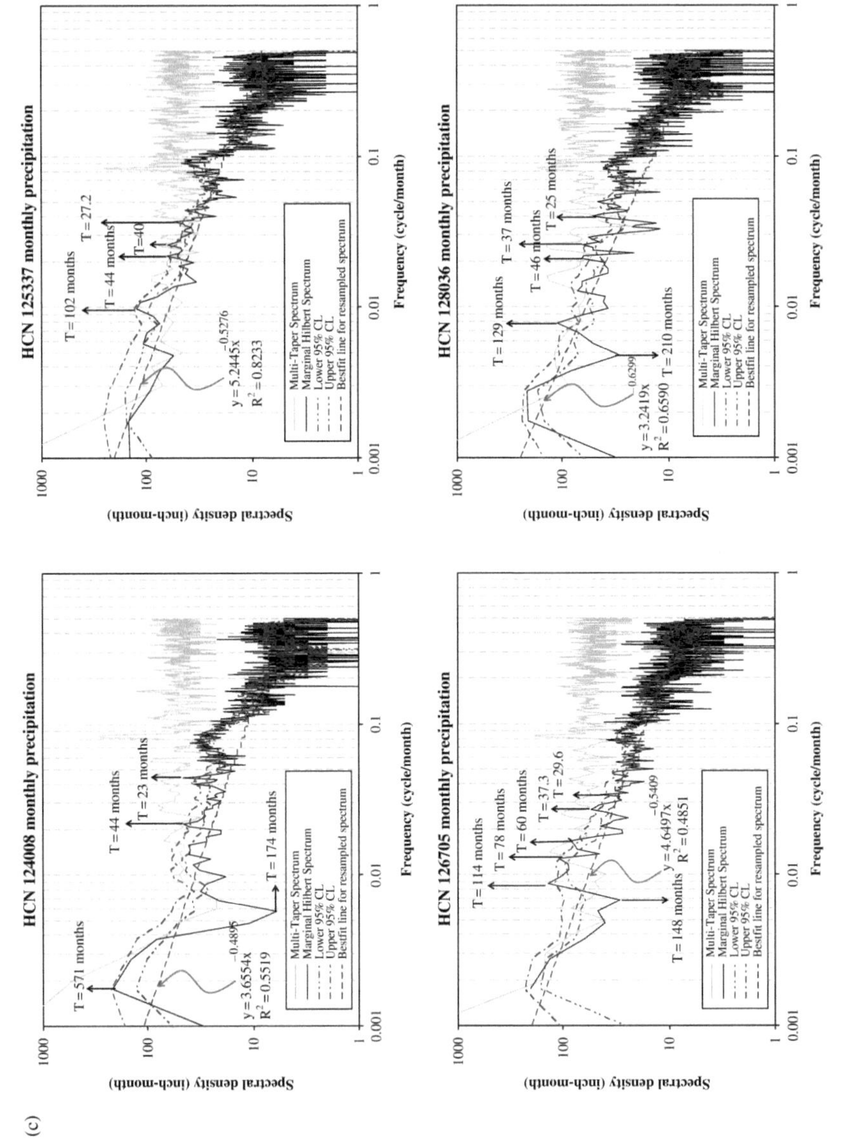

Figure 4.2.7. (Continued)

The degrees of stationarity (DS) for the nine HCN stations are plotted in Figure 4.2.8. The amplitudes of the degree of stationarity are quite close and almost overlay with each other. Theoretically the smaller DS means more stationary at that frequency. Hence, it is more stationary in low frequency region than in high frequency region. A dip-down in the degree of stationarity around the frequency 0.083 cycles/month indicates that the persistent annual cycle makes the spectrum more stationary at that frequency than in the adjacent frequencies.

The marginal Hilbert spectrum is fitted by the equation $y = af^b$, where y is the marginal Hilbert spectrum and f is the frequency. The parameter a, power law decay rates b and R-square value for the best fit line are summarized in Table 4.2.4. The regression is performed by considering both the entire spectra and the low frequency segment (frequency from 0.003 to 0.09 cycle/month). Whether entire spectrum or low frequency segment are fitted, they are resampled in the log-scale axis. For the entire spectrum, the power law decay rate is fairly steady around −0.57 with standard deviation of 0.07. The coefficient of variation is 0.12. For the low frequency resampled spectrum, the decay rate is around −0.37 with standard deviation of 0.19. It makes the coefficient of variation as high as 0.51. There is considerable variation in spectra in low frequency as seen from Figure 4.2.7. The monthly rainfall data does not yield a clear feature of power law because there is considerable leakage in the spectrum. Fitting the entire data is more acceptable if the power law is used.

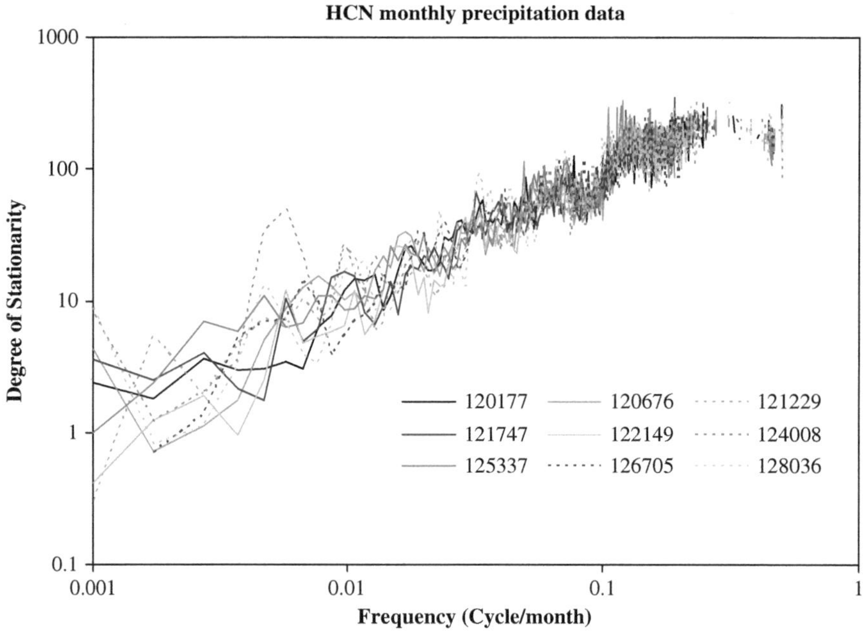

Figure 4.2.8. Degree of stationarity for HCN monthly rainfall data

Table 4.2.4. Parameters of the best fit lines to the marginal Hilbert spectra of HCN monthly rainfall time series

	Entire resampled spectrum			Low frequency resampled spectrum		
Station No	a	b	R^2	a	b	R^2
12-0177	2.919	−0.691	0.356	4.692	−0.584	0.800
12-0676	2.380	−0.654	0.637	5.797	−0.417	0.613
12-1229	5.299	−0.540	0.806	12.257	−0.340	0.539
12-1747	5.564	−0.500	0.622	5.807	−0.536	0.730
12-2149	4.373	−0.574	0.735	6.574	−0.501	0.513
12-4008	3.655	−0.490	0.552	31.345	0.068	0.016
12-5337	5.245	−0.528	0.823	8.016	−0.448	0.622
12-6705	4.650	−0.541	0.485	10.913	−0.332	0.350
12-8036	3.242	−0.630	0.659	12.080	−0.296	0.317
Mean	4.147	−0.572	0.631	10.831	−0.376	0.500
Std. Dev.	1.147	0.071	0.151	8.206	0.193	0.240

($y = af^b$, y is the marginal Hilbert spectrum; f is the frequency)

4.3. NCDC RAINFALL DATA

4.3.1 Long-Term Oscillations

The intrinsic mode functions of the NCDC monthly rainfall data are shown in Figure 4.3.1. There are nine subdivisions in the state of Indiana. They are identified as 01–09 by their location of from west to east and from north to south. The modified Mann-Kendall method is applied to the time series based on the segmentation of the last IMF component. The results are summarized in Table 4.3.1. The z values computed from Mann-Kendall test are also indicated in Table 4.3.1.

Unlike the high variability in HCN rainfall data, NCDC rainfall data have similar IMF components among the nine divisions. For example, all the residuals show an upward trend while the residuals of HCN data are not consistent at all. Also, the empirical mode decomposition yields similar lengths of oscillations. The segments of the last IMF components are plotted in Figure 4.3.2 with the positive/negative signs obtained from modified Mann-Kendall test. For NCDC monthly rainfall, the number of consistent results is 26 out of 42. This corresponds to a 62% consistency. This is smaller number than in HCN data. This may be due to the small amplitude of the last IMF component compared to the original data; hence it cannot accurately represent the trend in the time series.

4.3.2 Time-Frequency Distribution

The Hilbert spectrum of NCDC monthly rainfall data is shown in Figure 4.3.3. Along the time axis, apart from the Hilbert spectrum, the original data, volatility and instantaneous energy (IE) are plotted. For the instantaneous energy, the bold gray line indicates the average energy at the corresponding time while the dashed

(a)

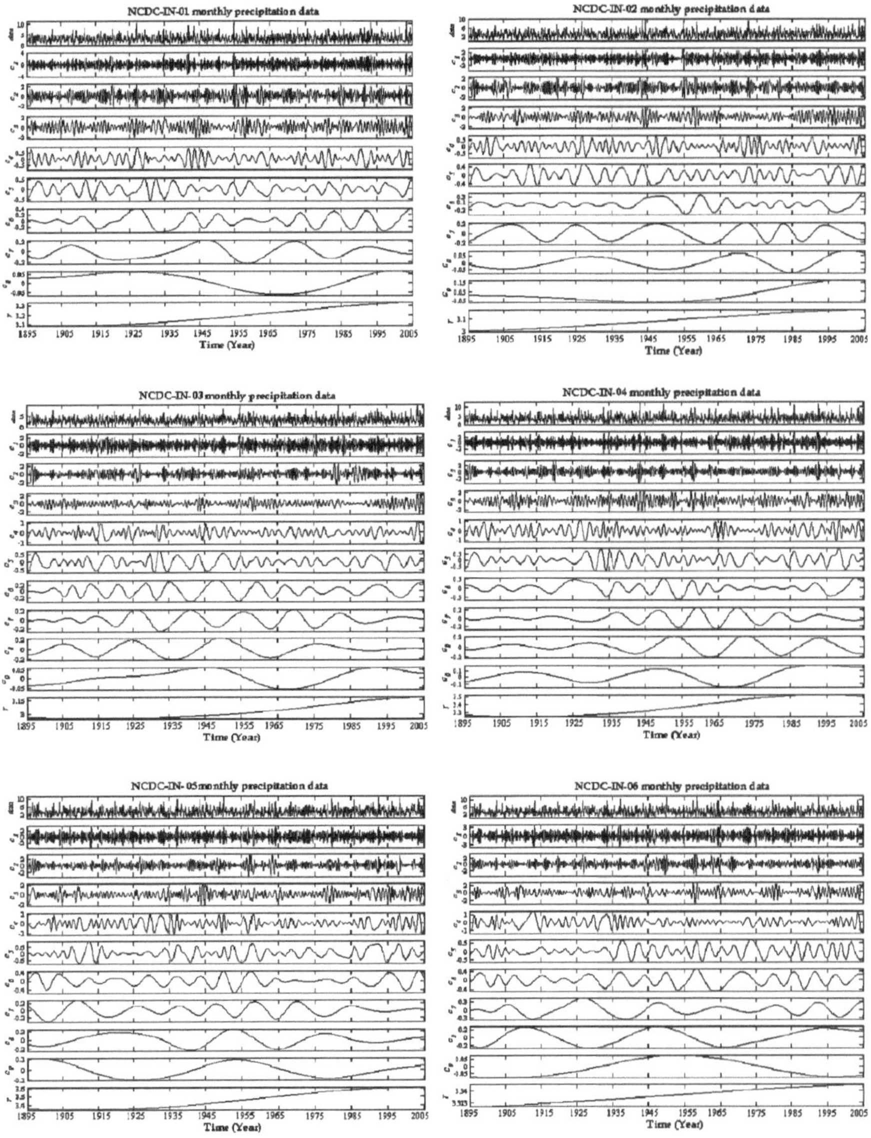

Figure 4.3.1. Intrinsic mode functions of Indiana NCDC monthly rainfall data

(b)

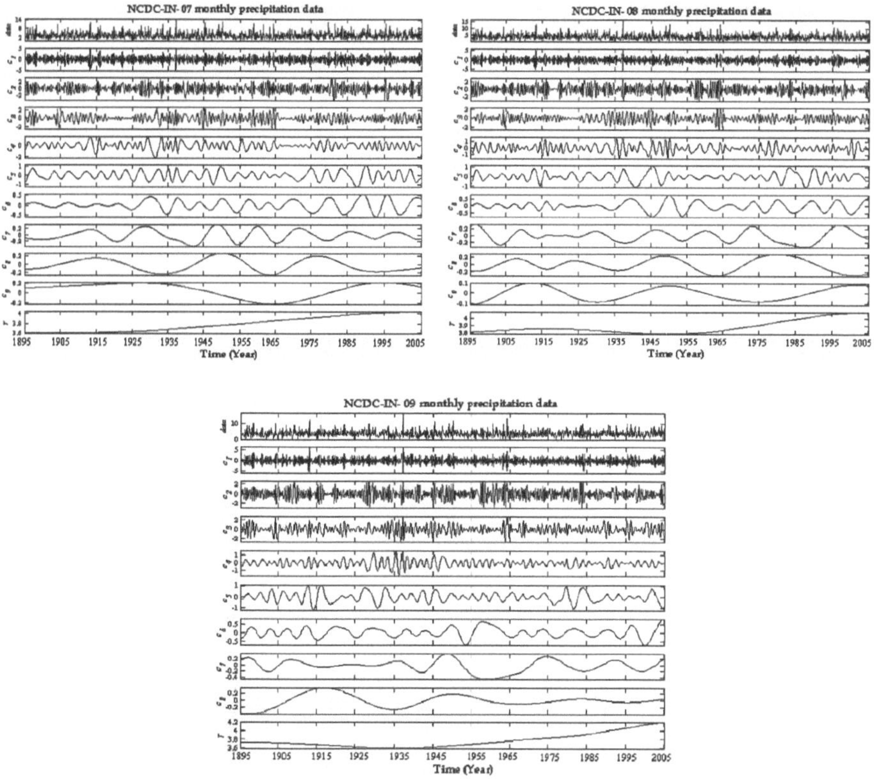

Figure 4.3.1. (*Continued*)

black line refers to the standard deviation of the instantaneous energy. It is of interest to observe the time with high variation and investigate the relationships among these plots (data, volatility, IE and Hilbert spectrum). Along the frequency axis, the degree of stationarity corresponding to each frequency is shown.

There are transparently shaded boxes added to each plot. They indicate some strong peaks or dips in the time series. Based on these highlighted time segments, the corresponding volatility, instantaneous energy, standard deviation of instantaneous energy, and Hilbert spectrum are easily compared. For example, the data in 1935 and 1947 have a stronger peak than the adjacent years. Looking through the nine divisions, there is a strong variation in either the corresponding volatility (derived from intrinsic mode functions) or the standard deviation of the instantaneous energy (derived from Hilbert spectrum) or both of them. It means that the abnormal events influence both empirical mode decomposition and the instantaneous spectrum. In order to investigate the details of these segments, the segments are zoomed in as shown in Figure 4.3.4.

Table 4.3.1. Trend test results for NCDC monthly rainfall time series

Region	Segment	Modified Mann-Kendall
01	1	Stationary (z = −0.363)
	2	Stationary (z = −1.183)
	3	Stationary (z = −0.017)
	4	Stationary (z = 0.597)
02	1	Stationary (z = −0.685)
	2	Stationary (z = 0.916)
	3	Stationary (z = −0.489)
03	1	Stationary (z = −0.636)
	2	Stationary (z = −1.951)
	3	Stationary (z = 0.679)
	4	Stationary (z = 1.181)
04	1	Stationary (z = 1.312)
	2	Stationary (z = 1.148)
	3	Stationary (z = 1.426)
	4	Stationary (z = −1.773)
	5	Stationary (z = −0.475)
	6	Stationary (z = 0.575)
05	1	Upward trend (z = 1.963)
	2	Stationary (z = 0.589)
	3	Stationary (z = 1.257)
	4	Stationary (z = 1.380)
	5	Stationary (z = 1.268)
06	1	Stationary (z = 1.344)
	2	Stationary (z = 1.189)
	3	Upward trend (z = 1.726)
	4	Stationary (z = 1.260)
07	1	Stationary (z = 1.617)
	2	Stationary (z = −0.197)
	3	Stationary (z = 0.434)
	4	Stationary (z = 0.875)
08	1	Stationary (z = 0.000)
	2	Stationary (z = 1.612)
	3	Stationary (z = 1.303)
	4	Upward trend (z = 1.817)
	5	Stationary (z = −0.189)
	6	Stationary (z = −0.441)
	7	Stationary (z = 0.515)
09	1	Stationary (z = 1.639)
	2	Stationary (z = 1.054)
	3	Stationary (z = −1.436)
	4	Stationary (z = 1.368)
	5	Downward trend (z = −1.920)
	6	Stationary (z = 0.398)
	7	Stationary (z = −0.105)
	8	Stationary (z = 1.628)

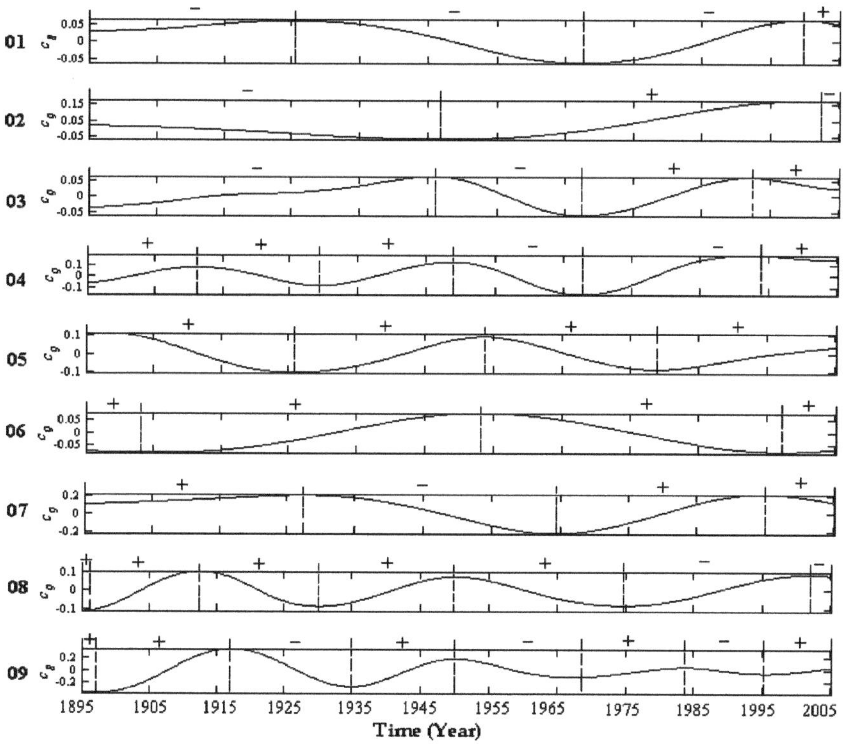

Figure 4.3.2. The last IMF component of NCDC monthly rainfall and the results of Mann-Kendall test for each segment

The segments are marked as ①②③, which corresponds the segments in Figure 4.3.3 from left to right. Similar to the results in HCN rainfall data, the peaks or dips of the volatility do not exactly correspond to the peaks or dips in the data, there are some lags between the volatility and data. Some volatility plots even do not have strong variation while there is a strong peak in the time series. The standard deviation of instantaneous energy is more useful in investigation of the variation in the time series. It identifies the occurrence of variation of time series better than volatility in these cases (Figure 4.3.4).

Studying the spatial and temporal variation and correlation of the spectra is of interest. The volatility and instantaneous energy is shown in Figure 4.3.5a, b, respectively. Similar to the results for HCN data, the difference in the results of these stations is not clear in the instantaneous energy plot. Therefore, volatility is used in time domain analysis. It is easier to investigate the spatial variation of NCDC data than HCN data, because the NCDC subdivisions are spread from west to east and from north to south (Figure 4.1.3). In Figure 4.3.5a, IN-01 and IN-02 have similar volatilities and IN-05 and IN-06 is the other pair. Also, IN-07 and IN-08 have similar volatility. It can be concluded that the adjacent subdivisions

(a)

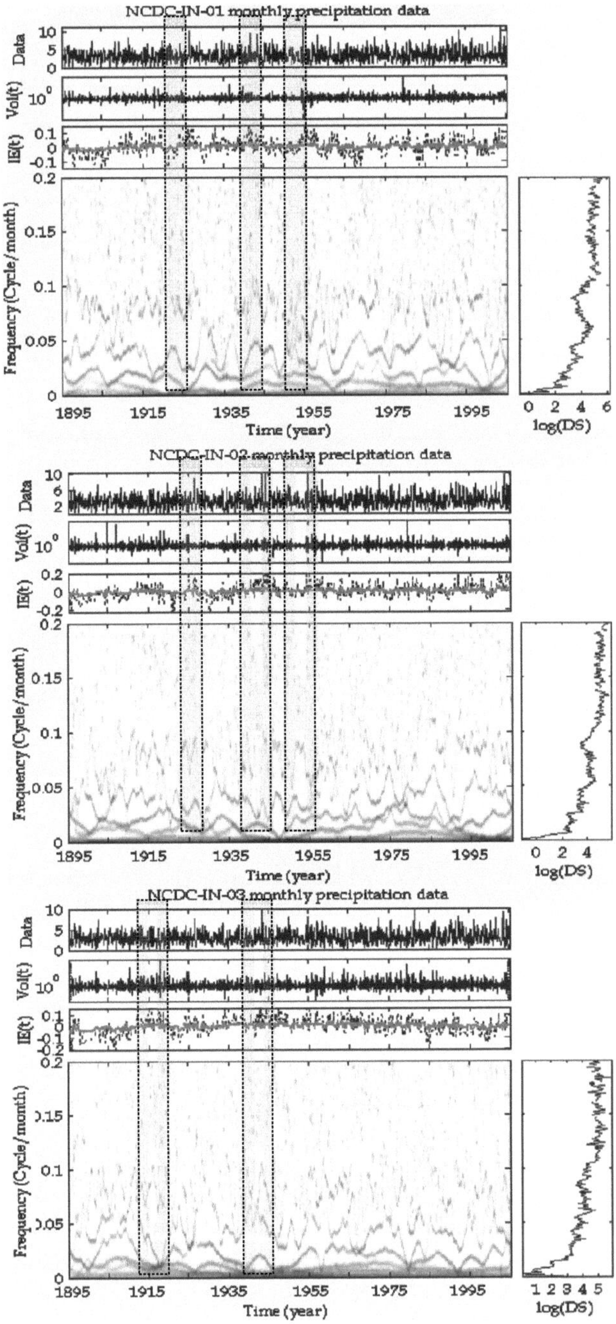

Figure 4.3.3. Time-frequency distributions of NCDC monthly rainfall data

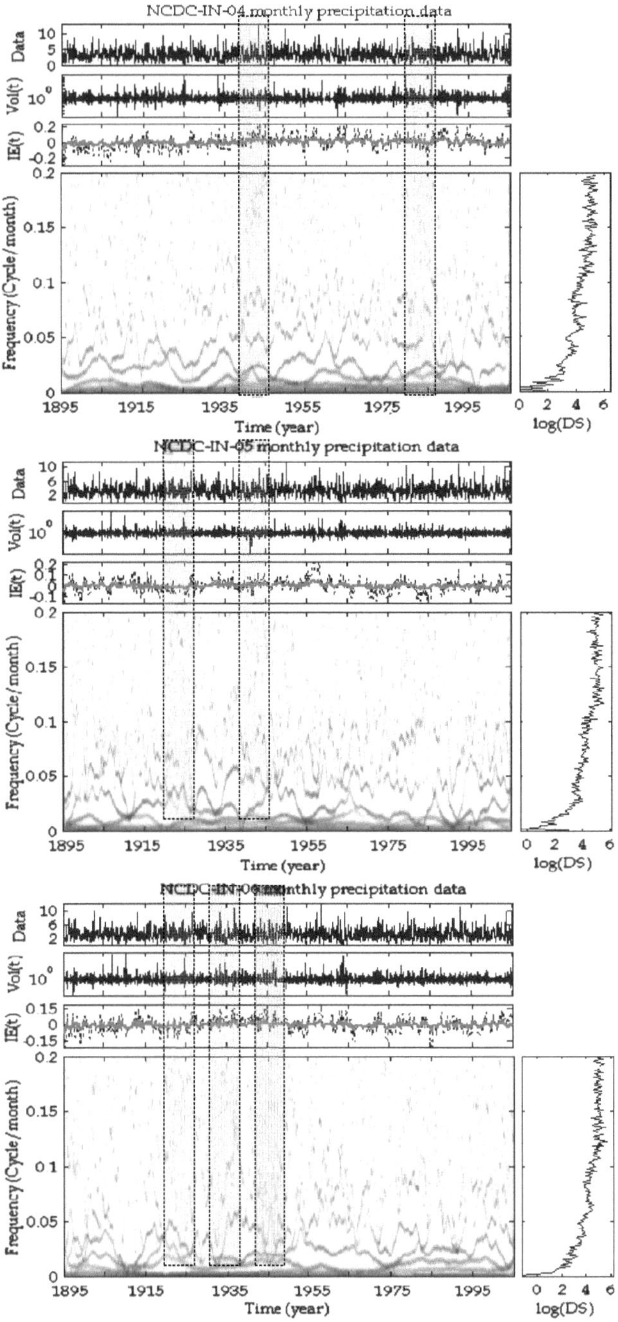

Figure 4.3.3. (*Continued*)

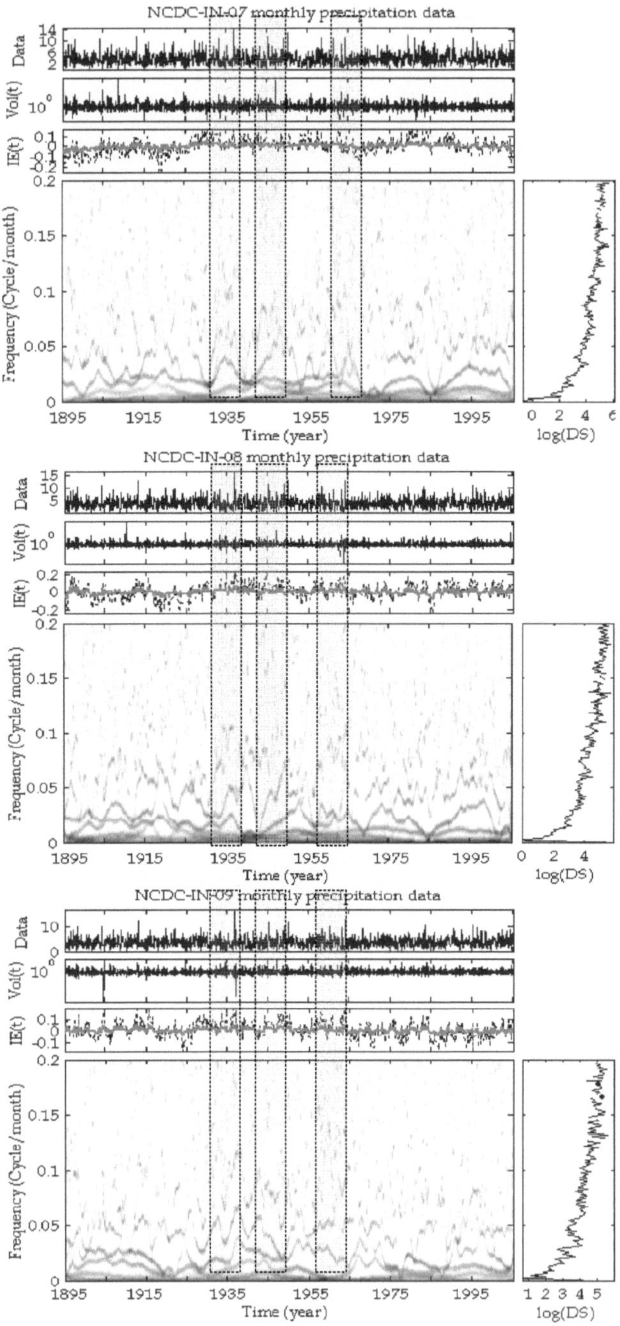

Figure 4.3.3. (*Continued*)

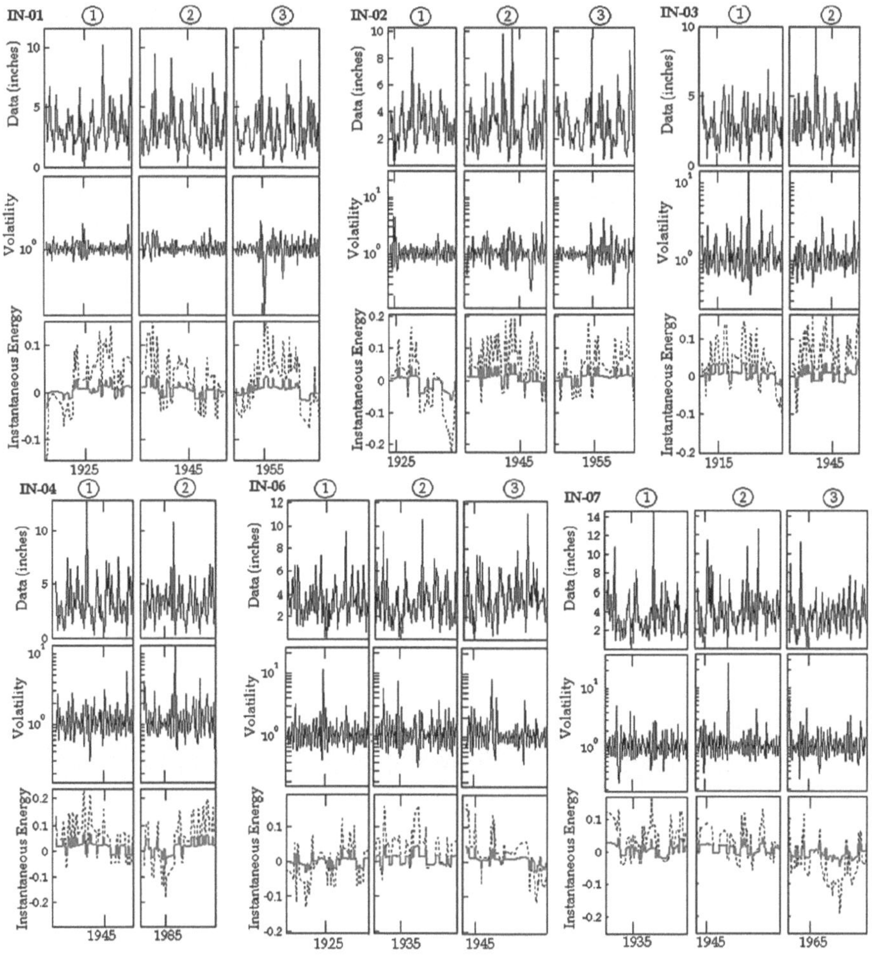

Figure 4.3.4. Time segments of NCDC rainfall data

have similar responses in volatility. Thus, the spatial correlations of the monthly rainfall data are revealed by the volatility.

The other way to look at the volatility is to find out the consistency of several peaks detected in Figure 4.3.5a. Six major peaks are compared. They occur in 1907, 1925, 1935, 1947, 1964 and 1980. In Figure 4.3.6, the shadowed areas are the subdivisions with significant volatilities. The instantaneous energy is too flat to tell and standard deviation of the instantaneous energy varies a lot within one segment. These results are not very useful for spatial analysis.

Most of the plots in Figure 4.3.6 show the spatial continuity except in year 1907. The shadowed areas gather in southern Indiana in year 1947 and 1964 while they appear in north in the year 1980. These results are quite consistent with the results

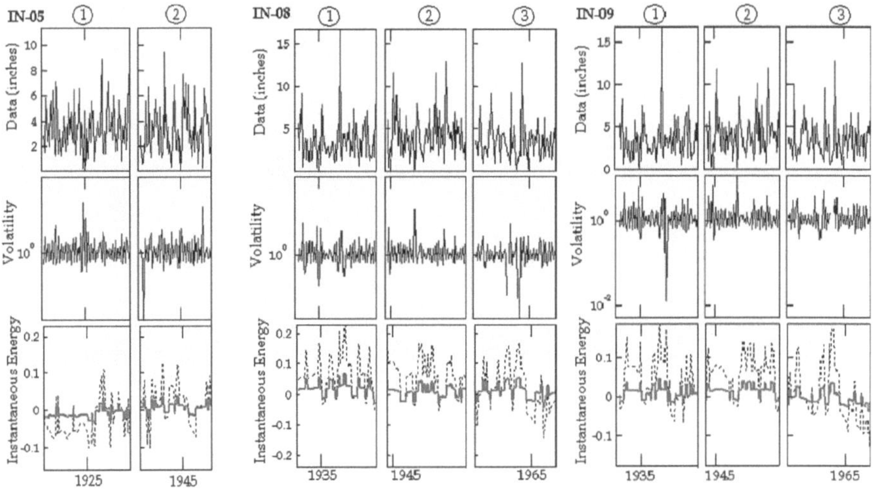

Figure 4.3.4. (*Continued*)

of HCN rainfall data. The volatility is detected for all regions in year 1925 and 1935, and this is the same conclusion from the analysis of HCN rainfall data.

4.3.3 Frequency Domain Analysis

The marginal Hilbert spectra of the NCDC monthly rainfall data are shown in Figure 4.3.7. The 95% confidence limits, best-fit line for the entire resampled spectrum and multi-taper spectrum are indicated as well. The significant periods are suggested based on the 95% confidence limits. They are summarized in Table 4.3.3. The multi-taper spectra are fairly flat and do not decay away in the high frequency regions.

In previous studies of HCN monthly rainfall data, the patterns of the marginal Hilbert spectra vary from station to station. For example, the annual cycle varies from 10 to 13 months (Table 4.3.2) and sometimes two peaks are detected. NCDC data is obtained by giving equal weight to stations reporting rainfall within a division. Therefore, the data are more stable and consistent than the data from

Table 4.3.2. Consistent volatilities in NCDC monthly rainfall time series

year	IN-01	IN-02	IN-03	IN-04	IN-05	IN-06	IN-07	IN-08	IN-09
1980	O	O	O	O	X	X	X	X	X
1964	X	X	X	X	O	O	O	O	O
1947	X	X	X	O	O	O	O	O	O
1935	O	O	O	O	O	O	O	O	O
1925	O	O	O	O	O	O	O	O	O
1907	O	O	X	X	O	O	O	O	X

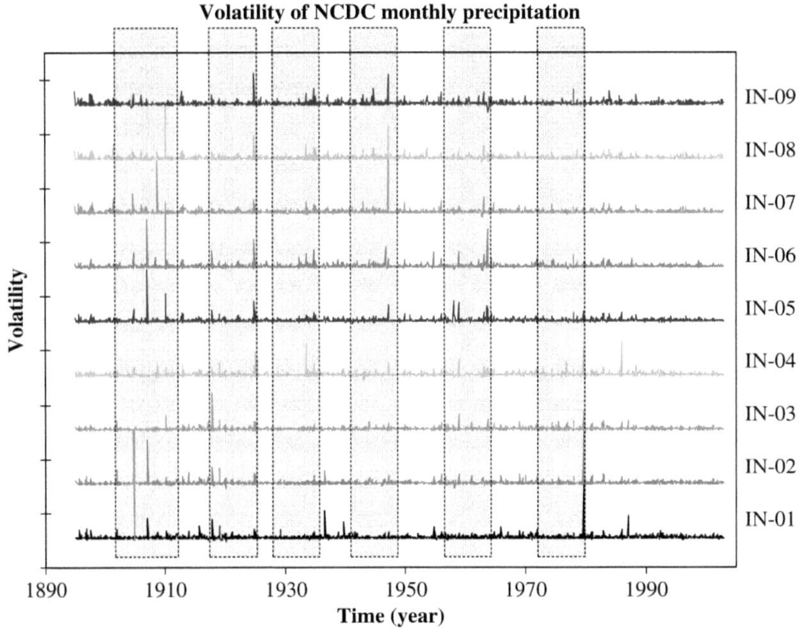

(a)

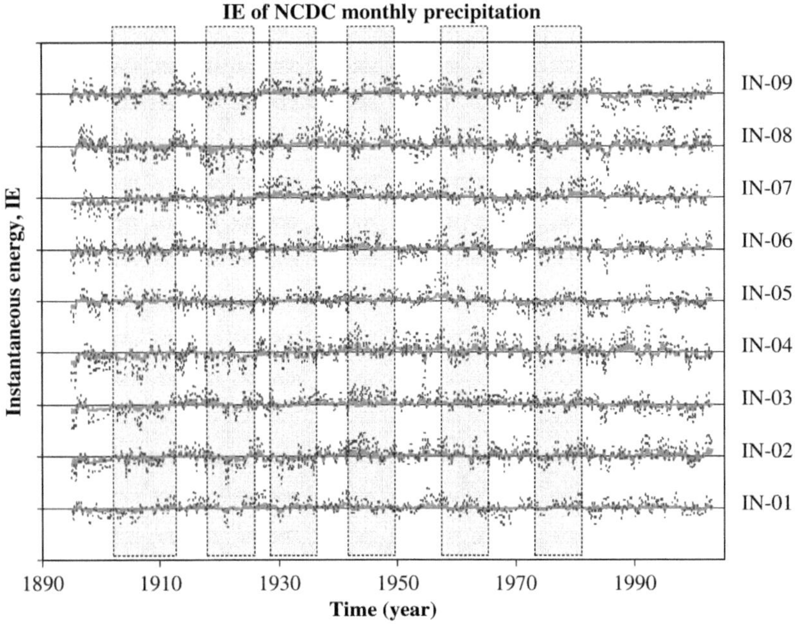

(b)

Figure 4.3.5. Volatility and instantaneous energy of NCDC rainfall data

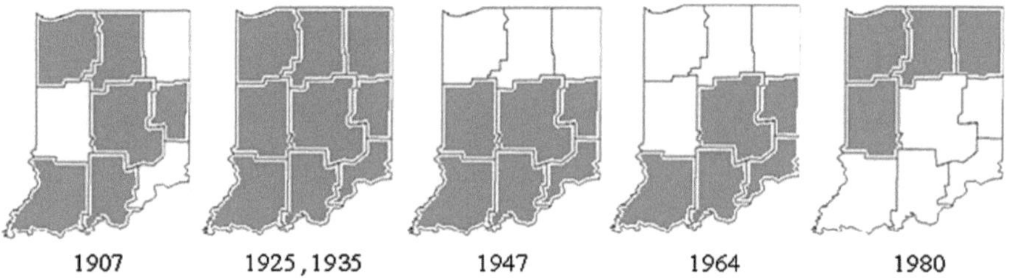

Figure 4.3.6. Significant volatilities detected in NCDC rainfall data

individual stations. Data from a single gauging station are affected by some local effects and by bias in the measurements.

The marginal Hilbert spectra for NCDC data have more consistent response in nine regions. Except the subdivision 8 (13.2 months), the annual peak is around 12 ± 0.6 months for the other eight regions. Multi-taper method identifies the annual peak better than the marginal Hilbert spectra. Marginal Hilbert spectra show the energy surrounding the annual peak but fail to resolve it exactly.

The 2 year period and the 3~7 year periods are detected in all regions. The 2 year period ranges around 2 ± 0.4 years. The 10 year period is detected in subdivision 3, 4, 6, 7 and 8. The longest period detected for NCDC data is a 22 year period from subdivision 4.

For the low frequency, some periodicities are indicated by both multi-taper and Hilbert spectra. In the high frequency, multi-taper spectra consistently have more energy than the marginal Hilbert spectra. Marginal Hilbert spectra obviously decay in this region while multi-taper spectrum is always constant. Multi-taper spectrum is fairly flat but has a well centered annual cycle with very strong energy. To investigate the long-term oscillations, the results from multi-taper method may be problematic if there is a trend embedded in the time series. The other advantage of using Hilbert transform is that it can better indicate the time-frequency characteristics than other methods.

Table 4.3.3. Detected periodicities in NCDC monthly rainfall time series (unit: years)

No.\ Period	1 year	2 years	3~7 years			10 years	> 15 years
1	1.0	2.1	3.4				
2	1.0	2.1	3.5	4.0			
3	1.0	2.0	2.9	7.1		14.5	
4	1.0	1.8	3.0	7.1		10.8	22.2
5	1.0	2.4	4.4	6.1			
6	1.0	2.4	3.1	4.3		8.5	
7	1.0	2.0	3.7	5.0		10.8	
8	1.1	1.8	4.4			12.3	
9	1.0	1.6	3.0	4.3	7.1		

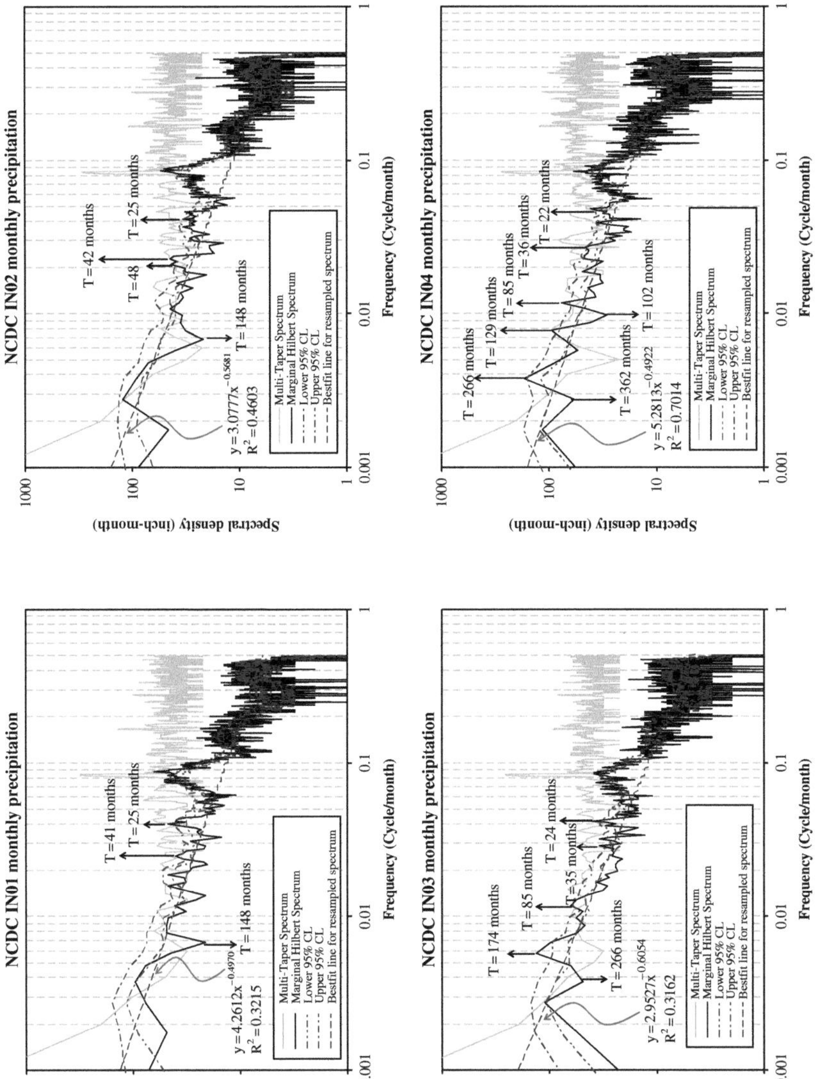

Figure 4.3.7. Marginal Hilbert spectrum and its confidence limits for NCDC monthly rainfall data

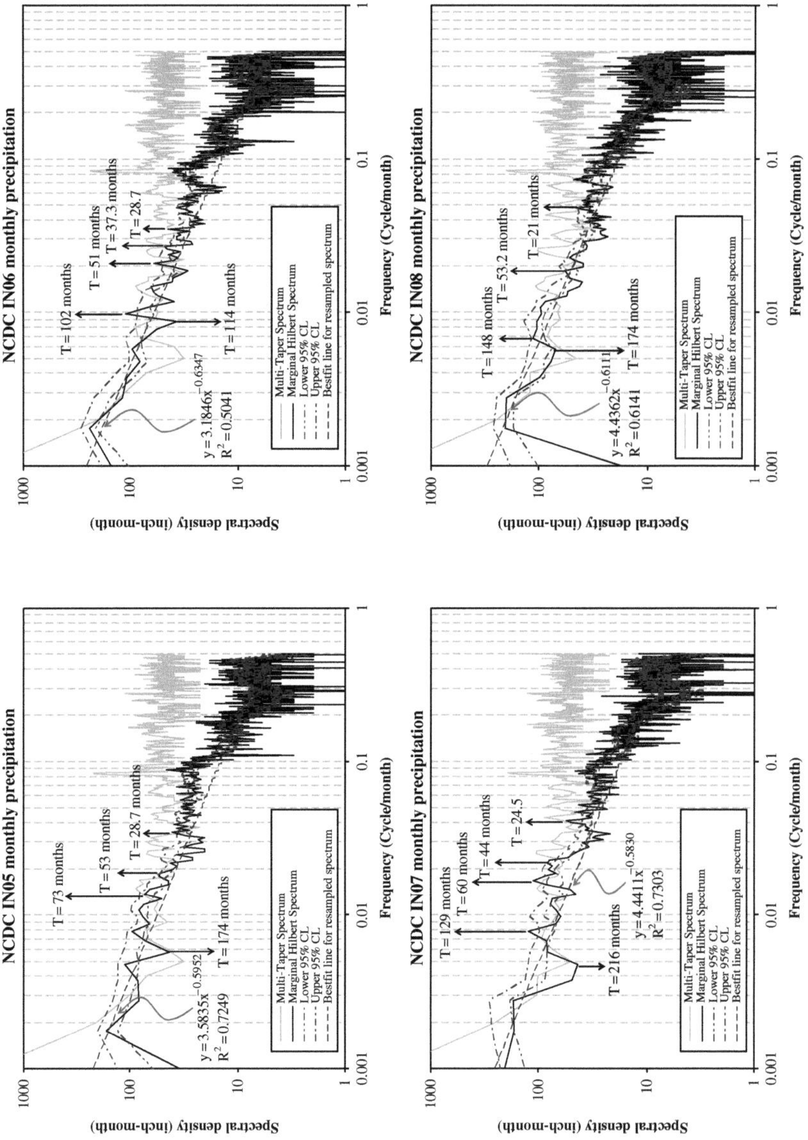

Figure 4.3.7. (Continued)

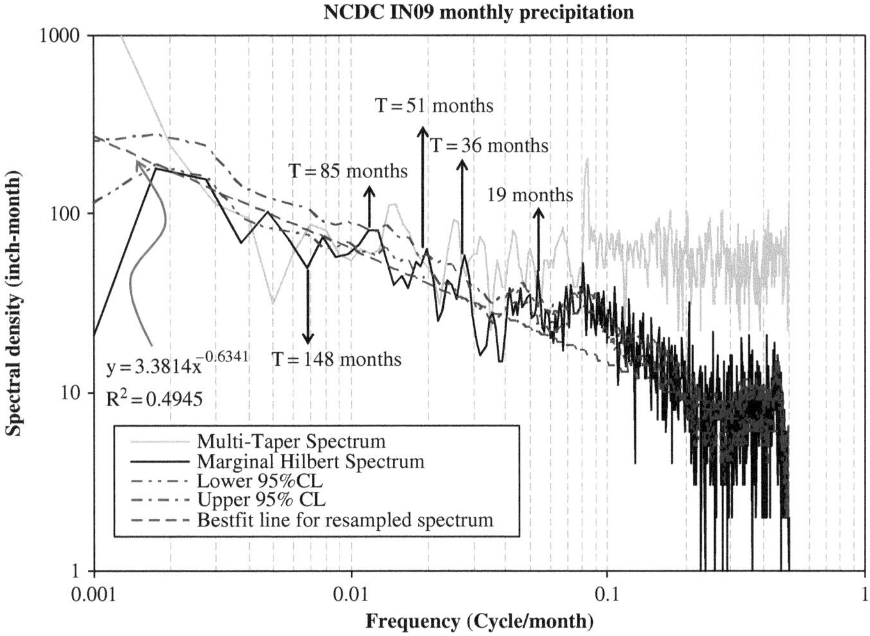

Figure 4.3.7. (Continued)

The degree of stationarity for the NCDC monthly rainfall data are shown in Figure 4.3.8. The amplitudes of the degree of stationarity are quite close and overlay with each other. As a smaller DS value means that the series is more stationary at that frequency, the NCDC data is more stationary in low frequency region than in high frequency region. A dip-down in the degree of stationarity around frequency 0.083 cycle/month indicates that the persistent annual cycle makes the spectrum more stationary at that frequency than in the adjacent frequencies. This reduction in degree of stationarity is stronger in NCDC data than in HCN data (Figure 4.2.8).

The marginal Hilbert spectrum is fitted by the power law equation $y = af^b$, where y is the marginal Hilbert spectrum and f is the frequency. The parameter a, power law decay rates b and R-square value for the best fit line are summarized in Table 4.3.4. The regression is performed by considering either the entire spectrum or the low frequency segment (frequency from 0.003 to 0.09 cycle/month). For the entire spectrum, the power law decay rate is fairly steady around −0.58 with standard deviation of 0.053. The coefficient of variation is 0.09. The decay rate is close to the value for HCN data (−0.57) and the coefficient of variation is smaller than that in the HCN data. Consequently the estimate from NCDC data is more stable than the one from HCN data. For the resampled spectrum in the low frequency, the power law decay rate is around −0.41 with standard deviation of 0.083. It makes the coefficient of variation as high as 0.20, which is again lower than the corresponding value from HCN data. However, if the power law is studied, the result from fitting the entire data is more acceptable than only using part of the spectrum.

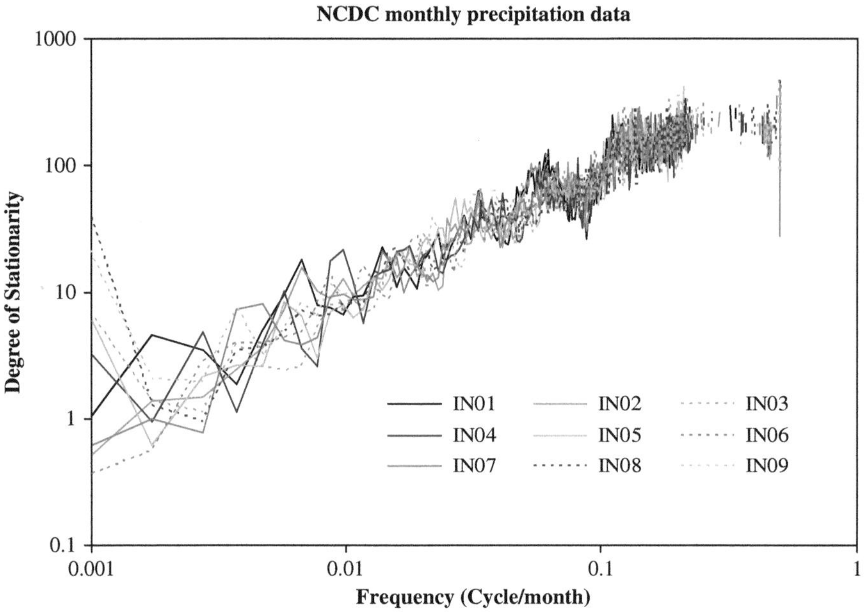

Figure 4.3.8. Degree of stationarity for NCDC monthly rainfall data

Table 4.3.4. Parameters of the best fit lines to the marginal Hilbert spectra of NCDC monthly rainfall time series ($y = af^b$, y is the marginal Hilbert spectrum; f is the frequency)

	Entire resampled spectrum			Low frequency resampled spectrum		
NCDC	a	b	R^2	a	b	R^2
01	4.261	−0.497	0.322	9.975	−0.307	0.383
02	3.078	−0.568	0.460	9.833	−0.294	0.399
03	2.953	−0.605	0.316	5.527	−0.472	0.630
04	5.281	−0.492	0.701	8.342	−0.413	0.584
05	3.584	−0.595	0.725	6.065	−0.508	0.743
06	3.185	−0.635	0.504	7.537	−0.444	0.733
07	4.441	−0.583	0.730	14.668	−0.306	0.380
08	4.436	−0.611	0.614	7.386	−0.483	0.751
09	3.381	−0.634	0.495	7.922	−0.433	0.684
Mean	3.844	−0.580	0.541	8.584	−0.407	0.587
Std. Dev.	0.794	0.053	0.162	2.721	0.083	0.160

4.4. CONCLUDING COMMENTS

For the rainfall data, the 12 month period in the marginal Hilbert spectra is not as strong as the annual cycle in the multi-taper method spectra. The annual cycle varies from 11 months to 13 months so the 12 month period cannot be clearly specified. The 1 and 2 year periods are detected. Approximate cycles of 4 years and 10 years are also detected. However, the periodicities greater than 15 years are detected only in a few stations. These periods are based on the 95% confidence intervals defined by the simulated data. Although the identification is based on the sample spectra, these periods are fairly strong and are detected in other approaches as well (Figure 4.2.7 and Figure 4.3.7).

HCN rainfall data are affected more by the local events than the NCDC rainfall data. Therefore, the results are more consistent in NCDC data than in HCN data. The residuals of HCN data vary a lot from one site to another; however, the NCDC data consistently yield upward trends for nine regions. Through investigation by the measures, the results in rainfall data show that the standard deviation of the instantaneous energy is more sensitive to the variation of time series than the measures of volatility and instantaneous energy.

In the study of spatial correlation, it is not surprising to see that the spatial distribution of the monthly rainfall data is highly correlated. This spatial continuity is seen in this study. The patterns of correlations exist over the state, events affecting either north or south region, or the northwest or the southeast regions.

CHAPTER 5

STREAMFLOW DATA ANALYSIS

5.1. INTRODUCTION AND DATA USED

There are several sets of streamflow data that are studied here. First is the data from USGS gauging stations in the state of Indiana. Others are the series from Warta River (Poland), Godavari and Krishna Rivers (India).

5.1.1 USGS Streamflow Data from Indiana

USGS surface-water data include more than 850,000 station years of time-series data that describe stream levels, streamflow (discharge), reservoir and lake levels, surface-water quality, and rainfall. The data are collected by automatic recorders and manual measurements at field installations across the U.S. Data are collected by field personnel or relayed through telephones or satellites to offices where they are stored and processed. Once a complete day of readings are received from a site, daily summary data are generated and stored in the data base. For the State of Indiana, the stations used in this study are listed in Table 5.1.1 and their geographical locations are shown in Figure 5.1.1. The annually averaged time series are shown in Figure 5.1.2.

5.1.2 Streamflow Data from Warta, Godavari and Krishna Rivers

The previous six flow data series have around 80-years length and are under similar meteorological condition. An additional streamflow series discussed here is that from Warta (1822–1990), which is a good candidate to investigate the long-term variation in streamflow. In order to study the characteristics of streamflows affected by monsoons, monthly streamflow from Krishna (1901–1979) and Godavari (1902–1960) are investigated.

1) Warta River

Warta is a river in west-central Poland, a tributary of the Oder River. With a length of approximately 808 kilometers it is Poland's third longest river. Its geographical location is shown in Figure 5.1.3. Warta has a basin area of 54,529 km^2. It rises in

Table 5.1.1. USGS stations for monthly streamflow data from Indiana

No.	State	USGS No	Station Name	Record Year	Length
1	IN	3276500	Whitewater River at Brookville	1924–2004	82
2		3324500	Salamonie River at Dora	1924–2001	78
3		3326500	Mississinewa River at Marion	1924–2004	81
4		3335500	Wabash River at Lafayette	1924–2004	81
5		3373500	East Fork White River at Shoals	1923–2004	82
6		5518000	Kankakee River at Shelby	1923–2004	82

the Jura Krakowska, S Poland, and flows northwest past Czjstochowa and Poznaq to the Oder River at Kostrzyn.

The daily data are available from 1822 to 1990 in the document edited by Olejnik (1991). The original unit in their record is m^3/s but converted to cubic feet per second (cfs) in this study. Data of both daily and monthly scales are investigated in this study. It is a rare streamflow time series with as long as 170 years length of record and a daily recording interval.

2) Godavari River

Godavari River is about 1450 km (900 miles) long. It rises at Triambakeshwar, in Nashik, near Bombay, flows southeast across south-central India into Andhra Pradesh, and empties into the Bay of Bengal. At Dhavaleswaram the river is nearly

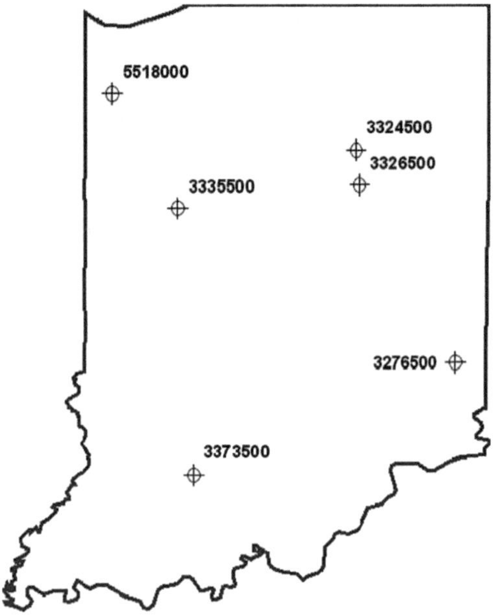

Figure 5.1.1. Location of the USGS flow stations

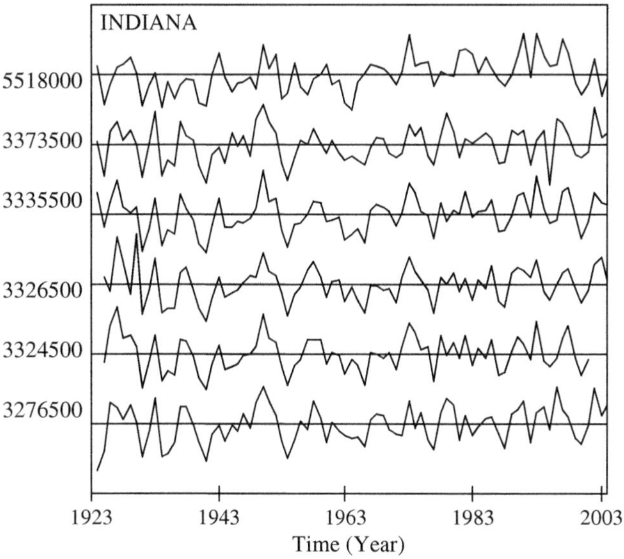

Figure 5.1.2. USGS annual average flow time series

Figure 5.1.3. Warta river, Poland (http://www.lib.utexas.edu/maps/poland.html)

4 miles wide. It is a seasonal river, approximately 80% of its discharge into the Bay of Bengal occurs during the monsoons between July and October. The color of the river is turbid yellow during the monsoons, while the water is clean and greenish the rest of the year. The geographical image of Godavari River is shown in Figure 5.1.4. The studied data have a record of length from 1902 to 1960. The unit is converted to cubic feet per second (cfs).

3) Krishna River

The Krishna is one of the longest rivers of India (about 1300 km in length). It originates at Mahabaleswar, passes through Sangli and meets the sea in the Bay of Bengal at Hamasaladeevi in Andhra Pradesh. The Krishna River flows through the states of Maharashtra, Karnataka and Andhra Pradesh (Figure 5.1.4). The data is available from Global River Discharge Database (RivDIS v1.1; http://www.rivdis.sr.unh.edu/) and the length of record is from 1901 to 1979. The gauging station is at Vijayawada and has an upstream area of 251355 km^2. The longitude and latitude of Vijayawada are 80.62° and 16.52°, respectively.

The two Indian rivers have several common characteristics. They are quite wide and have conspicuous deltas; both have heavy discharges of water during the monsoon season followed by low discharges during the dry season. Both discharge sediment into the Bay of Bengal and both are, like all rivers in India, sacred to Hindu religion.

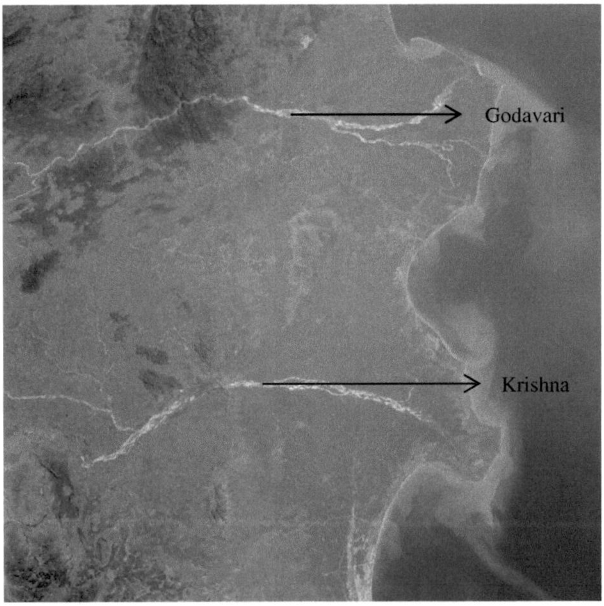

Figure 5.1.4. Godavari and Krishna River Deltas, India, October 1989 (NASA, downloaded from http://earth.jsc.nasa.gov/)

5.2. USGS STREAMFLOW DATA

5.2.1 Long-Term Oscillations

The intrinsic mode functions for the monthly streamflow data of the six USGS stations are shown in Figure 5.2.1. These components range from high frequency to low frequency, which refer to the scale of couple months to yearly or decadal spans.

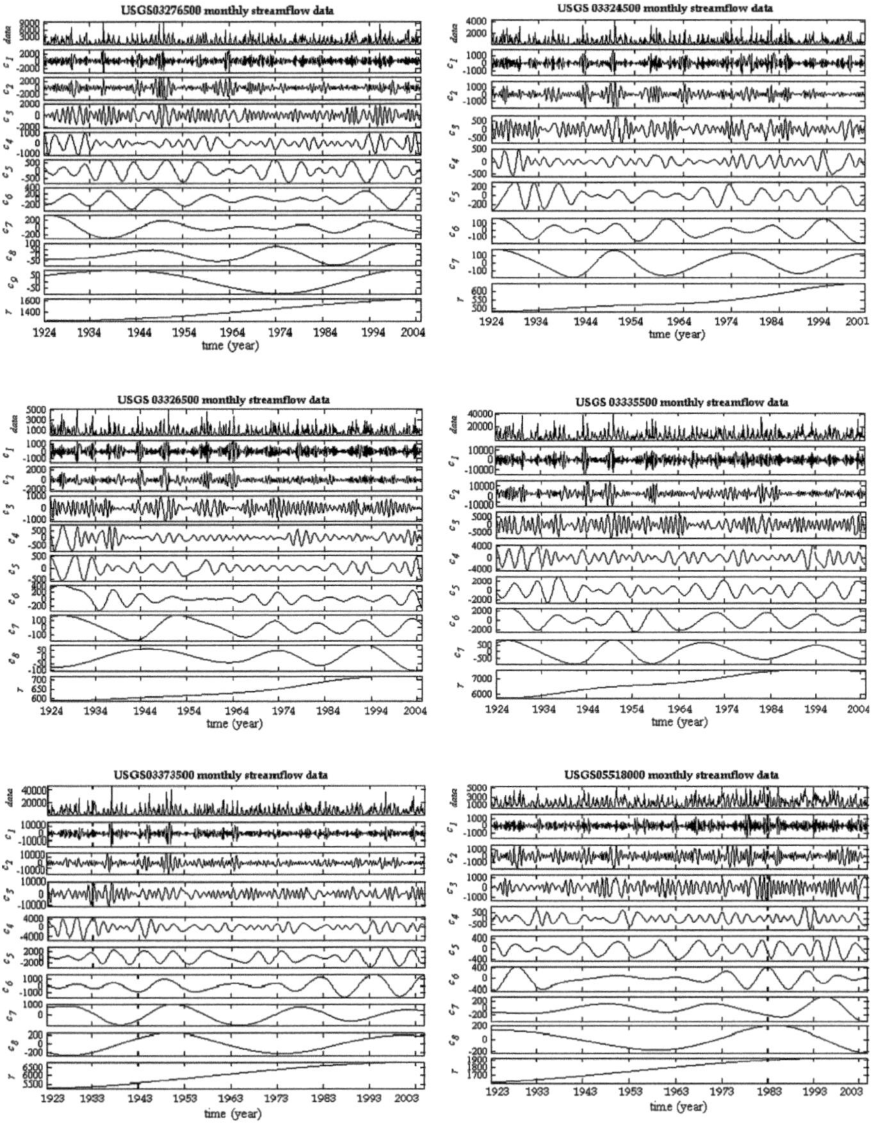

Figure 5.2.1. Intrinsic mode functions of USGS monthly streamflow data (unit: cfs)

For these six series, an upward residual is seen in all of them and the amplitudes of the residuals represent around 1~2% of the variation of data.

The results of trend test performed by using the modified Mann-Kendall tests are shown in Table 5.2.1. The segments of the last IMF components are compared to the z values of modified Mann-Kendall test in Figure 5.2.2. The ratio of consistent to inconsistent is 17:14 which means 55% of these segments are consistent. This is a poor result compared to the results from rainfall and temperature data.

5.2.2 Time-Frequency Distribution

The Hilbert spectra for USGS monthly streamflow data are shown in Figure 5.2.3. Along the time axis, the original data, volatility and instantaneous energy (IE)

Table 5.2.1. Trend test results for USGS monthly streamflow time series

Station	Segment	Modified Mann-Kendall
3276500	1	Stationary (z = −1.347)
	2	Stationary (z = 1.472)
	3	Stationary (z = 0.291)
	4	Stationary (z = −1.712)
3324500	1	Downward trend (z = −2.023)
	2	Upward trend (z = 1.881)
	3	Stationary (z = −0.394)
	4	Upward trend (z = 1.758)
	5	Stationary (z = −0.06809)
	6	Stationary (z = −1.147)
3326500	1	Downward trend (z = −1.819)
	2	Downward trend (z = −2.136)
	3	Stationary (z = −1.286)
	4	Upward trend (z = 1.802)
	5	Stationary (z = −0.337)
	6	Stationary (z = 1.549)
	7	Downward trend (−2.015)
3335500	1	Stationary (z = −0.575)
	2	Downward trend (z = −1.939)
	3	Upward trend (z = 1.753)
	4	Downward trend (z = −1.886)
	5	Stationary (z = −0.389)
	6	Stationary (z = −0.460)
	7	Stationary (z = 1.523)
	8	Stationary (z = 0.796)
3373500	1	Stationary (z = 1.045)
	2	Stationary (z = 1.522)
	3	Stationary (z = 0.960)
	4	Stationary (z = −1.094)
5518000	1	Stationary (z = 0.946)
	2	Upward trend (z = 2.064)
	3	Downward trend (z = −1.805)

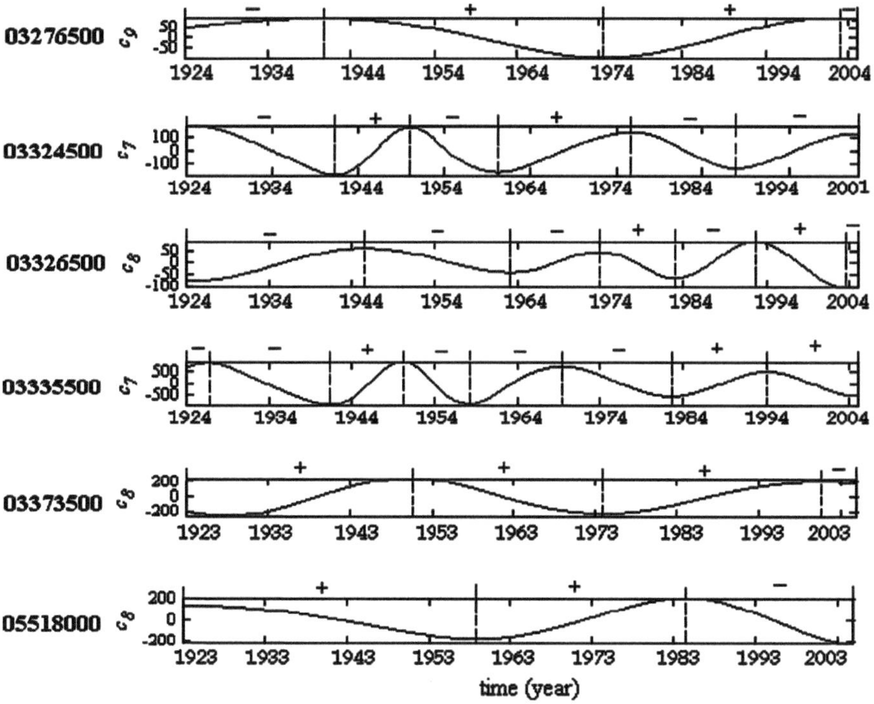

Figure 5.2.2. The last IMF component of USGS monthly streamflow and the results of Mann-Kendall test for each segment

are also plotted. The times at which high variation in these plots may be used to investigate the relationships among data, volatility, IE and Hilbert spectrum are also identified. Along the frequency axis, the degree of stationarity corresponding to each frequency is shown. The transparently shaded boxes ( ) in each plot indicate some strong peaks or dips in the time series.

In these highlighted time segments, the volatility, instantaneous energy, standard deviation of instantaneous energy, and Hilbert spectrum are easily compared. In order to investigate the details in these segments, the segments are zoomed in as shown in Figure 5.2.4. They are marked as (①②③), which correspond to the segments in Figure 5.2.3 from left to right. As seen in monthly rainfall data, it shows that the peaks or dips of volatility do not exactly correspond to the peaks or dips in the data, there are some lags between the volatility and data. In some instances volatility does not even have strong variation while there is a strong peak in the time series.

The standard deviation of instantaneous energy is quite consistent with the variation of data while the instantaneous energy does not clearly indicate these variations. A large standard deviation of IE refers to higher streamflows and a low value indicates a continued low flow or drought. The volatility and instantaneous energy of the six stations are extracted and summarized in Figure 5.2.5. There are consistencies in volatilities in the years of 1935, 1947 and 1964 as shown in the shaded areas of Figure 5.2.5.

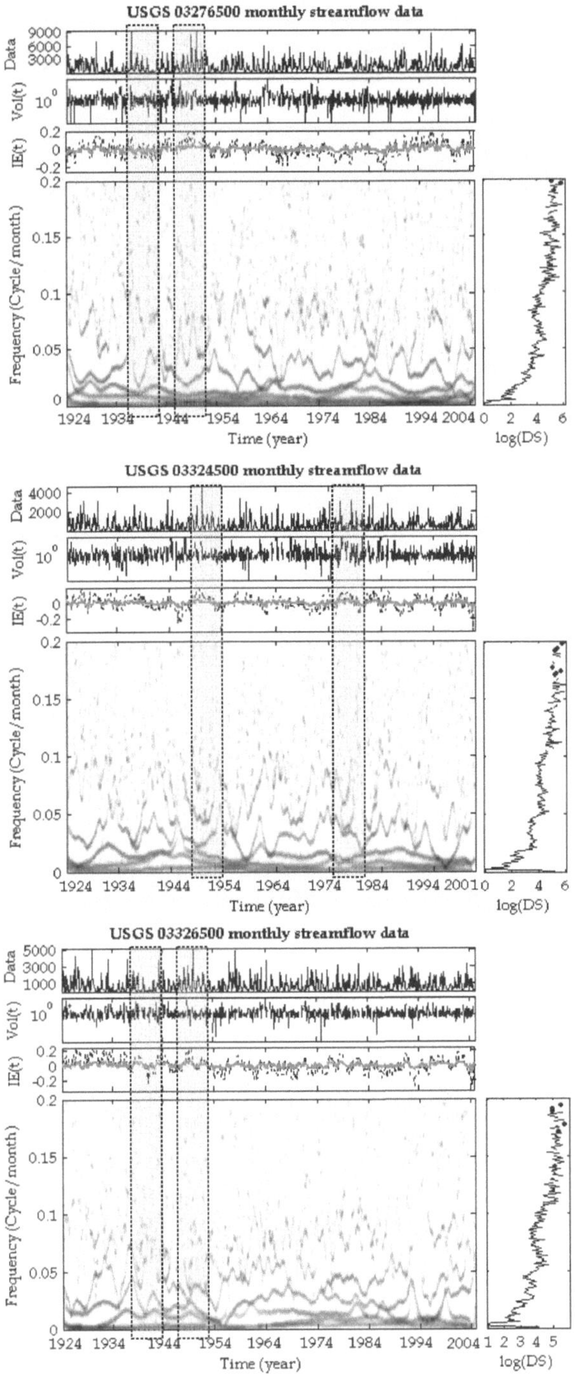

Figure 5.2.3. Time-frequency distributions of USGS monthly streamflow data

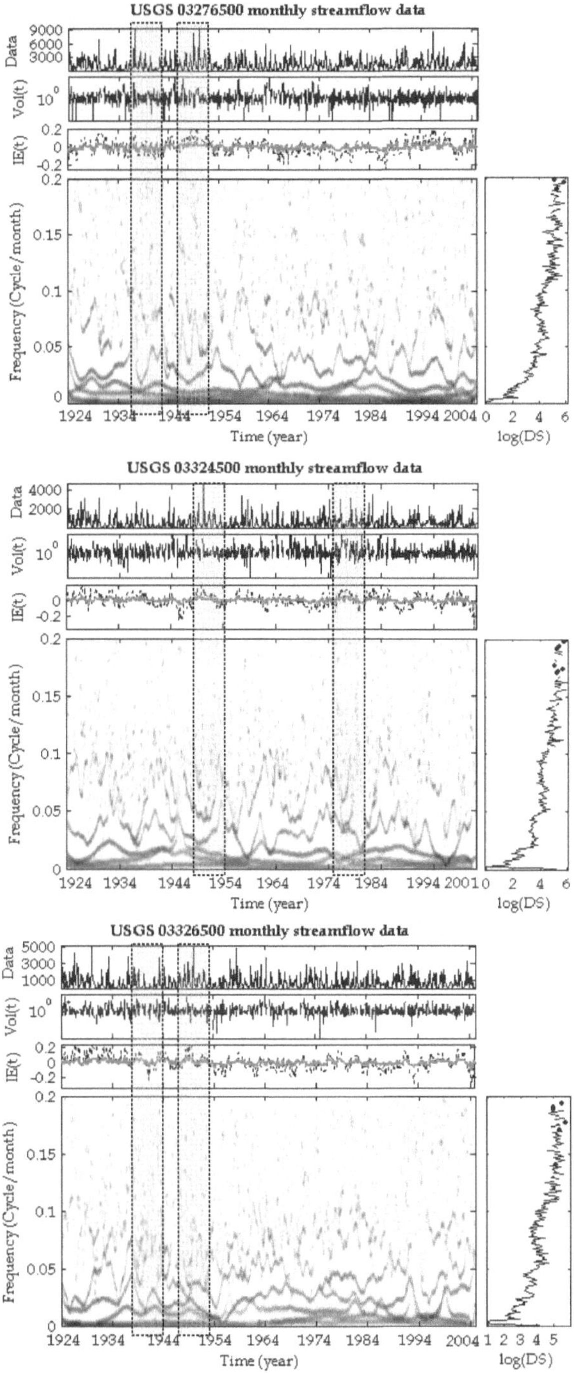

Figure 5.2.3. (*Continued*)

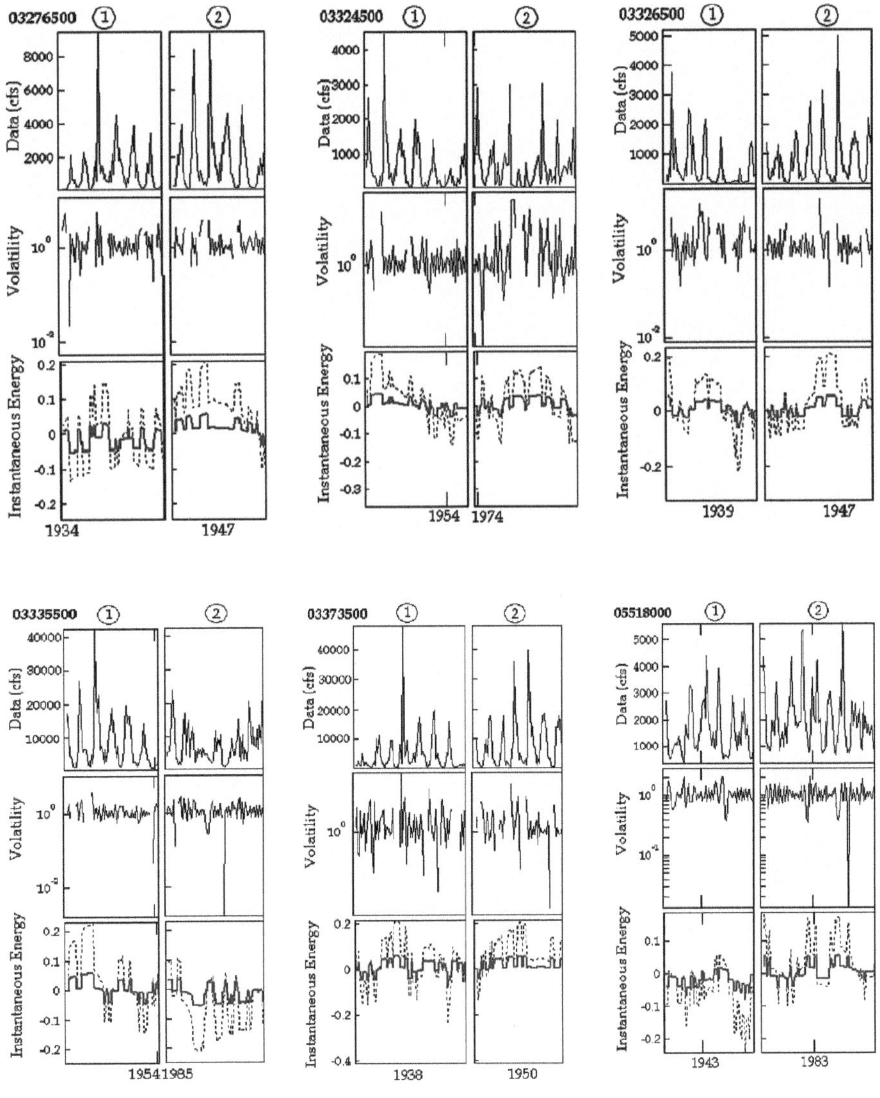

Figure 5.2.4. Time segments of USGS monthly streamflow data

5.2.3 Comparison with MTM Spectra

The marginal Hilbert spectra of runoff data and their 95% confidence limits are plotted in Figure 5.2.6. The best fit line for resampled marginal Hilbert spectra and the multi-taper spectra are also shown in Figure 5.2.6. The significant periods summarized in Table 5.2.2 are detected based on the 95% confidence limits.

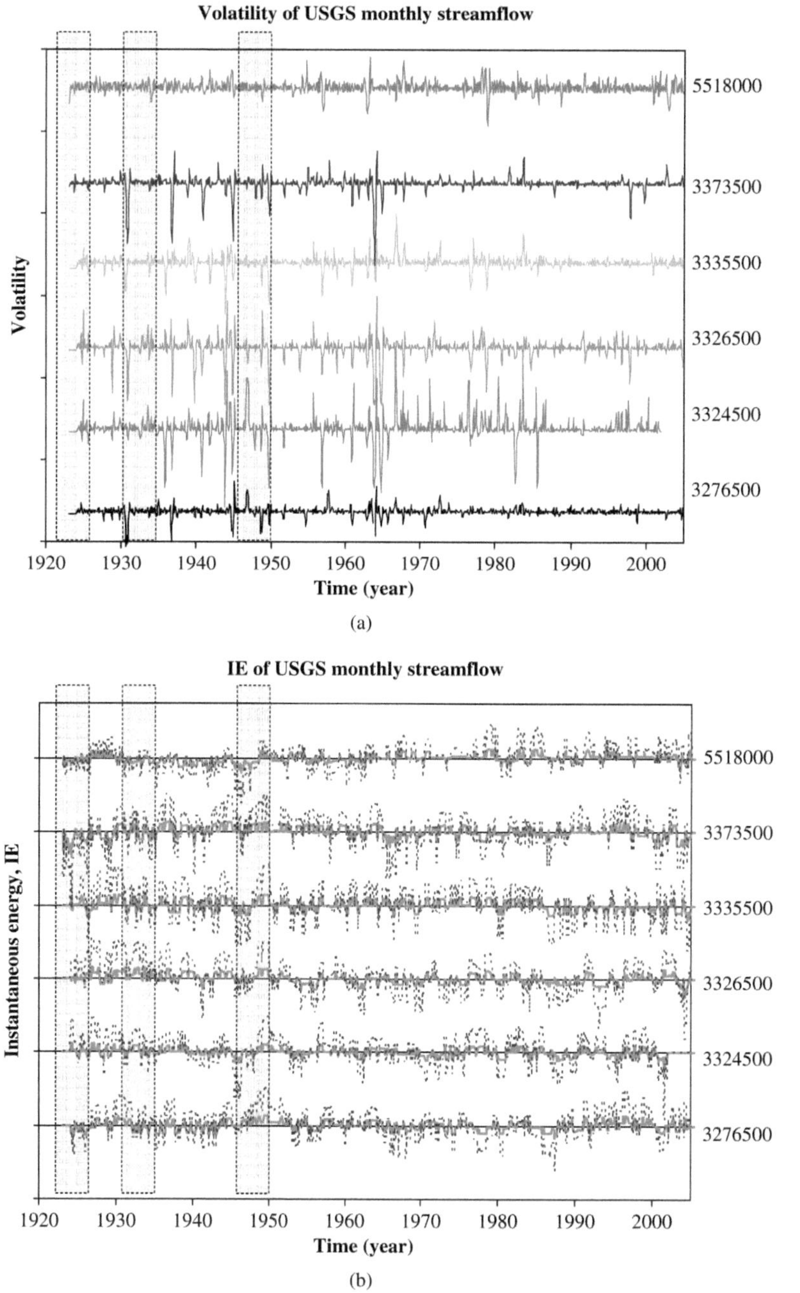

Figure 5.2.5. (a) Volatility and (b) instantaneous energy of USGS monthly streamflow data

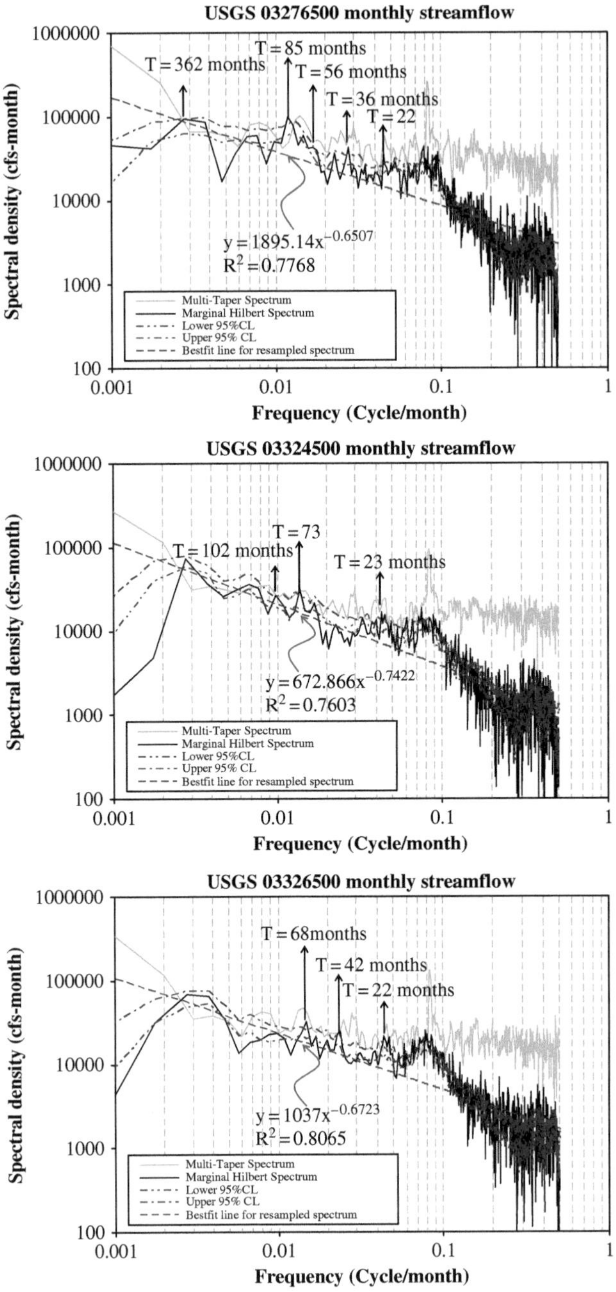

Figure 5.2.6. Marginal Hilbert spectra and their confidence limits for USGS monthly streamflow data

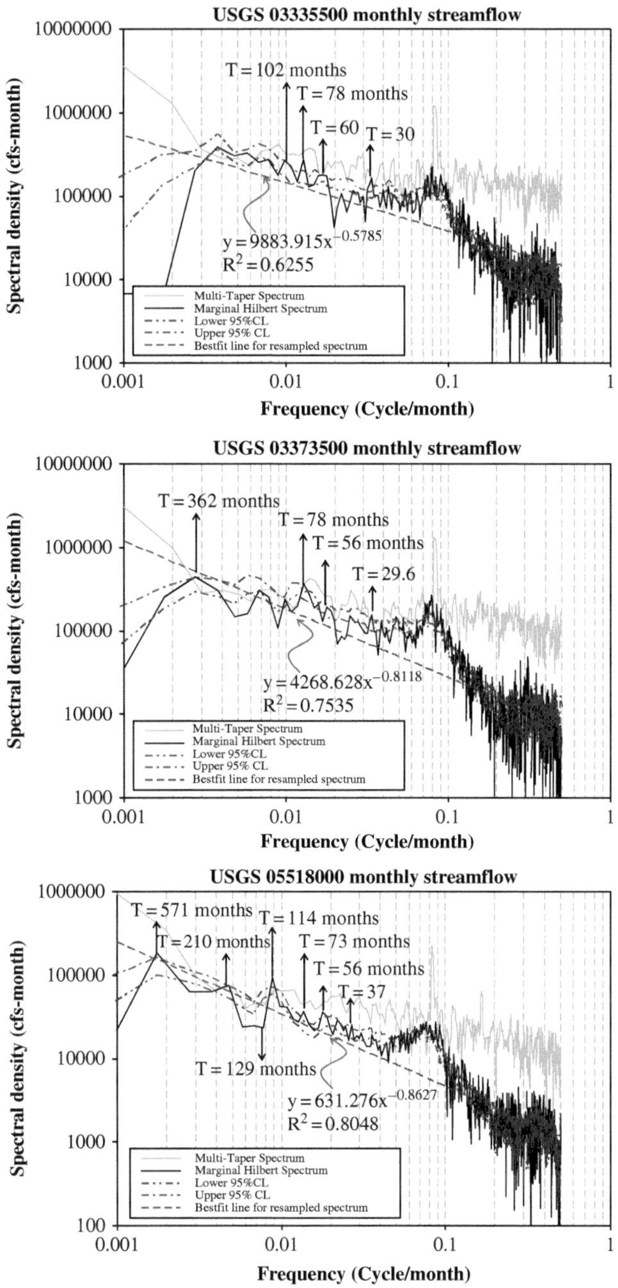

Figure 5.2.6. (*Continued*)

Table 5.2.2. Periodicities detected in USGS monthly streamflow time series (Unit: years)

No. Period	1 year	2 years	3~7 years			10 years	>15 years
03276500	1.1	1.8	3.0	4.7	7.1		30.2
03324500	1.0	1.9	6.2			8.5	
03326500	1.0	1.8	3.5	5.7			
03335500	1.0	2.5	5.0	6.5		8.5	
3373500	1.0	2.5	4.7	6.5			30.2
05518000	1.0		3.1	4.7	6.1	9.5	17.5

Based on the R-square values (Figure 5.2.6), the marginal Hilbert spectra do not exhibit a strongly log-linear relationship. Also, the high variation in low frequency makes the fitting more problematic. The decay rates of the power law equation for these six USGS stations are −0.6507, −0.7422, −0.6723, −0.5785, −0.8118 and −0.8627, respectively. The mean, standard deviation and coefficient of variation for these decay rates are −0.720, 0.106 and 0.148. Hence, the power law of best fit line to the spectra is not consistent in the streamflow data.

Apart from the annual cycle (Table 5.2.2), 2 year period is detected in all the series except for station No. 5518000. The 3~7 year period is also detected in all stations. Ten year period is detected in station No. 03324500, 03335500 and

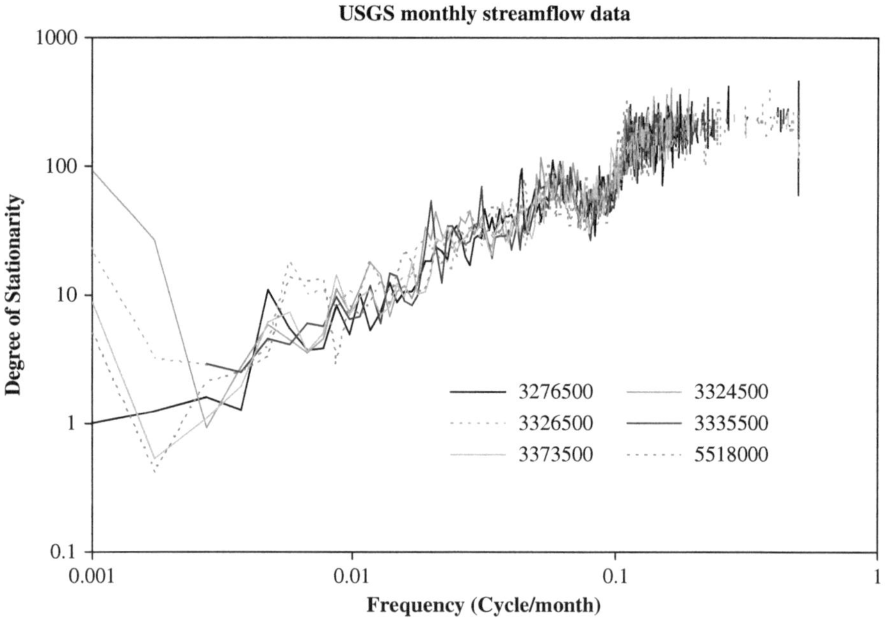

Figure 5.2.7. Degree of stationarity of USGS monthly streamflow data

055180000. The decadal long term oscillations are detected in station No. 03276500, 03373500 and 05518000. In Figure 5.2.6, the degree of stationarity becomes high and abnormal in station 03324500, 03326500, 03373500, 05518000 at frequencies close to zero. This situation indicates that the marginal Hilbert spectrum in that frequency is problematic.

The degrees of stationarity for six USGS stations are plotted in Figure 5.2.7. These lines are very close to each other. As usual, a dip-down in the annual cycle is seen. The strong annual cycle makes the Hilbert spectrum keeping constant around this frequency and it is the reason to pull down the degree of stationarity. The high degree of stationary in low frequency region indicates that these frequencies vary with time. This can be also seen from Figure 5.2.3. At the frequency band close to zero, the energy of the spectrum changes considerably over the time span. This variation results in high value of degree of stationarity at low frequencies. In addition, the marginal Hilbert spectra (Figure 5.2.6) do not behave as steadily at that frequency either. Since the marginal Hilbert spectrum is presenting the integral energy in that frequency, the results are not reliable. It may be better to investigate these oscillations by using the time-frequency distribution than by simply using the marginal Hilbert spectra.

5.3. ANALYSIS OF WARTA, GODAVARI AND KRISHNA RIVER FLOW DATA

5.3.1 Warta River Daily Streamflow Data

The Warta River flow data are used to construct intrinsic mode functions as shown in Figure 5.3.1a. It requires 22 IMFs to successfully decompose the data. The residuals have a small upward trend although it is not monotonic. The Hilbert spectrum of daily streamflow data in Warta River is shown in Figure 5.3.1b. Along the time axis, apart from the Hilbert spectrum, the original data, volatility and instantaneous energy (IE) are plotted. The bold gray line in the plot of instantaneous energy indicates the average energy while the dashed black line refers to the standard deviation of the instantaneous energy. There are three peaks shaded in boxes as examples. They occur in year 1870, 1887 and 1922.

The consistency among the peak flows, volatility and the standard deviation of instantaneous energy is obvious. In Figure 5.3.1b, the degree of stationarity gradually increases from low frequency to high frequency. This is mainly because of the larger number of data points in this case. Hence, each data point has less contribution and smaller effect on nonstationarity. This situation leads to a smooth degree of stationarity values. This also results in a smooth marginal Hilbert spectrum as shown in Figure 5.3.2.

Although most parts of the spectra are significant at 95% confidence level, only one peak, 0.00276 cycle/day, can be clearly recognized in low frequency. A large part of the spectrum is also outside the 95% confidence level. This frequency corresponds to a 362 days period and that refers to the annual cycle. There are some

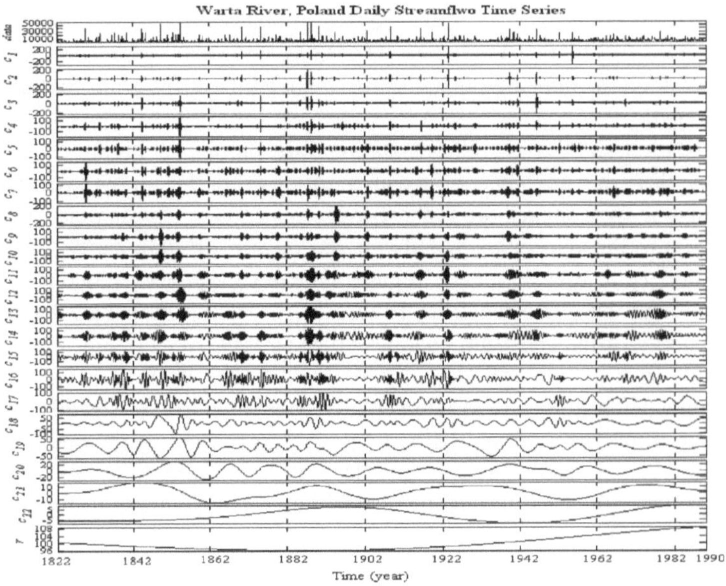

(a)

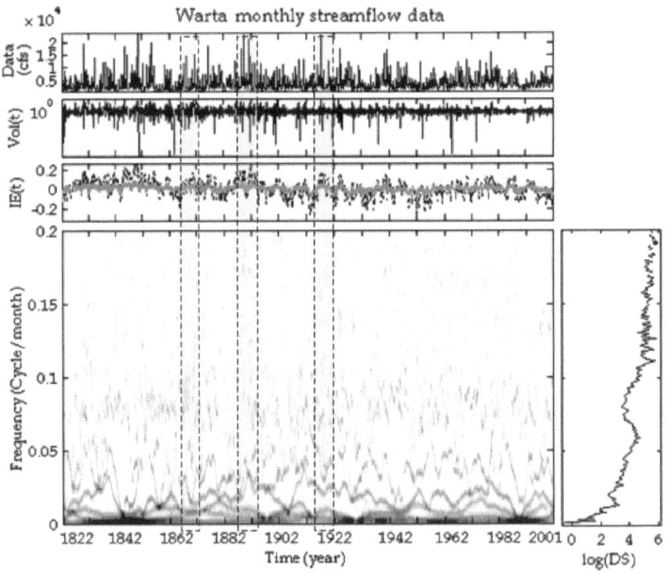

(b)

Figure 5.3.1. (a) Intrinsic mode functions and (b) time-frequency distribution of Warta River daily streamflow data

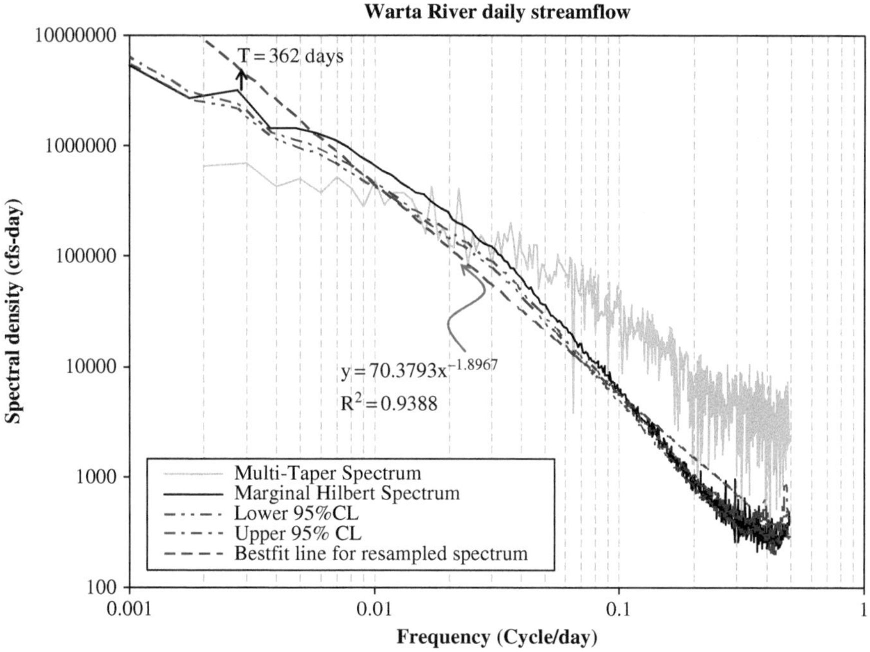

Figure 5.3.2. Marginal Hilbert spectrum of Warta River daily streamflow data

spikes in multi-taper spectrum. These spurious peaks make it not easy to detect the annual cycle from multi-taper method. As for the best fit line for the resampled marginal Hilbert spectrum, it reveals a good power law relationship. The decay rate is -1.897 with a fairly high R-square value of 0.939. Multi-taper spectrum also shows the energy decay in high frequency.

A large amount of data make the results from Hilbert spectra too smooth. An experiment is performed here to investigate the changes by studying shorter time series. The entire series is divided into 11 segments and each segment is 15 years long. The means and standard deviations of these segments are computed. The result showed that the means of these segments are within the 95% confidence limits.

The marginal Hilbert spectra are computed for each segment. The results of for these eleven segments are shown in Figure 5.3.3. These spectra still follow a power law well with an average decay rate -1.928 and standard deviation 0.046. The coefficient of variation is 2.4%. However, there are more peaks and dips in the spectra in Figure 5.3.3 than those in Figure 5.3.2. This is more helpful in detecting periodicities than using the entire length of data. The annual cycle is detected in all these spectra. In high frequency, there is a half year period is detected. Also 2, 3, and 6 year periods are detected.

However, this analysis depends on the length of segments. If these segments are not strictly stationary, it is difficult to expect the spectrum from a segment to

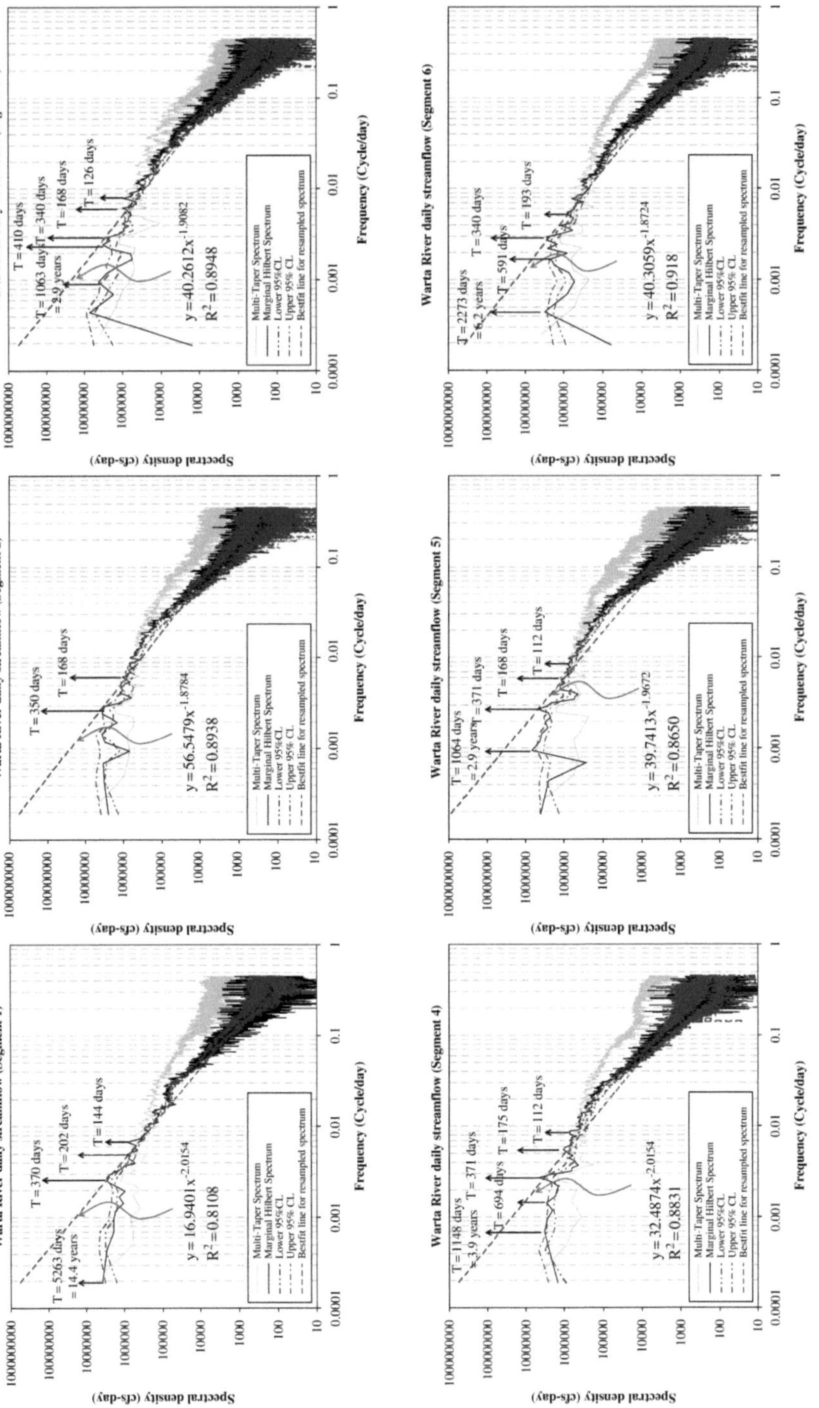

Figure 5.3.3. Marginal Hilbert spectra of segments of Warta daily streamflow data

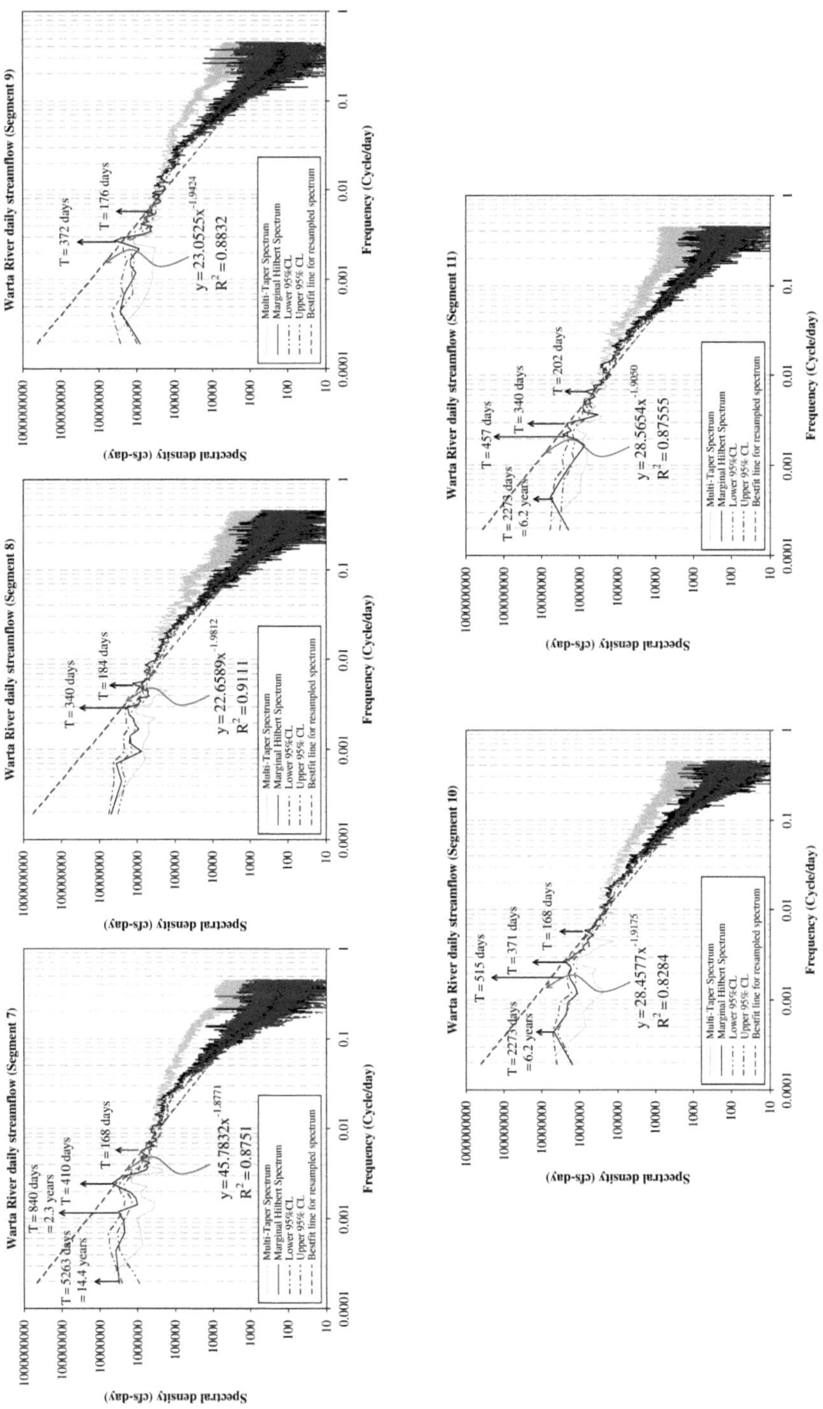

Figure 5.3.3. (Continued)

represent the entire time series. An average spectrum from these eleven spectra is computed and shown in Figure 5.3.4. The standard deviation of the average spectrum is also shown. Since the data are not strictly stationary the spectra vary from one segment to another. These variations are smoothed in the average spectrum. In the averaged spectra there are only two peaks, one corresponding to 1 year and the other to 6 years.

The effect of long time series in smoothing the spectrum is clearly seen. The HHT and MTM spectra are fairly close for the results of these eleven segments (Figure 5.3.3). However, from the results in Figure 5.3.2, the HHT spectrum is quite smooth but the MTM spectrum has quite a few spikes and valleys.

5.3.2 Warta River Monthly Streamflow Data

It is also of interest to study the Warta river streamflow data with a different time scale. Hence, the monthly data obtained by averaging the daily data are studied. The intrinsic mode functions are shown in Figure 5.3.5a. Compared to the results in Figure 5.3.1a, it is interesting to see that both the data have similar residuals. Although the values are different, the trends are the same.

The Hilbert spectrum of monthly streamflow data from Warta River is shown in Figure 5.3.5b. Apart from the Hilbert spectrum, the original data, volatility, instantaneous energy and its standard deviation are plotted. There are three peaks

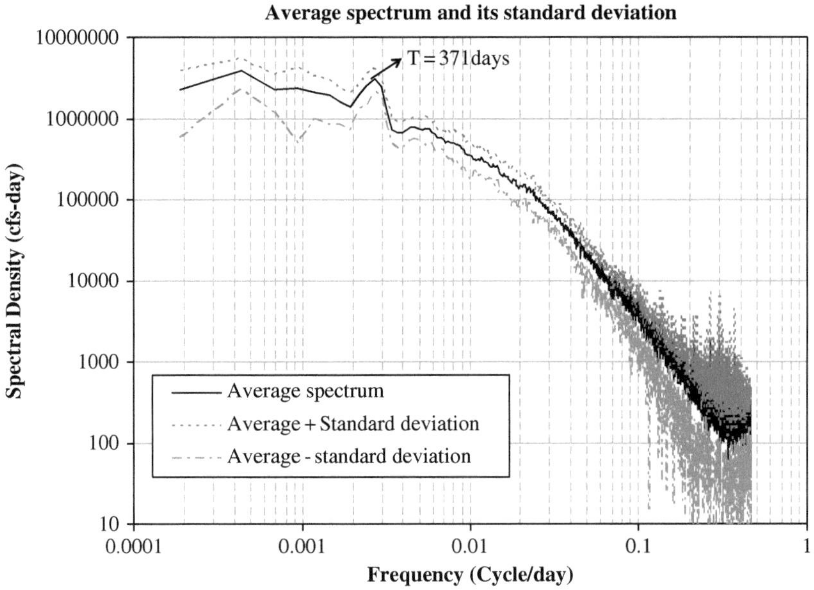

Figure 5.3.4. Average marginal Hilbert spectra of segments obtained from Warta River daily stream flow data

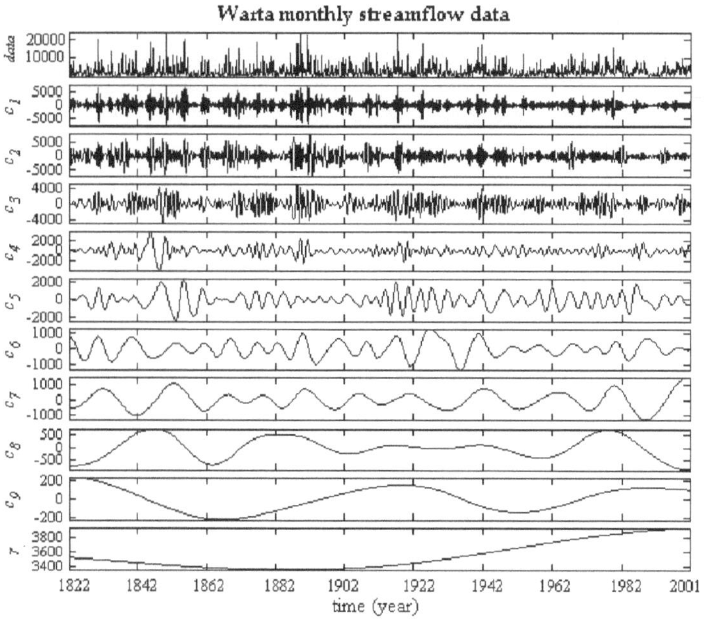

(a)

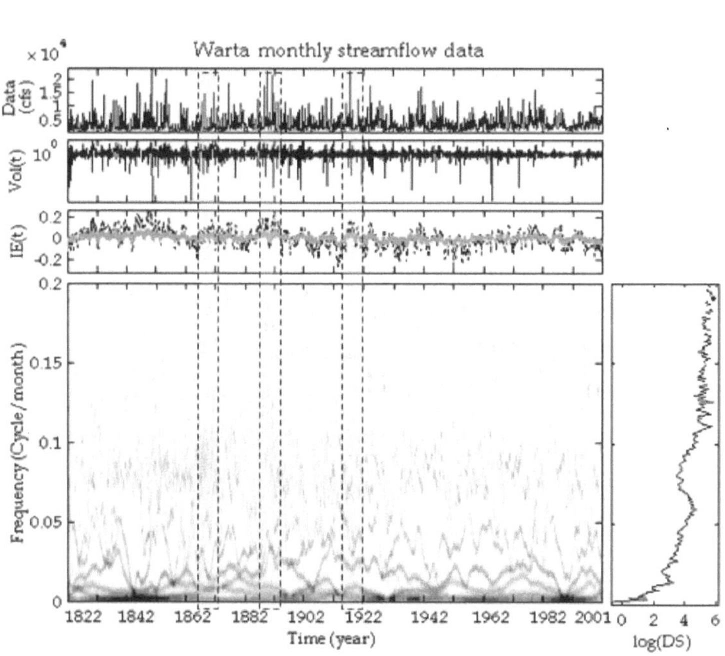

(b)

Figure 5.3.5. (a) Intrinsic mode functions and (b) time-frequency distribution of monthly streamflow data in Warta River shaded in boxes as examples. They occur in year 1870, 1887 and 1923. These peaks are the same as in daily data. For the years 1887 and 1923, the consistency among the peak, volatility and the standard deviation of instantaneous energy are very clearly seen. However, this consistency is not clear in year 1870. This may be due to the averaged monthly series smoothing the strong variations in daily data.

The degree of stationarity is similar to the results of USGS monthly streamflow data. The degree of stationarity is fairly low in low frequency and gradually increases when the frequency increases. There are some fluctuations and also a dip-down at the annual cycle. The marginal Hilbert spectrum and the 95% confidence intervals are shown in Figure 5.3.6. Besides the annual cycle, the long term oscillations are detected at 2 year, 8.5 year and 22 year periods. Multi-taper spectrum does not indicate these long term oscillations well. The energy of HHT spectrum is splitting around the annual cycle so it is not easy to clearly identify the 12 month period.

The high frequency region is very noisy and is not easy to interpret. However, it is interesting to see there are 6 month and 4 month periods occurring in this monthly streamflow data. They are inter-annual variations. The power law is not a good fit for these data because the annual peak introduces errors when fitting that segment. This is a common feature of monthly streamflows investigated in this study.

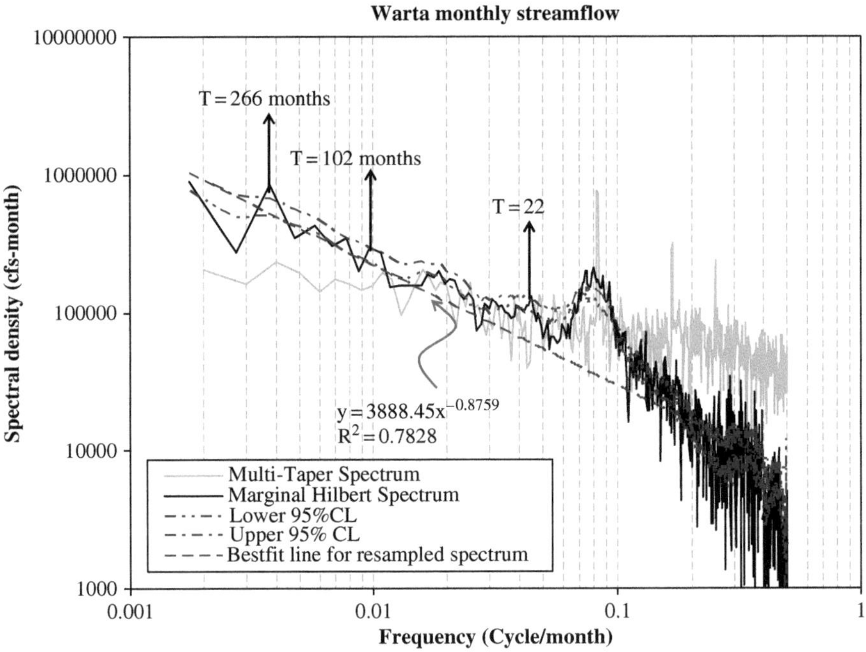

Figure 5.3.6. Marginal Hilbert spectrum and its confidence limits for Warta monthly streamflow data

5.3.3 Godavari River Monthly Streamflow Data

The intrinsic mode functions for the Godavari monthly streamflow are shown in Figure 5.3.7a. The data, volatility, instantaneous frequency and its standard deviation, Hilbert spectrum and degree of stationarity are shown in Figure 5.3.7b. The marginal Hilbert spectrum with its 95% confidence limits is shown in Figure 5.3.8.

There is an upward residual obtained from the IMF analysis (Figure 5.3.7a). The amplitude of the trend is about 3000 cfs, which is 2.5% the amplitude of original data. In Figure 5.3.7b, two peaks in year 1907 and 1953 and one low flow in year 1930. It is quite consistent from the data, volatility and the standard deviation of instantaneous energy. For the continuous low flow around year 1930, it also affects the instantaneous energy and clearly decreases the energy in that time span.

As for the degree of stationarity, it shows high nonstationary at the frequency close to zero. After that frequency, it goes down and then gradually increases as frequency increases. There are three dip-downs in degree of stationarity and they occur at frequencies 0.083, 0.05 and 0.025 cycle/month corresponding to 1, 1.7 and

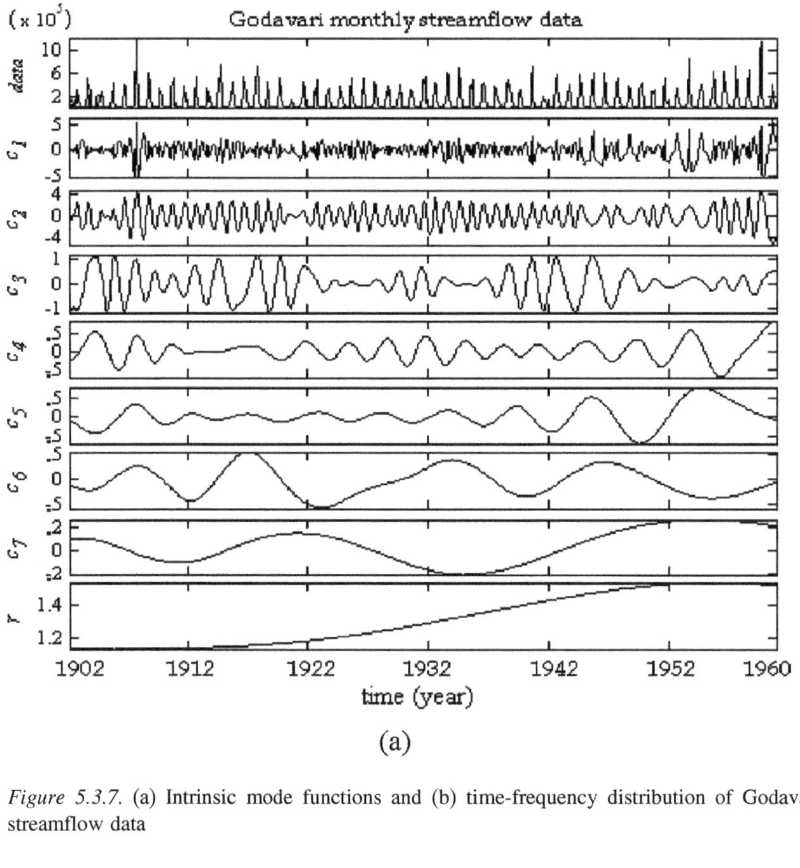

(a)

Figure 5.3.7. (a) Intrinsic mode functions and (b) time-frequency distribution of Godavari monthly streamflow data

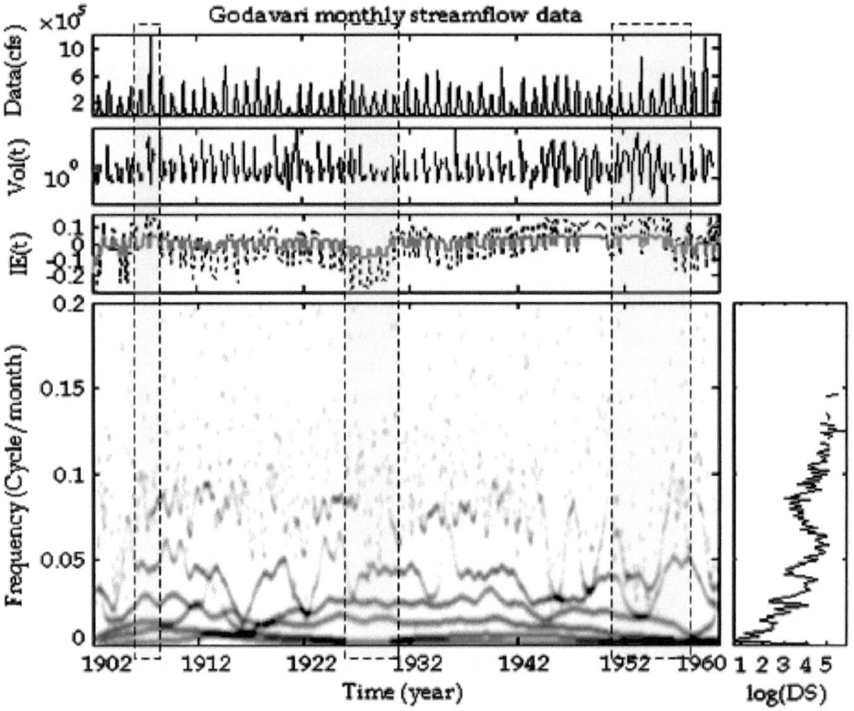

(b)

Figure 5.3.7. (Continued)

3.3 years. In Figure 5.3.8, besides the strong annual cycle, 2 year, 2.5 year, 3 year and 12 year periods are detected by using the 95% confidence intervals.

Also, the energy is quite high and persistent around the 2–3 year cycle, and this situation corresponds to the dip-downs of degree of stationarity at the frequency 0.05 and 0.025 cycle/month, corresponding to 20 and 40 month periods, respectively.

In the high frequency region, there are two peaks corresponding to 4 month and 6 month periods observed from the multi-taper spectrum. These are not present in marginal Hilbert spectrum. These components have strong energy. These interannual frequencies may mislead the interpretation. This result is also present in the Bayesian spectral analysis in Figure 5.3.9 (Hsu and Rao, 2005; Tirtotjondro, 1992). The results from Bayesian spectral analysis for the Godavari monthly streamflow is shown in Figure 5.3.9. The 4 month period is very strong. This is not reasonable because its strength is higher even than that of the annual cycle.

5.3.4 Krishna River Monthly Streamflow Data

The intrinsic mode functions for Krishna river monthly streamflow are shown in Figure 5.3.10a and the Hilbert spectrum, volatility, instantaneous energy and degree of stationarity are shown in Figure 5.3.10b.

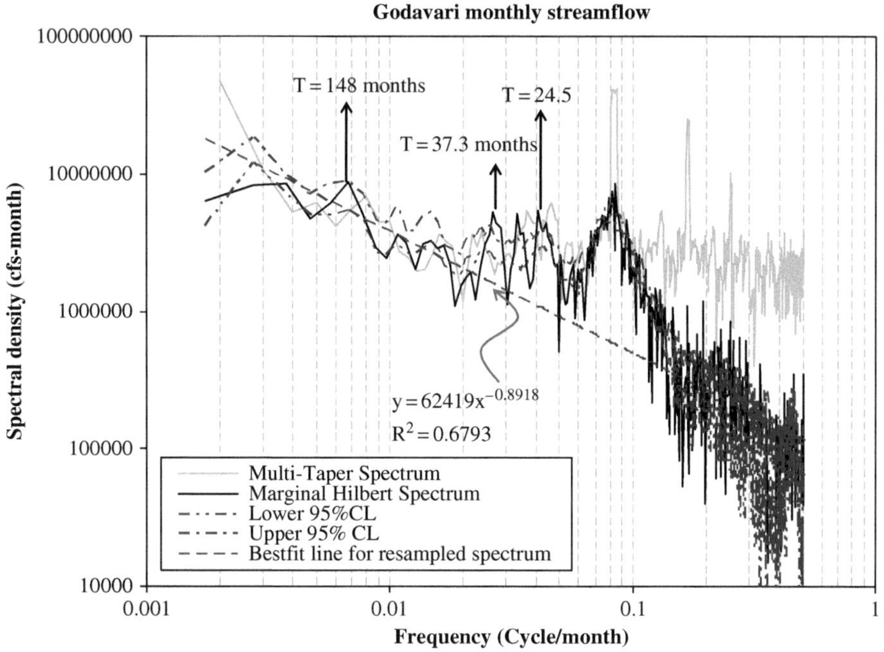

Figure 5.3.8. Marginal Hilbert spectrum of Godavari monthly streamflow data

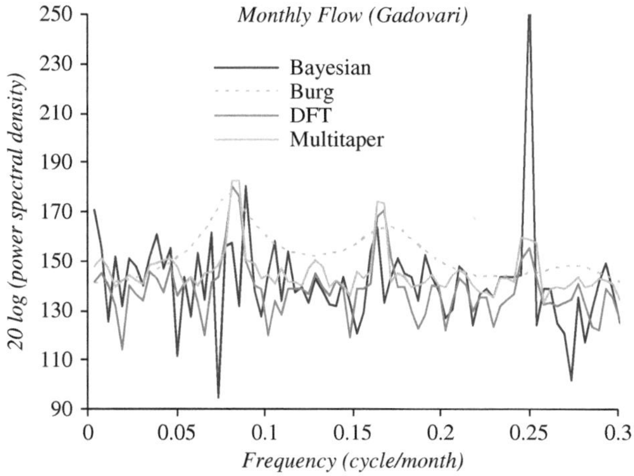

Figure 5.3.9. Bayesian, DFT, MTM and Burg spectra for Godavari monthly runoff data (Hsu and Rao, 2004)

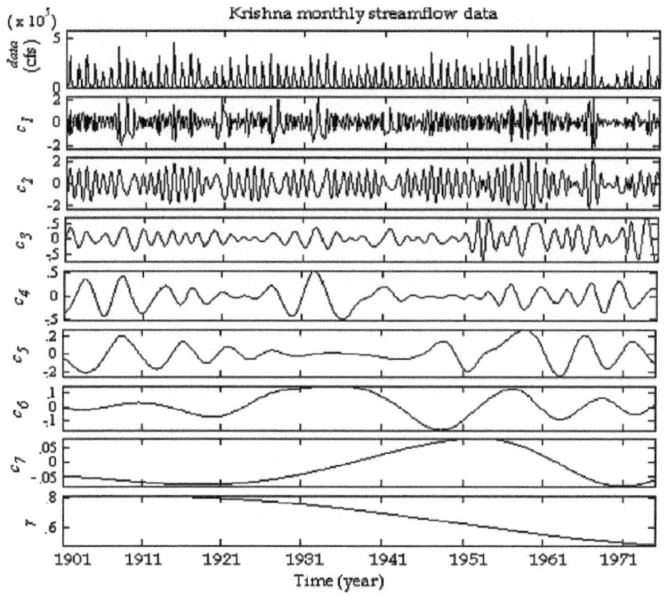

(a)

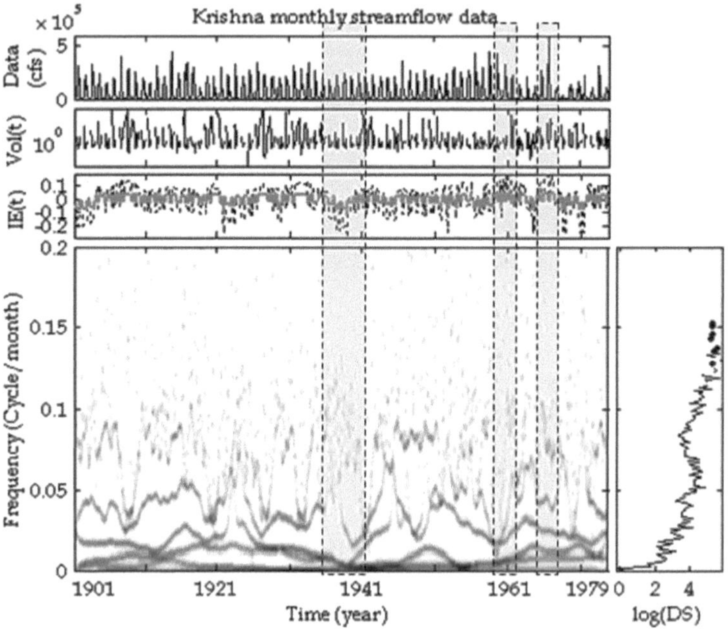

(b)

Figure 5.3.10. (a) Intrinsic mode functions and (b) time-frequency distribution of Krishna monthly streamflow data

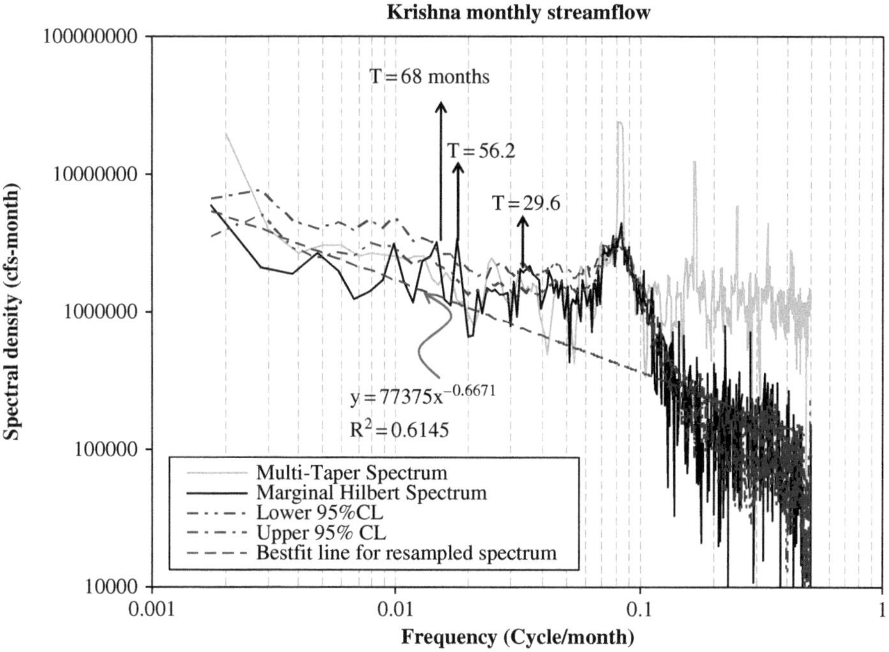

Figure 5.3.11. Marginal Hilbert spectrum and its confidence limits for Krishna monthly streamflow data

There is a downward residual which is around 5% of the streamflow amplitude (Figure 5.3.10a). This may be caused by the lasting low flows from year 1966. The reason may relate to the construction of Nagarjuna Sagar Dam, which started in 1956 and was completed in 1969. It is a masonry dam built across Krishna River in Nagarjuna Sagar, Andhra Pradesh, India. Three peaks are marked in the shaded box in Figure 5.3.10b. The segment in year 1941 corresponds to a low flow series and it lowers the instantaneous energy and makes a dip in standard deviation.

The marginal Hilbert spectrum and its confidence intervals are shown in Figure 5.3.11. Besides the annual cycle, 2.5, 5.7 and 6.7 year periods are detected. The power law is not a good fit in this case. The interannual periods are seen from multi-taper spectrum in Krishna River data. They are 6, 4, 3 and 2 month spikes in the high frequency region. The strong energy of these pseudo frequencies lead to misinterpretation of periodicities. These peaks are not present in the HHT spectra. This is the significant difference between the two estimates.

5.4. CONCLUDING COMMENTS

Several conclusions are made from the study of streamflow data. All the six USGS stations in the state of Indiana have upward residuals. Applying the Mann-Kendall test to each time series, the z values are 1.703 (U: a significant upward trend),

1.047, 1.581, 1.699 (U), 1,706 (U) and 1.984 (U) for the six stations. It shows that all of these series have an upward trend. Looking at the data in Figure 5.1.2, it is observed that the average streamflow data increases after year 1970. Quantitatively, the means before and after year 1970 are computed for station No. 03276500 (before: 1253 cfs; after: 1427cfs; difference: 12.2%), 03324500 (499 cfs; 548 cfs; 8.9%), 03326500 (624 cfs; 676 cfs; 7.7%), 03335500 (6250 cfs; 7380 cfs; 15.3%), 03373500 (5335 cfs; 6097 cfs; 12.5%), and 05518000 (1536 cfs; 1902 cfs; 19.2%). However, although there is an upward trend in NCDC rainfall data also, the percentage differences for the nine divisions are 3.1%, 6%, 5.9%, 5.2%, 6.4%, 2.3%, 8.8%, 7.3% and 7.7%, which are smaller than the differences in streamflows. In the frequency domain, the 12 month period in the marginal Hilbert spectra is not as strong as the annual cycle shown in multi-taper method spectra. This is similar to the results from monthly rainfall analysis for the state of Indiana. The energy spread around annual cycle indicates that the annual cycle may not exactly be 12 months, but vary between 11 and 13 months.

The results for the daily streamflow data of Warta River bring out the issue of smooth spectra caused by the large amount of data with high resolution. The marginal Hilbert spectrum is computed by integrating the Hilbert spectrum over time span. Unless there are fairly strong periodicities, the smoothness is quite easily produced by integrating tens of thousands variations in spectral estimates. However, the smooth spectra obtained from the high frequency signals may be useful to investigate the power law. In order to investigate the long-term oscillation and to identify the significant periodicities, the monthly streamflow data may be better. The annual cycle in marginal Hilbert spectra for Warta, Godavari, and Krishna rivers are stronger than in the spectra of USGS streamflow data. These results indicate these rivers have a clear division of flood and drought seasons.

CHAPTER 6

TEMPERATURE DATA ANALYSIS

6.1. INTRODUCTION AND DATA USED

Long-term monthly temperature time series in Europe and the state of Indiana are studied in this chapter. The long-term observations from Europe are daily climatic series (temperature and pressure) from instrumental measurements taken nearly continuously since the 18th century. Several papers (Camuffo and Jones, 2002) have been devoted to the correction of data derived from instruments in long time series. They describe the backgrounds of the instrument features, calibration methodologies, operational procedures, maintenance, relocation and instrument replacements. Many of these series have been corrected for systematic errors. The critical work of debugging, correcting, validating and homogenizing the series is essential, so that a reliable climate signal is available for study. The available record length from these meteorological observation sites are shown in Table 6.1.1. The time series are shown in Figure 6.1.1. The original temperature data are recorded daily, and the monthly time series have been derived from them.

For the state of Indiana, the data from HCN and NCDC are used. The locations are the same as the sites used in rainfall data (Figure 4.1.1 and Figure 4.1.3). The lengths of record for HCN data are listed in Table 6.1.2. The annual average time series for HCN and NCDC data are shown in Figure 6.1.2 and Figure 6.1.3, respectively.

6.2. EUROPEAN LONG-TERM MONTHLY TEMPERATURE TIME SERIES

A record of monthly temperatures longer than 200 years in European meteorological stations is a valuable source for studying the long-term climatic behavior. There are data from six stations investigated in this chapter. Also, the analysis is performed on original, linear detrended and annual-cycle removed time series.

Table 6.1.1. European long-term temperature time series

Station	Country	Record period	Length (yrs)
Cadiz	Spain	1786–2000	215
Milan	Italy	1763–1998	236
Padova	Italy	1725–1997	273
Stockholm	Sweden	1756–2000	245
St. Petersburg	Russia	1743–1997	255
Uppsala	Sweden	1722–2000	279

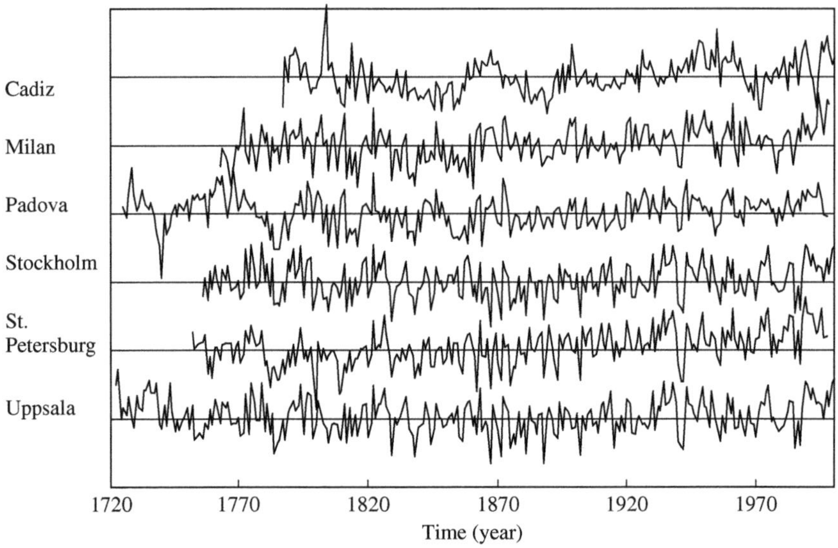

Figure 6.1.1. European monthly temperature time series

Table 6.1.2. HCN stations for temperature time series

No.	State	Station No.	Station Name	Record Year	Length
1		120177–05	Anderson Sewage Plant	1895–2002	108
2		120676–03	Berne	1910–2002	93
3		121229–06	Cambridge City	1892–2002	111
4		121747–05	Columbus	1885–2002	118
5	IN	122149–02	Delphi 3NNE	1885–2002	118
6		124008–01	Hobart 2WNW	1919–2002	84
7		125337–05	Marion 2N	1885–2002	118
8		126705–08	Paoli	1898–2002	105
9		128036–07	Shoals Highway 50 Bridge	1912–2002	91

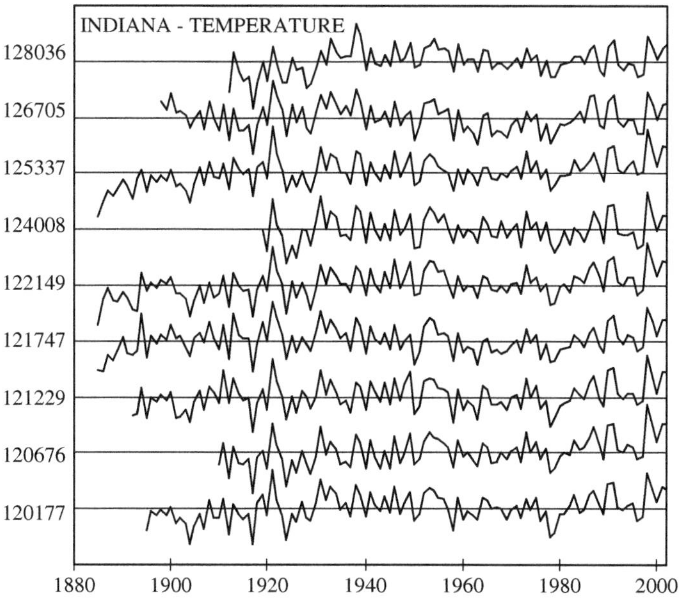

Figure 6.1.2. Annual HCN temperature time series

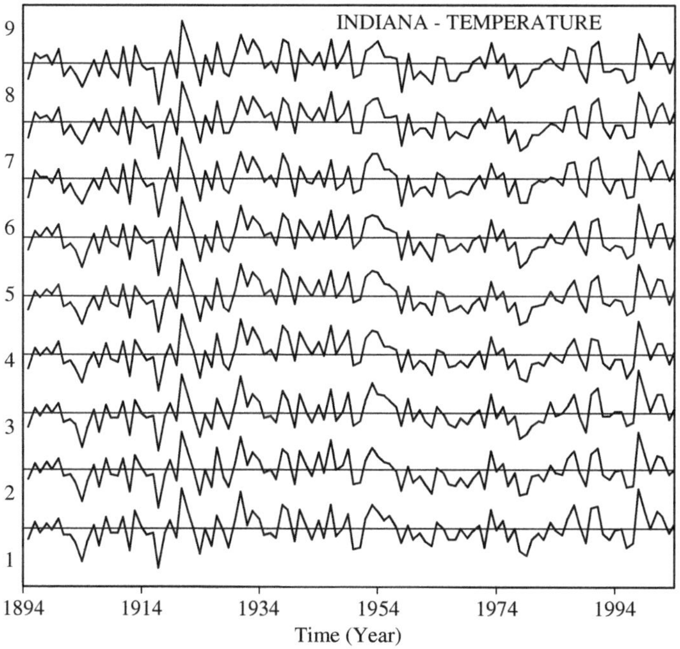

Figure 6.1.3. Annual NCDC temperature time series

6.2.1 Original Data

Through performing empirical mode decomposition, the original time series, IMF components ($c_1, c_2, c_3, c_4 \ldots$) and residual ($r$) for each data series is shown in Figure 6.2.1. In order to study how the last IMF components relate to the

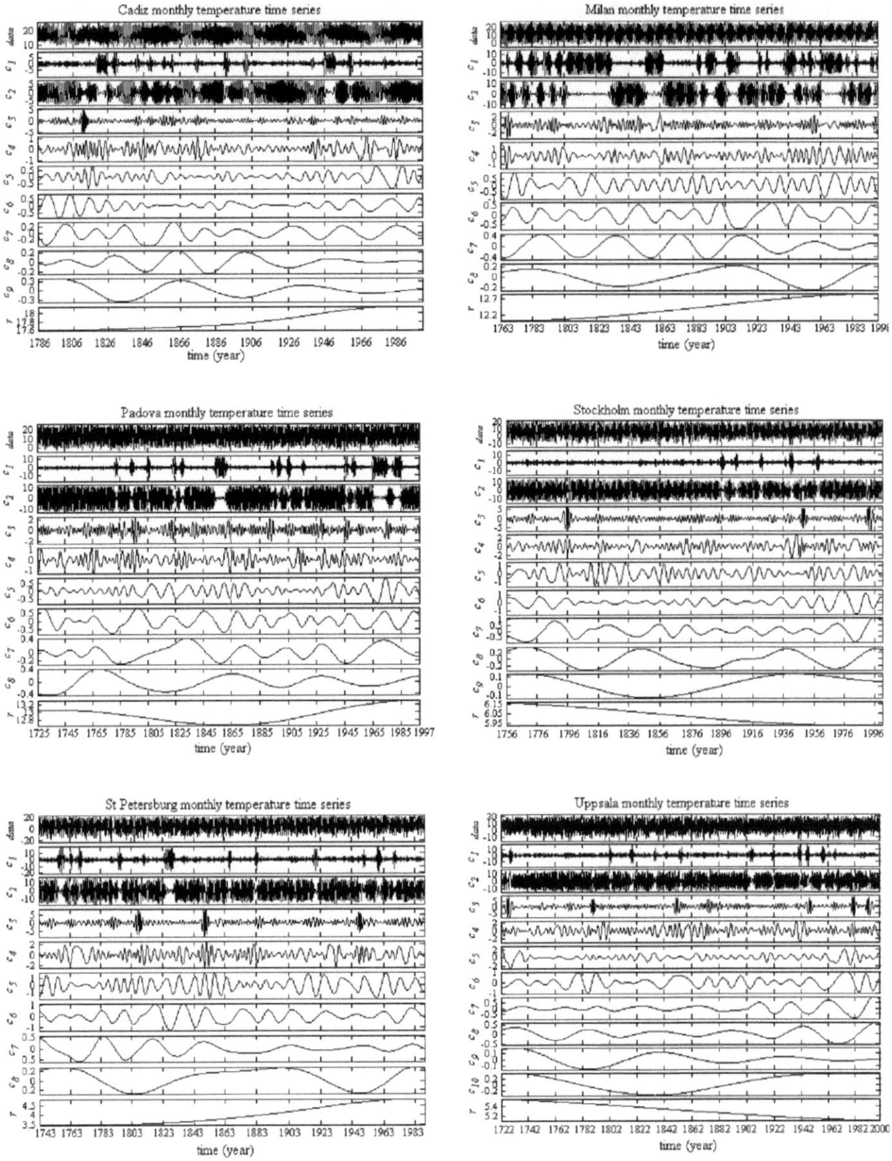

Figure 6.2.1. Intrinsic mode function components of European monthly temperature data

real time series, the modified Mann-Kendall test is used. The result is shown in Table 6.2.1. The piecewise trends from Mann-Kendall test compared to the last IMF component is shown in Figure 6.2.2 with a positive or negative sign. Also, the results of fitting a linear equation to the entire time series are shown in Figure 6.2.2.

First of all, the linear trend fitted for the entire time series is discussed. The general slope in the linear equation in Figure 6.2.2 relates to the residual (r) in Figure 6.2.1. The straight-solid line in Figure 6.2.2 is the linear regression for the entire data, which is the trend of the entire series. The fitted equation is shown in the

Table 6.2.1. Trend test results for monthly temperature time series in Europe

Station	Segment	Modified Mann-Kendall
Cadiz	1	Stationary (z = 0.7400)
	2	Stationary (z = −1.2893)
	3	Upward trend (z = 1.754)
	4	Stationary (z = −0.7844)
	5	Stationary (z = 0.5727)
	6	Downward trend (z = −1.7083)
	7	Stationary (z = 1.4982)
Milan	1	Stationary (z = 1.386)
	2	Stationary (z = −0.6103)
	3	Stationary (z = 0.7562)
	4	Stationary (z = 0.1589)
	5	Stationary (z = 1.189)
Padova	1	Stationary (z = 1.3163)
	2	Upward trend (z = 1.9048)
	3	Stationary (z = −0.9431)
	4	Stationary (z = −1.5911)
	5	Stationary (z = −1.2087)
	6	Stationary (z = 0.7629)
	7	Stationary (z = −0.2319)
	8	Stationary (z = 1.0896)
Stockholm	1	Stationary (z = −0.1013)
	2	Stationary (z = −0.8983)
	3	Stationary (z = 0.7441)
	4	Stationary (z = 0.6480)
St Petersburg	1	Stationary (z = −0.4300)
	2	Stationary (z = −1.0074)
	3	Stationary (z = 1.2887)
	4	Stationary (z = −0.2959)
	5	Stationary (z = 0.7762)
	6	Stationary (z = −0.4978)
Uppsala	1	Stationary (z = −1.3485)
	2	Stationary (z = −1.1082)
	3	Stationary (z = 1.1717)
	4	Stationary (z = 1.2278)

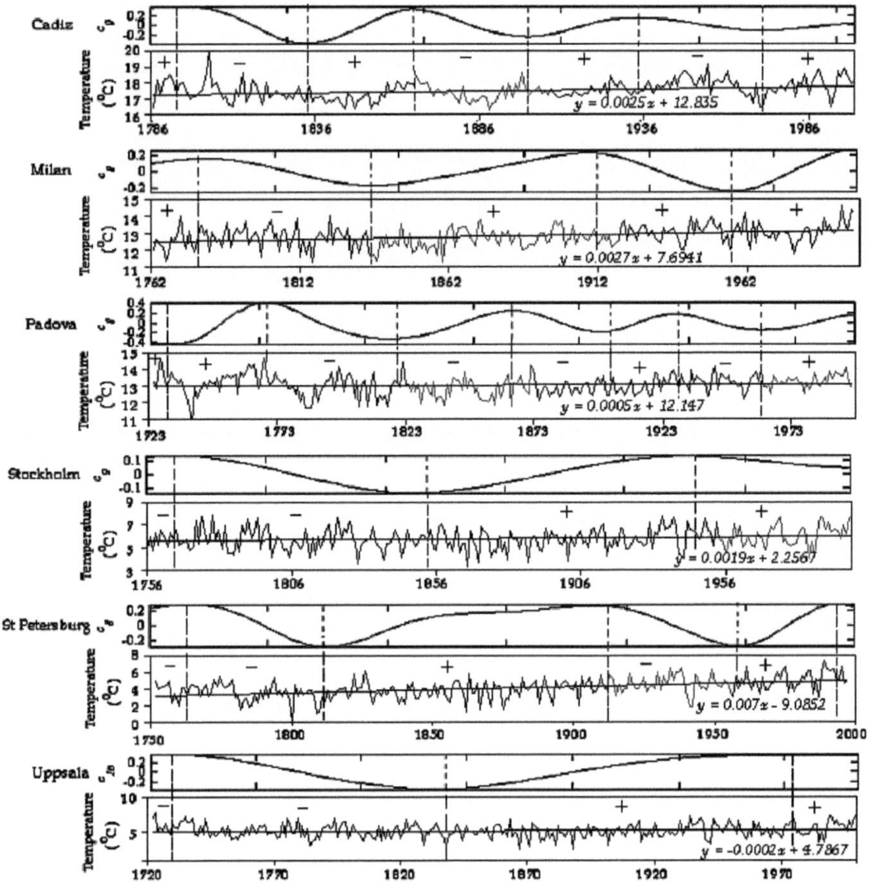

Figure 6.2.2. Linear-trend lines fitted to the European long-term temperature data. Solid straight line is regressed from the entire time series. The positive and negative signs are obtained from segments corresponding to the last IMF component by modified Mann-Kendall test

plot. The slopes read from the equations for Cadiz, Milan, Padova, Stockholm, St Petersburg and Uppsala are 0.0025, 0.0027, 0.0005, 0.0019, 0.0007 and −0.0002, respectively. Thus data from Uppsala is the only one having a decreasing overall trend.

The results of empirical mode decomposition show that (Figure 6.2.1), Stockholm and Uppsala data have negative trends (r) while the other data have positive trends. Most of the results match the characteristics shown by linear regression analysis (Figure 6.2.2). In Figure 6.2.1, data from Cadiz, Milan and St Petersburg have a positive slope and that from Uppsala has a negative slope, which is consistent with

Figure 6.2.2. The residual for Padova has two peaks, but regression of the entire data yields a positive slope.

Stockholm data does not have a consistent result with linear trend fitting. It has a positive slope in Figure 6.2.2 (+0.0019) but a decaying trend in EMD analysis. Therefore the signal itself is investigated further. By putting a breakpoint around year 1940, the entire signal can be shown to have two significant trends. Before 1940, a length of 180-year record has a negative trend, and after that a length of 60-year record has a positive trend. Although the persistence of negative trend is longer, the amplitude of the positive trend is larger, and hence it pulls up the trend of the overall record and ends up with a positive trend if it is fitted as a simple line. Hence, the residual in IMFs reveals the trend for most segments and also its variation with time. The residuals in Figure 6.2.1 give more detailed information about the trend.

As for the amplitude of the trend or the temperature difference for the entire span in Figure 6.2.1, Cadiz increases 0.5°C, Milan increases 0.6°C, Padova increases 0.2°C, Stockholm decreases 0.2°C, St Petersburg increases 1.4° and Uppsala decreases 0.4°C. The corresponding amplitudes of the trends read from Figure 6.2.2 are +0.5°C, +0.6°C, +0.14°C, +0.46°C, +1.6°C and −0.1°C. They are comparable to those in Figure 6.2.1.

The trends in segments of the data are considered next. The segments are separated depending on the last IMF components, i.e., c_9 for Cadiz and c_8 for Milan. In Figure 6.2.2, there are two plots for each station, one is the last IMF and the other is the corresponding time series. The time series are divided into several segments based on the maxima and minima of the last IMF curve to investigate the trends. The positive and negative signs on the top of each segment are obtained from Mann-Kendall test (Table 6.2.1). It is interesting to analyze the long-term oscillation from the intrinsic mode functions. Most of them have consistent results. Consequently the long-term oscillations can be investigated through the empirical mode decomposition without making a priori assumptions. Thus the empirical mode decomposition appears to work well for these data. Also, since the nonstationary phenomenon is investigated from Figure 6.2.2 for temperature data, it is not reasonable to use Fourier and other linear spectral analysis methods, which would produce spurious estimates in low frequencies and mislead the interpretation of spectra.

Hilbert transform is applied to all IMF components and then constructing the time-frequency distribution. The results for Cadiz, Milan and Padova data are shown in Figure 6.2.3 and the results for Stockholm, St Petersburg and Uppsala data are shown in Figure 6.2.4. The time-frequency distribution obtained by Short-time Fourier transform (STFT, spectrogram) and Multi-taper method are plotted as well for comparison.

The time frequency diagrams are significantly different for the HHT and the Fourier and MTM spectra. For the Cadiz data, only the low frequency component is seen in the Fourier and MTM spectra. The frequency around 0.09 cycle/month is

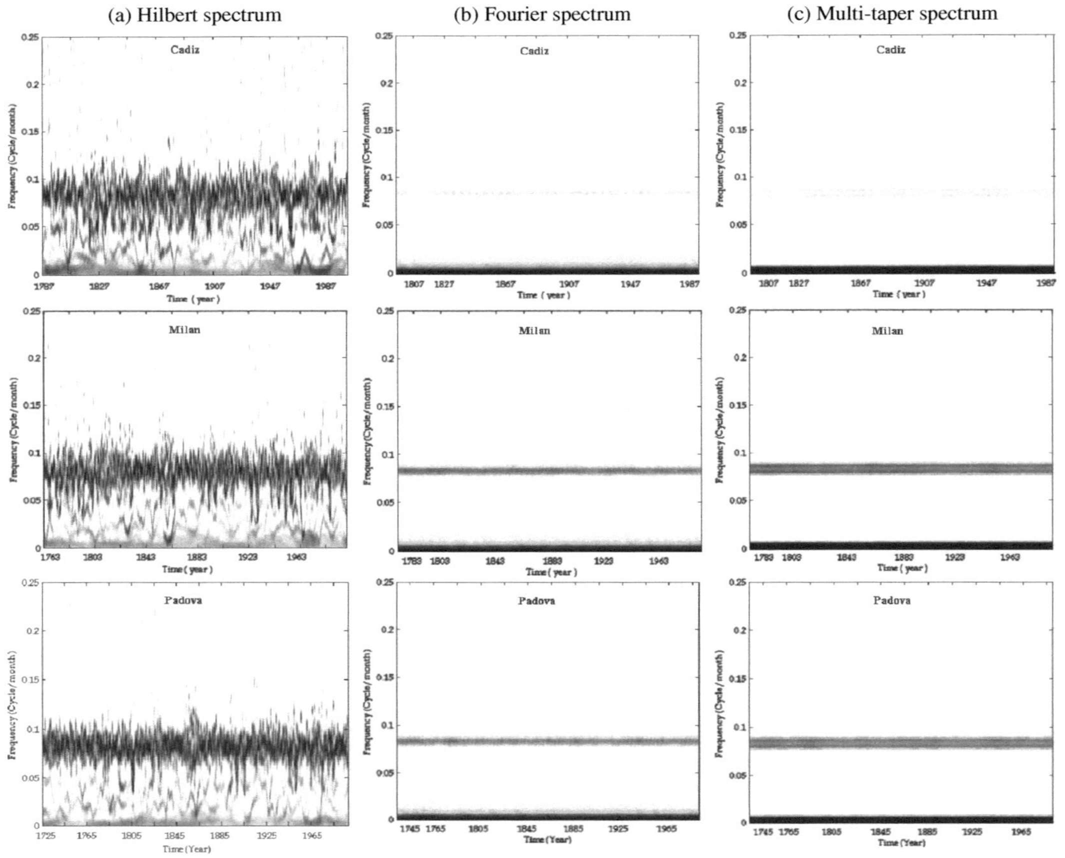

Figure 6.2.3. The time-frequency distribution of the European monthly temperature obtained by a 8 × 8 Gaussian filtered Hilbert spectrum, Fourier spectrum, and Multi-taper spectrum for data from Cadiz, Milan and Padova (a) Hilbert spectrum (b) Fourier spectrum (c) Multi-taper spectrum

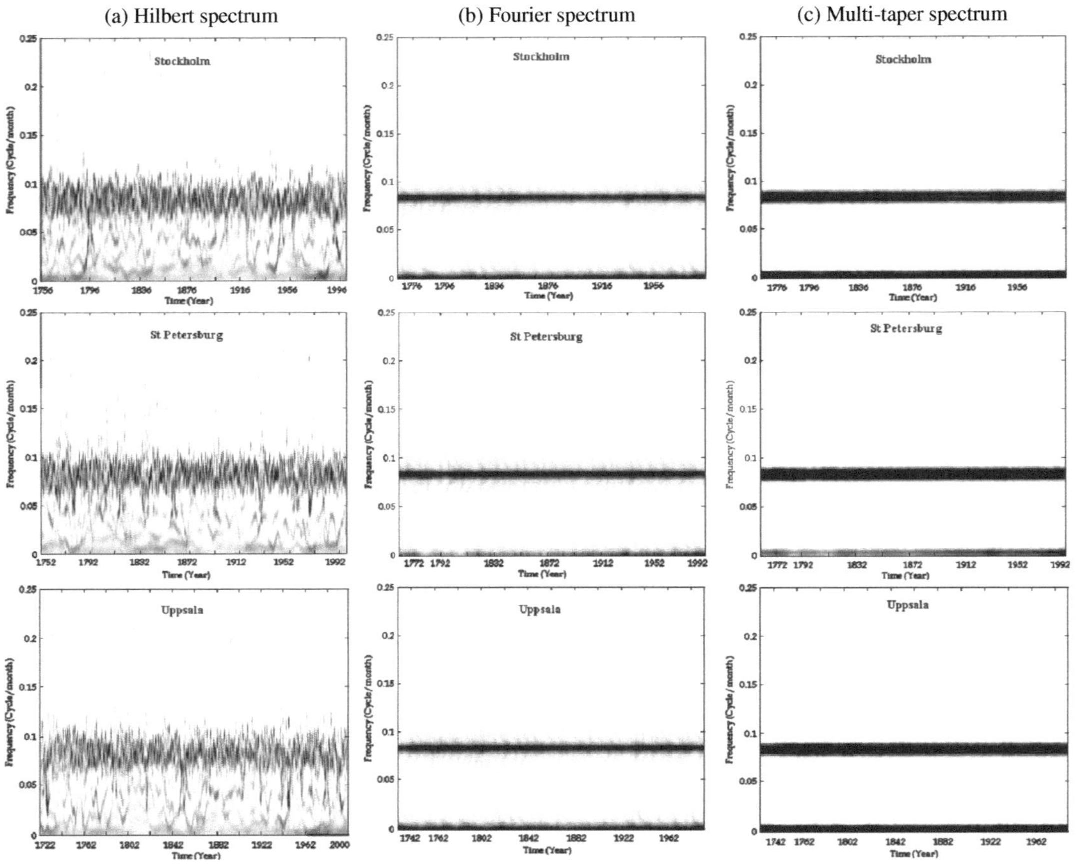

Figure 6.2.4. The time-frequency distribution of the European monthly temperature obtained by a 8 × 8 Gaussian filtered Hilbert spectrum, Fourier spectrum, and Multi-taper spectrum for data from Stockholm, St Petersburg and Uppsala (a) Hilbert spectrum (b) Fourier spectrum (c) Multi-taper spectrum

weak for the Fourier and MTM spectra whereas they are quite strong for the HHT spectra for Milan and Padova data. For these data, there is very little power in the low frequencies for the HHT spectra but it is quite strong in the Fourier and MTM spectra.

The marginal Hilbert, Fourier and Multi-taper spectra are shown in Figure 6.2.5. Also, a zoom-in version for Milan data is shown in Figure 6.2.6. From Figure 6.2.3,

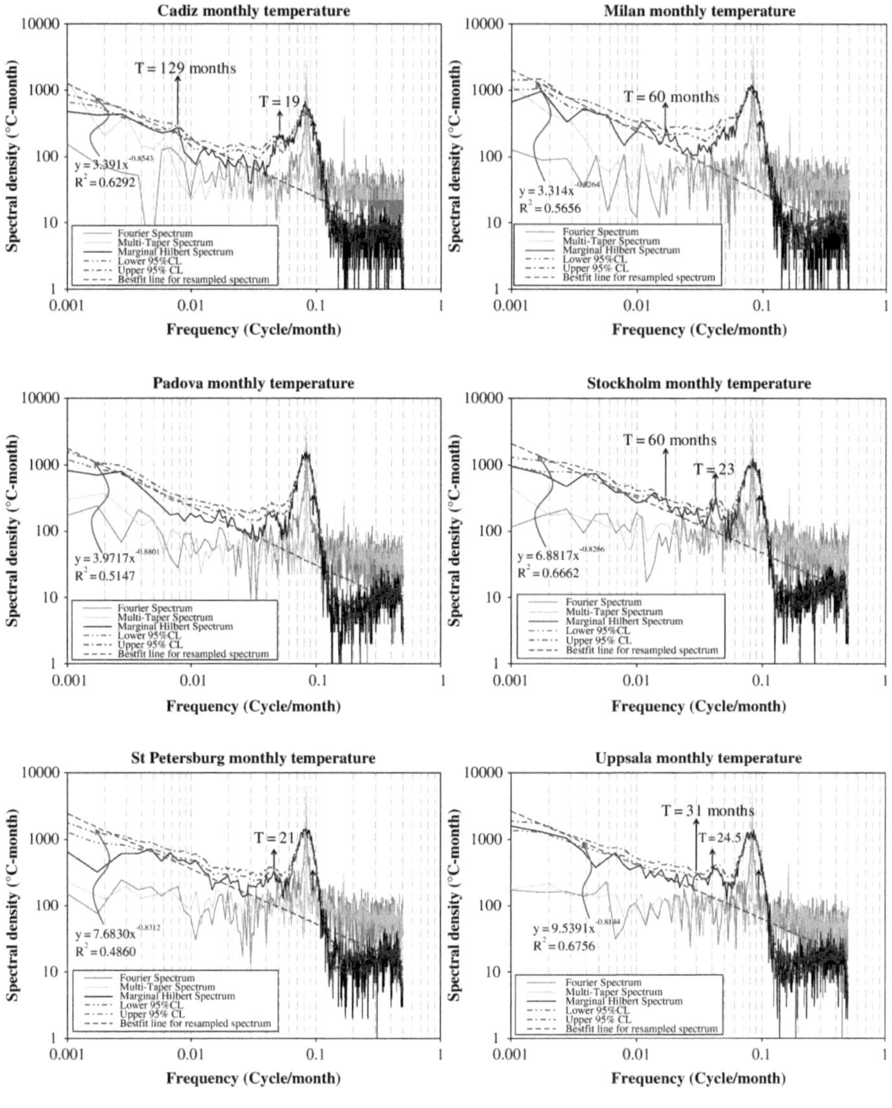

Figure 6.2.5. The Marginal Hilbert spectrum, Fourier spectrum and MTM spectrum of the European monthly temperature data

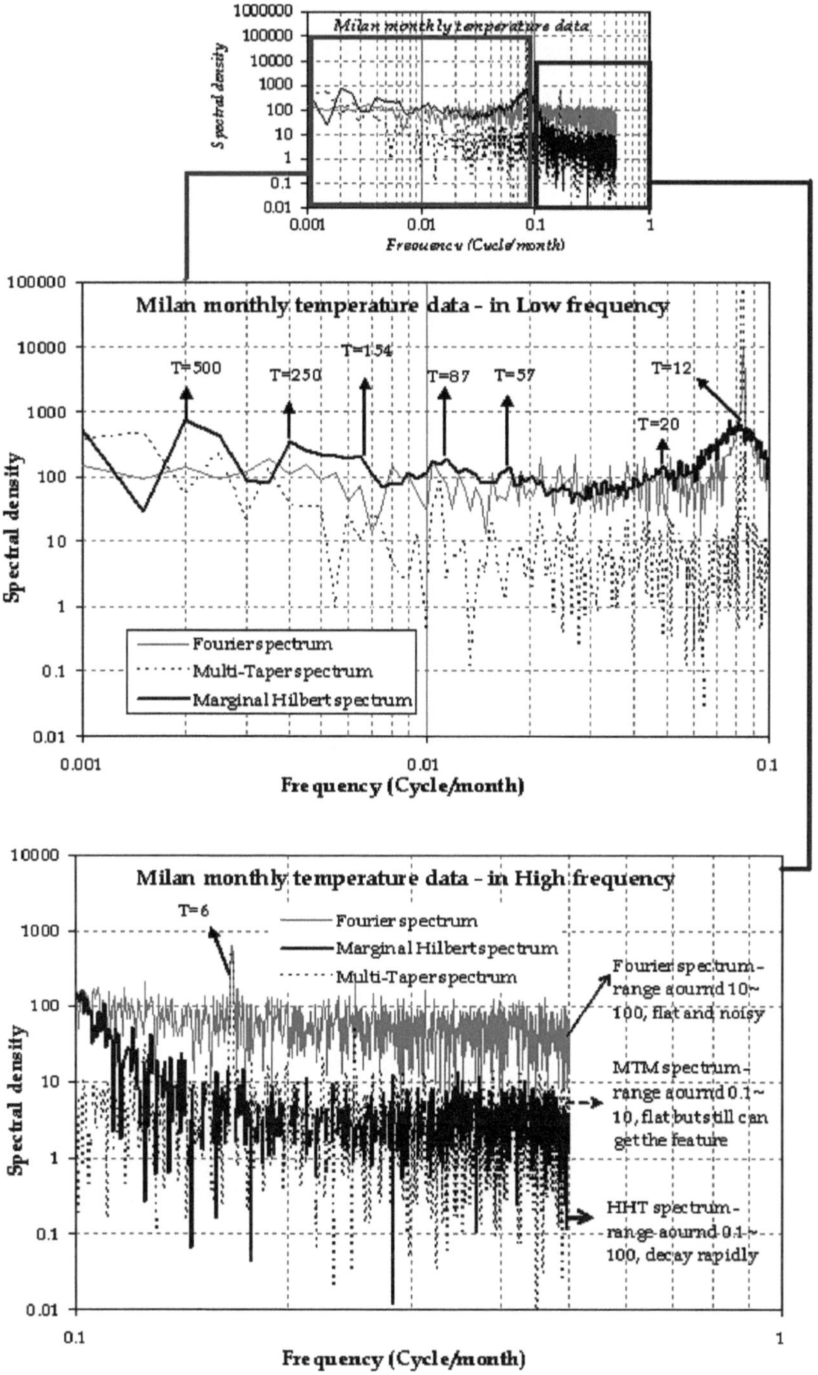

Figure 6.2.6. Zoom-in versions of the low frequency and high frequency spectra of Milan monthly temperature data

an annual cycle, which corresponds to frequency 0.083, is observable in all of them, but it is relatively weak in Fourier and MTM spectra. This might be caused by the nonstationarity and nonlinearity of the time series, since Fourier and Multi-taper spectra use linear schemes in their estimates. The long-term oscillations are stronger than the annual harmonic, and results in exaggerated power in very low frequencies and mislead the interpretation of significant periodicities. Hilbert-Huang spectra do not involve linear and stationary assumptions. The low-frequency variation in time is not prominent. Hilbert spectrum in Figure 6.2.4 still has a good representation of annual cycle. The low frequency estimate around zero has similar order of magnitude as the estimate of the annual frequency in Fourier and Multitaper spectra and hence they provide information which is erroneous.

From the time-frequency distribution (TFD) in Figure 6.2.3 and Figure 6.2.4, Multi-taper spectrum produces less leakage than the others but has the disadvantage that it yields a wider range of peaks when short time multi-taper spectra are estimated. From Figure 6.2.3, the MTM has a wide spectrum with less leakage. The center of the peak of multi-taper spectra matches the center of the peak obtained from DFT. However, if Figure 6.2.5 is examined, while the time span gets larger, the range of peak frequency obtained from Multi-taper spectrum gets narrower.

Ideally, the peak bandwidth should not have much difference between DFT and MTM. As the window length for each segment gets smaller, the frequency resolution gets coarser. By trials, it was found that the bandwidth calculated from MTM gets wider than DFT as the window length decreases. The time-frequency distributions of DFT and MTM shown here are calculated by using the same window length, length of overlapping and step of sliding window. An experiment may be made by trying different window lengths to see which one comes out to be the optimal length. However, during the trials, the loss of information in the beginning and ending portion of time series gets larger. It is the familiar trade-off situation in the sliding window analysis.

To illustrate the results from spectral analysis as shown in Figure 6.2.5, one plot is zoomed in as shown in Figure 6.2.6. It is separated to two regions, one is the low frequency and the other is the high frequency. The long term oscillation is investigated by detecting significant peaks in the low frequency regions. The high frequency contains too much noise and is impossible to interpret.

Comparing the three spectral analysis methods, Fourier transform is the poorest one in identifying the dominant frequencies because of energy leakage. Multi-taper spectra as well as Hilbert-Huang spectra have a good representation in frequency domain analysis especially in low frequency domain although multi-taper method still has leakage. The detected periods are also shown in Figure 6.2.5 and summarized in Table 6.2.2. Annual cycle is the obvious one detected for all of them while the 6 month period could be embedded in the annual cycle. Hilbert-Huang spectra fail to locate this 6-month period and that comes from

Table 6.2.2. Periods detected in European monthly temperature time series (unit: years)

Period	1 year	2 years	3~7 years	10 years	> 15 years
Casiz	1.0	1.6	–	10.8	–
Milan	1.0	–	5.0	–	–
Padova	1.0	–	–	–	–
Stockholm	1.0	1.9	5.0	–	–
St Petersburg	1.0	1.8	2.0 2.6	–	–
Uppsala	1.0	–	–	–	–

the integrating process in calculating the marginal Hilbert spectrum. A biannual cycle is detected but since the spectral amplitude is not concentrated at one frequency, the biannual cycle is not an exact value but ranges from 19~24 months. As for the ENSO-like cycle, which is a 4 year cycle, it ranges from 3 to 6 years. Quasidecadal (10~12 year period) cycles are also detected in Cadiz. The broad-band variations at low frequency relate to the large scale trends in the data.

6.2.2 Linear-Trend Removed Data

There is a linear trend in the data from six stations from both time domain analysis and frequency domain analysis (Figure 6.2.3). The trend makes the power spectra inflated near-zero frequency for Fourier and multi-taper spectra. Therefore the linear trend from the six time series are removed and the time-frequency distribution recomputed. Figure 6.2.7 shows the results for Cadiz, Milan and Padova and Figure 6.2.8 shows the results for Stockholm, St Petersburg and Uppsala. After the trend-removal, the power in low frequency is absent in the Fourier and MTM spectra. The Hilbert spectrum is not affected significantly by the removal of the linear trend. Therefore the Fourier and multitaper spectra are very sensitive to the non-stationarity of the time series.

The marginal spectra after removing the trend are shown in Figure 6.2.9. Compared to results in Figure 6.2.5, which are computed from the original time series, the spectra are different. From these results, we can see that removing a linear trend not only affects the near-zero frequency but also shifts other long-term oscillations. The leakage from Fourier and multi-taper spectra is reduced in detrended data than in the original data.

One year cycle is detected in all of them. Two years period is detected in Milan (1.5 year), Stockholm (2 years), St Petersburg (1.8 years) and Uppsala (2 years). Only one long term oscillation is observed in Cadiz and that is a 30 years period.

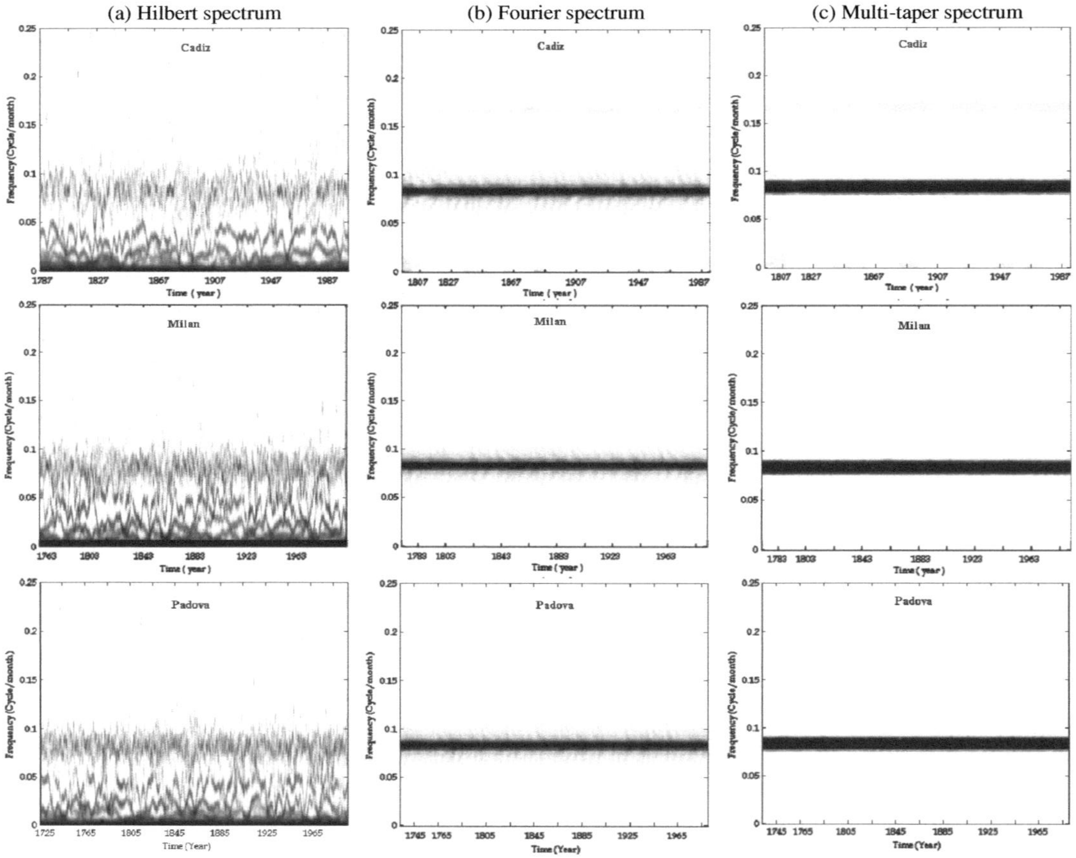

Figure 6.2.7. The time-frequency distribution of the linear-detrended European monthly temperature obtained by a 8 × 8 Gaussian filtered Hilbert spectrum, Fourier spectrum, and Multi-taper spectrum for data from Cadiz, Milan and Padova (a) Hilbert spectrum (b) Fourier spectrum (c) Multi-taper spectrum

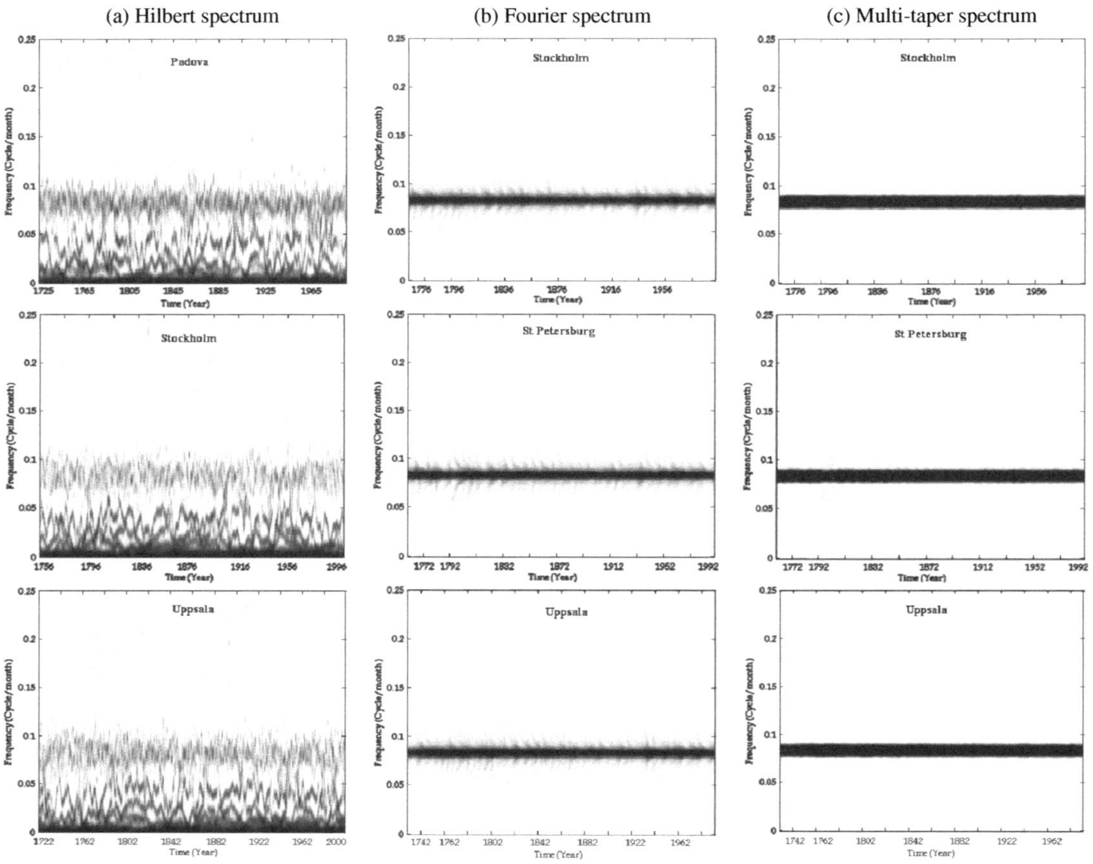

Figure 6.2.8. The time-frequency distribution of the linear-detrended European monthly temperature obtained by an 8 × 8 Gaussian filtered Hilbert spectrum, Fourier spectrum, and Multi-taper spectrum for data from Stockholm, St Petersburg and Uppsala (a) Hilbert spectrum (b) Fourier spectrum (c) Multi-taper spectrum

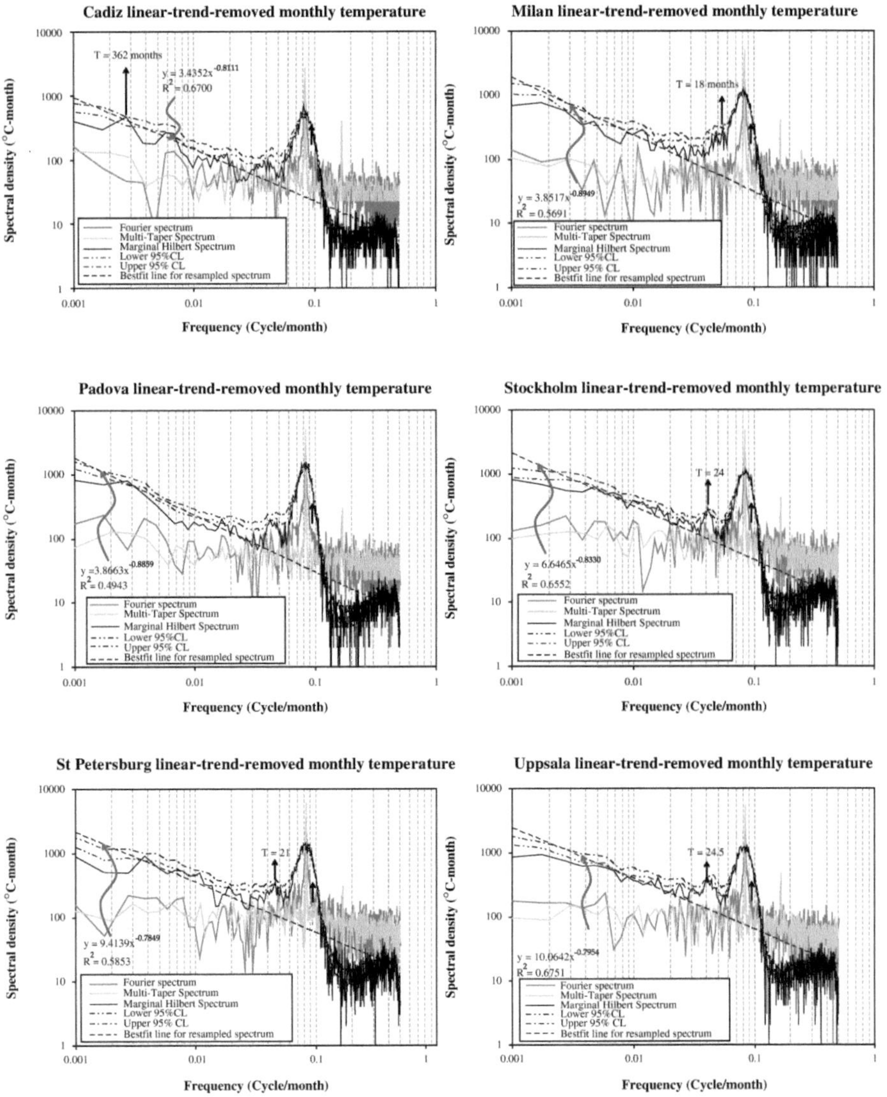

Figure 6.2.9. Marginal Hilbert spectrum, Fourier spectrum and MTM spectrum of the linear-trend-removed European monthly temperature data

6.2.3 Annual-Cycle Removed Data

The annual cycle is much stronger than other long-term oscillations so that the annual frequency dominates the spectrogram. The principal frequencies appear to remain the same over the time span, but their strength varies. Removing the annual cycle may enable better inspection of other cycles; hence, the annual period is removed by using Eq. (6.2.1) and a detrended series is considered. A linear trend is removed prior to removing the annual cycle in order to investigate the remaining oscillations.

$$y_{i,j} = \frac{x_{i,j} - m_j}{s_j}, i = 1, 2, \ldots, n(\text{year}); j = 1, 2, \ldots, 12(\text{month}) \tag{6.2.1}$$

where m_j and s_j are the mean and standard deviation for the jth month, respectively.

The time-frequency distributions for these detrended time series are shown in Figure 6.2.10 and Figure 6.2.11. The time-frequency distribution of the annual-cycle removed signal behaves like random noise without any harmonic waves. In the Stockholm, St Petersburg and Uppsala data (Figure 6.2.11), a smeared annual trend still exists in the beginning of the time series, which is the time before 1820 A.D. This may indicate that the earlier monthly temperature has more variation and bias than the whole series and a small oscillation remains even after removing the annual cycle. If we look at the first-half time series of Padova in the Figure 6.2.10, it has several low frequencies, which corresponds to 20~60 years long-term oscillations. If we look back to Figure 6.2.1, it indeed has the sinusoidal waves. Similarly, it is helpful to address the connection between the detected frequency in time-frequency distribution (TFD) and actual time series.

The analysis of the annual-cycle removed data shows that, the time-frequency representation looks more random and does not yield a unique pattern or frequency. That is because after removing the annual cycle, the data behaves similar to white noise except for several long-term oscillations embedded in the data. Because of the high randomness of the data, the frequency versus spectra representation is shown in Figure 6.2.12 cannot provide as much information as original series and linear-detrended series. The detected periods based on the 95 % confidence intervals are summarized in Table 6.2.3.

Removing a linear trend or annual cycle changes or shifts some periodicities detected in the original data. Hence, it is preferable to analyze the original data. From the time-frequency distribution, the nonstationarity embedded in the original data has strong effect upon the Fourier and Multi-taper spectra. Hilbert-Huang spectrum has less impact from trend and nonstationarity than the other two methods, so overall it provides stable and reasonable results.

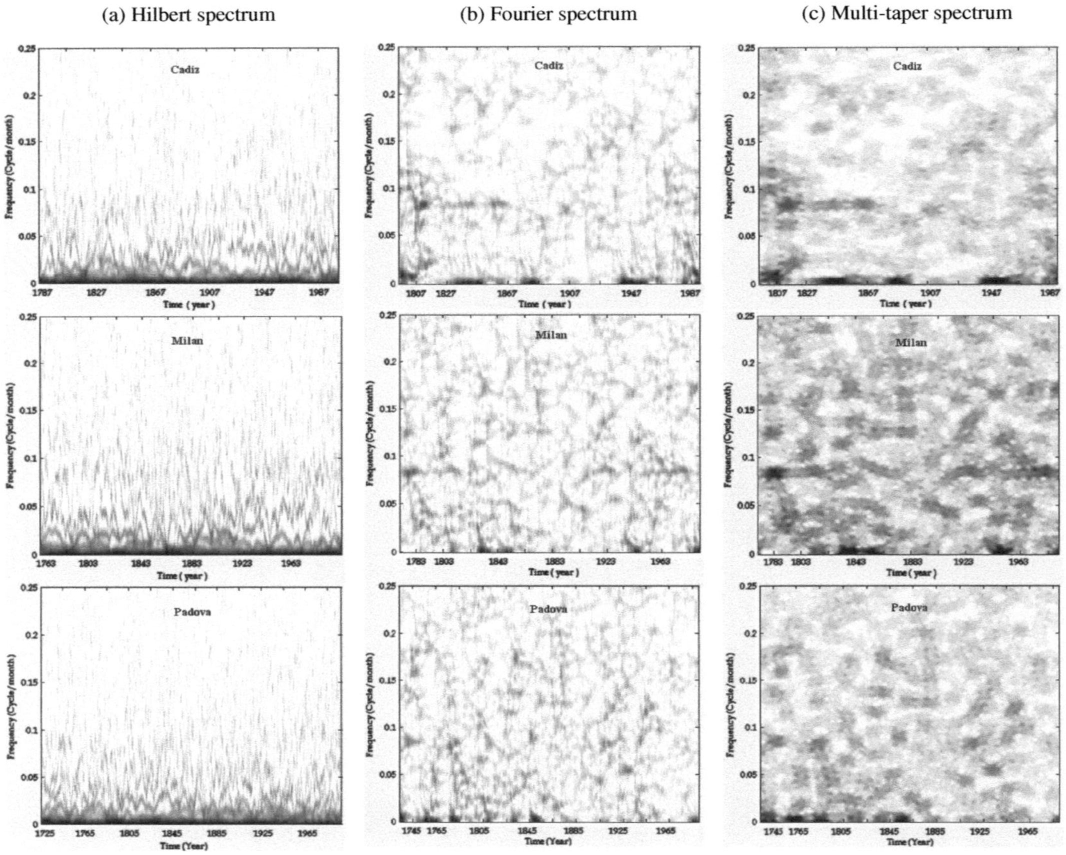

Figure 6.2.10. The time-frequency distribution of the annual-cycle removed European monthly temperature obtained by a 5 × 5 Gaussian filtered Hilbert spectrum, Fourier spectrum, and Multi-taper spectrum for data from Cadiz, Milan and Padova (a) Hilbert spectrum (b) Fourier spectrum (c) Multi-taper spectrum

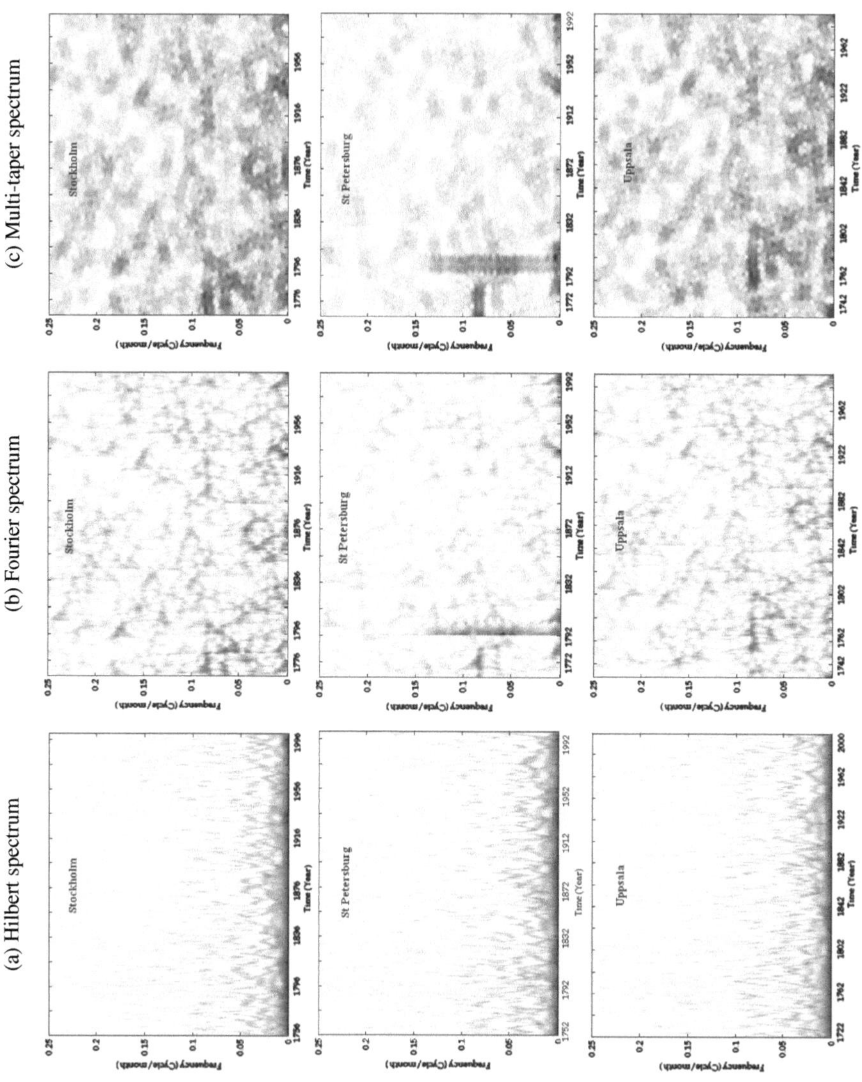

Figure 6.2.11. The time-frequency distribution of the annual-cycle removed European monthly temperature obtained by a 5 × 5 Gaussian filtered Hilbert spectrum, Fourier spectrum, and Multi-taper spectrum for data from Stockholm, St Petersburg and Uppsala (a) Hilbert spectrum (b) Fourier spectrum (c) Multi-taper spectrum

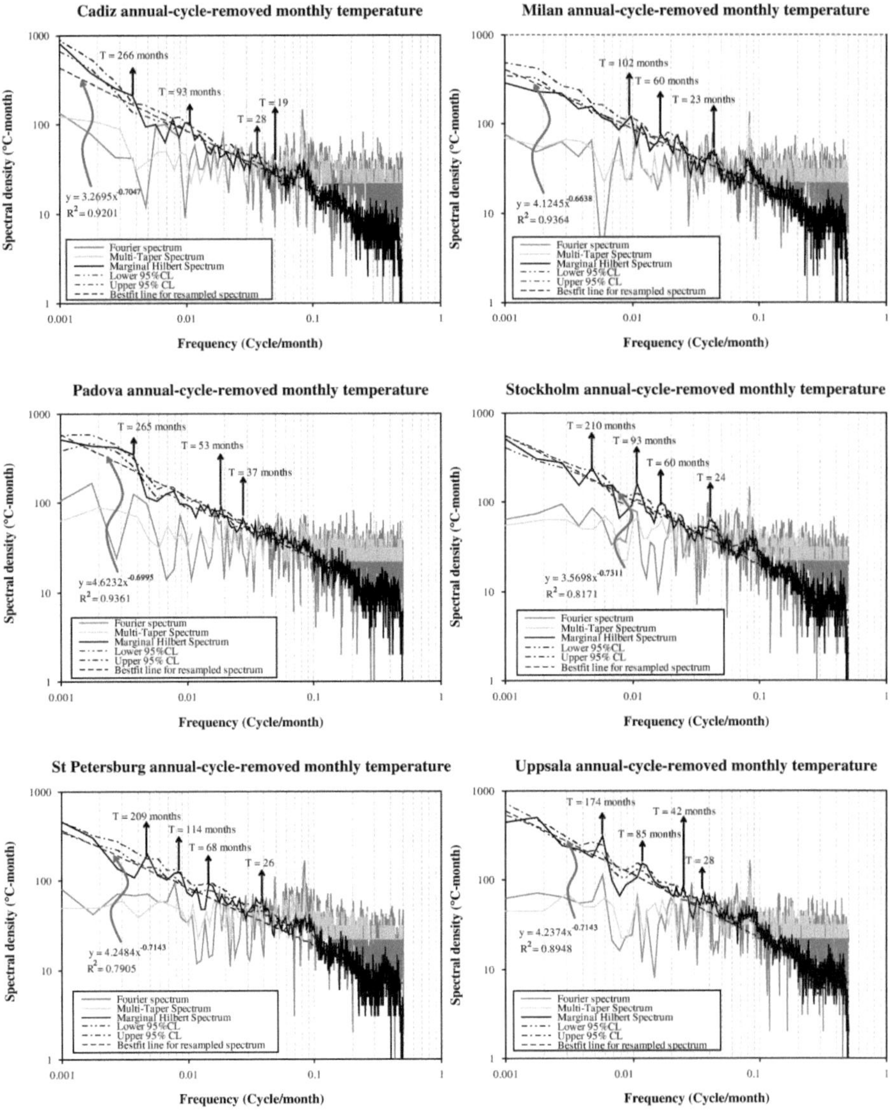

Figure 6.2.12. Marginal Hilbert spectrum, Fourier spectrum and MTM spectrum of the annual- cycle-removed European monthly temperature data

Table 6.2.3. Periods detected in annual-cycle removed European monthly temperature time series (unit: years)

Period	1 year	2 years	3~7 years	10 years	> 15 years
Casiz	1.0	1.6 2.3	7.8	–	22.2
Milan	1.0	1.9	5.0	8.5	–
Padova	1.0	–	3.1 4.4	–	22.1
Stockholm	1.0	2.0	5.0 7.8	–	17.5
St Petersburg	0.9	2.2	5.7	9.5	17.4
Uppsala	1.0	2.3	3.5 7.1	14.5	–

6.3. HCN AND NCDC MONTHLY TEMPERATURE TIME SERIES

6.3.1 HCN Monthly Temperature Time Series

The empirical mode decomposition is applied to the monthly average temperature from the nine HCN stations. The intrinsic mode function components are shown in Figure 6.3.1. The modified Mann-Kendall test is applied to the segmented data based on the last IMF components, which are c_7, c_8, c_7, c_8, c_7, c_7, c_8, c_7 and c_7, respectively. For the series considered, the stationarity or nonstationarity of those segments are summarized in Table 6.3.4. The consistency is examined in Figure 6.3.2. There are 28 consistent segments out of 36 segments.

The time-frequency distribution is studied by Hilbert spectra, which presents the information of time, frequency and the Hilbert spectrum amplitude in a two-dimensional plot. Also, the auxiliary time/frequency series, such as volatility, instantaneous energy, degree of stationarity and marginal Hilbert spectrum, are plotted in Figure 6.3.3. The shaded areas in Figure 6.3.3 refer to the strong variations. For the plot of the marginal Hilbert spectrum, the 95 % confidence intervals, best fit line for resampled data and multi-taper spectrum are also represented. The degrees of stationarity, the volatility and instantaneous energy for the nine HCN stations are extracted and plotted together in Figure 6.3.4, Figure 6.3.5a, b, respectively.

In Figure 6.3.4, similar to what has observed in rainfall and streamflow data, degree of stationarity is more prominent in low frequency than in high frequency; however, the dip-down in annual cycle is stronger than the results in rainfall and streamflow. It means that temperature time series have a strong and consistent 12 months period. The stronger 12 month period is also seen in the marginal Hilbert spectra. In addition to that, a dip-down with small amplitude occurs around 0.05 cycle/month (2 years period) is of interest. A 2 year period is detected in most of the HCN stations.

For the shaded areas in Figure 6.3.3, the volatility, instantaneous energy and standard deviation of instantaneous energy are quite consistent. Some of these segments can refer to variations in the original data but some of them are not easy to investigate. This is because temperature data is more stable compared to

(a)

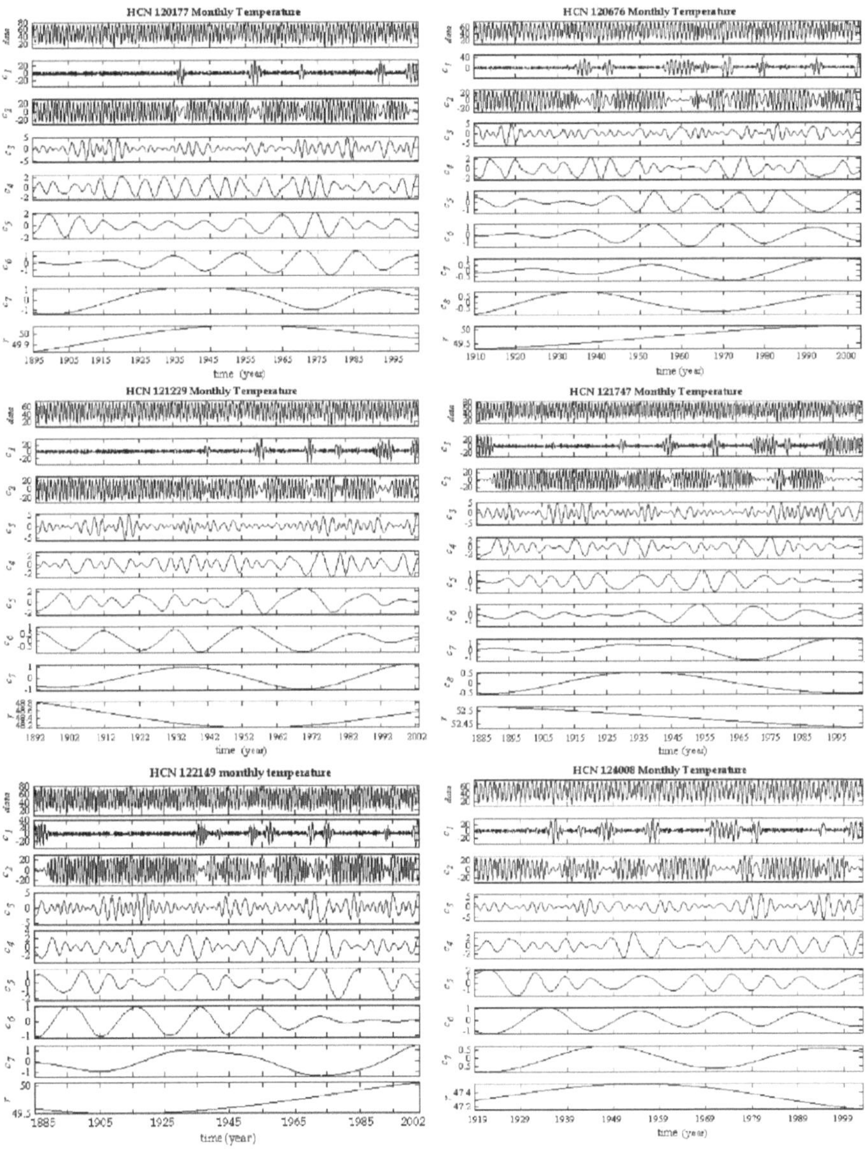

Figure 6.3.1. Intrinsic mode functions of HCN monthly temperature data

(b)

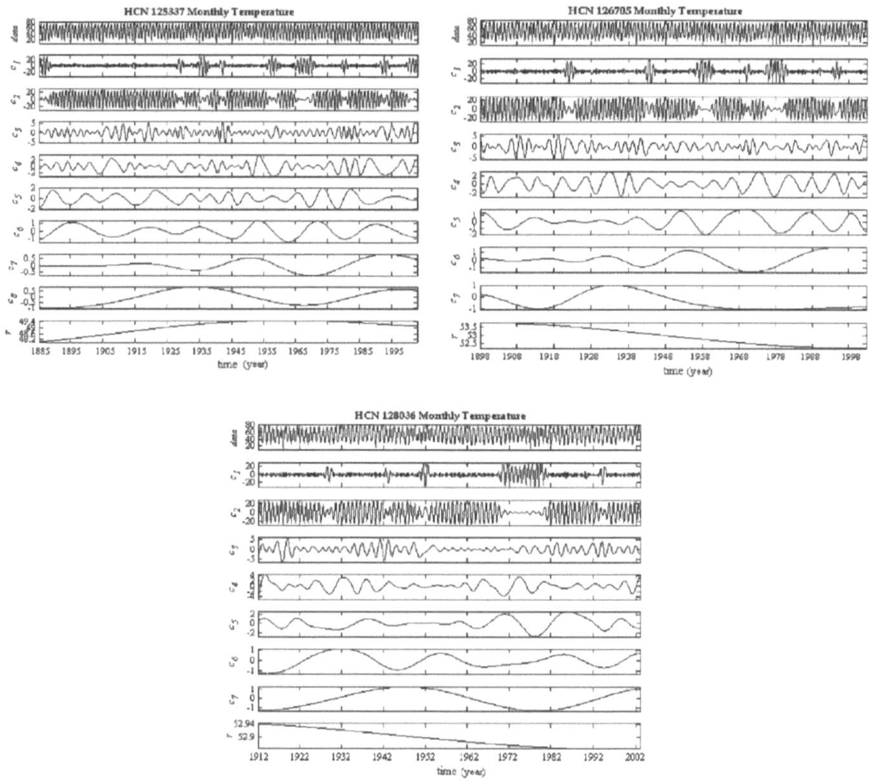

Figure 6.3.1. (*Continued*)

rainfall or runoff data. It has an obvious annual cycle and less variation from month to month, from year to year. On the other hand, rainfall and runoff are associated with global and local storm events. They are affected by flood or drought seasons.

To investigate the spatial correlation of the volatility and instantaneous energy, the results in Figure 6.3.5 are used. Data from years of 1936, 1958, 1971, 1980 and 1994 are used as the example. The results are shown in Table 6.3.5. If volatility is significant in a station, a circle mark ("O") is used to indicate it. Otherwise, a cross mark ("X") is used. For these 5 years, more than half the stations yield as high a variation of volatility as the others. For further comparison with NCDC data, the HCN stations are assigned to the corresponding NCDC divisions. From subdivision 1 to 9, the corresponding HCN stations are 124008, 125337, 120676, 122149, 120177, 121229, 128036, 126705 and 121747. The NCDC subdivision is marked shaded if the volatility is significant of the HCN station and left blank if not.

Table 6.3.4. Trend test results for HCN monthly temperature time series

Station	Segment	Modified Mann-Kendall
120177	1	Stationary (z = −0.162)
	2	Stationary (z = 1.568)
	3	Stationary (z = −0.859)
	4	Stationary (z = 1.550)
	5	Upward trend (z = 1.882)
120676	1	Stationary (z = 1.008)
	2	Stationary (z = −0.129)
	3	Stationary (z = 0.671)
	4	Stationary (z = −0.524)
121229	1	Stationary (z = 1.242)
	2	Stationary (z = 0.483)
	3	Stationary (z = 0.001)
	4	Stationary (z = 0.993)
	5	Stationary (z = −1.126)
121747	1	Stationary (z = 1.121)
	2	Stationary (z = 0.825)
	3	Stationary (z = −0.600)
122149	1	Stationary (z = −0.655)
	2	Stationary (z = 0.662)
	3	Stationary (z = −1.069)
	4	Stationary (z = 0.743)
124008	1	Stationary (z = 1.056)
	2	Stationary (z = −0.618)
	3	Stationary (z = −0.484)
	4	Stationary (z = 0.261)
125337	1	Stationary (z = 1.645)
	2	Stationary (z = −0.235)
	3	Stationary (z = 1.458)
	4	Stationary (z = 1.354)
126705	1	Stationary (z = −0.146)
	2	Upward trend (z = 1.959)
	3	Stationary (z = −0.677)
	4	Stationary (z = 1.192)
128036	1	Stationary (z = 1.594)
	2	Stationary (z = −0.729)
	3	Stationary (z = 0.980)

This is shown in Figure 6.3.6. The results indicate a consistent spatial continuity. For monthly temperature data, the volatility and standard deviation of instantaneous energy are quite consistent for most cases. Therefore, the spatial continuity study also represents the correlation of the instantaneous energy for the adjacent divisions.

The significant periodicities are investigated by using the marginal Hilbert spectra in Figure 6.3.3. These periods are summarized in Table 6.3.6. Annual

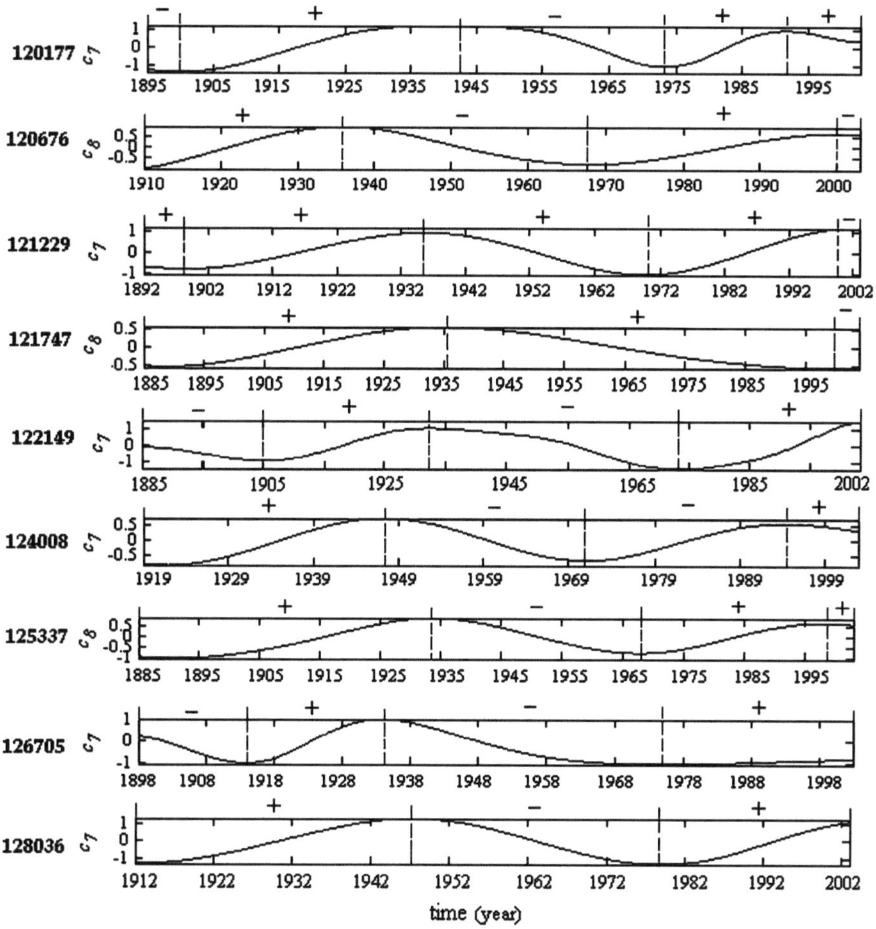

Figure 6.3.2. The last IMF components of HCN monthly temperature data and the results of modified Mann-Kendall test

cycle is commonly detected. Two years period is indicated in most stations. Four year, 10 year and 18 year periods are detected in some stations. The residuals are different from site to site as shown in Figure 6.3.1. The trends are $[+, +, -, -, +, -, +, -, -]$, which are corresponding to the station order used in this study, i.e. 120177, 120676, ..., 128036.

As for the best fit line for the resampled marginal Hilbert spectra, the average decay rate is –0.812 and standard deviation is 0.056. The coefficient of variation is 0.069. The fitting is poor based on the low R-square values. It is affected by the annual cycle and the low frequency spectra, which has lots of noise with low energy. A power law is not suitable for the monthly temperature data.

(a)

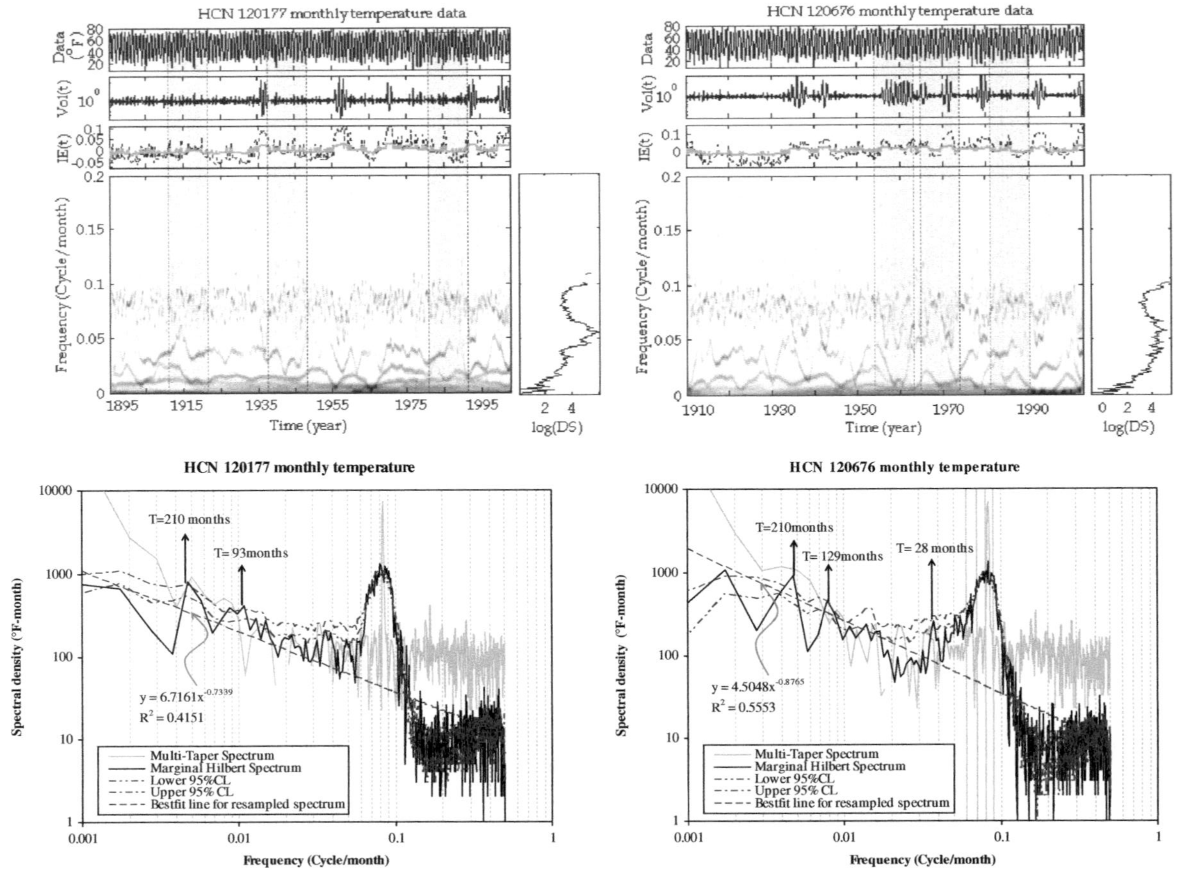

Figure 6.3.3. Time-frequency distribution and marginal Hilbert spectra of HCN monthly temperature data

(b)

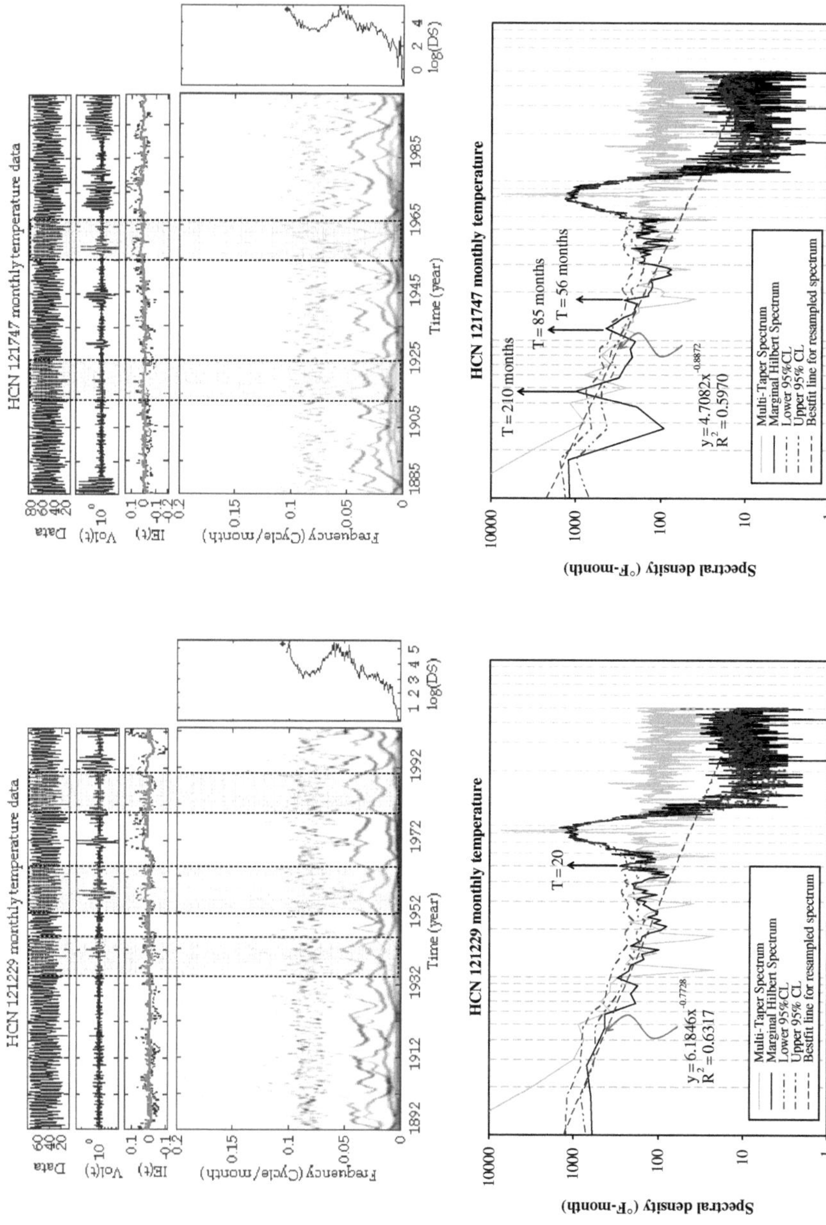

Figure 6.3.3. (*Continued*)

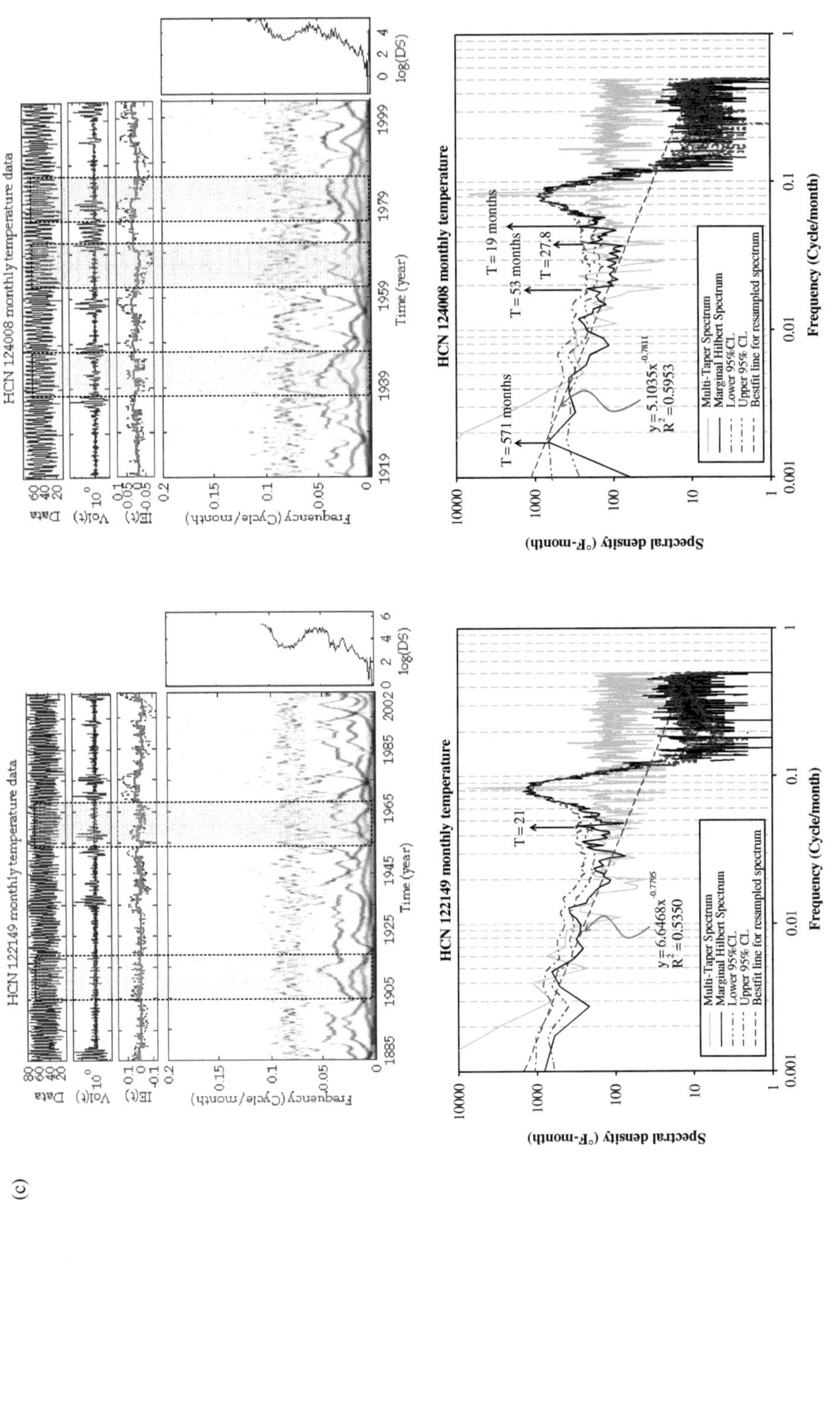

Figure 6.3.3. (Continued)

(d)

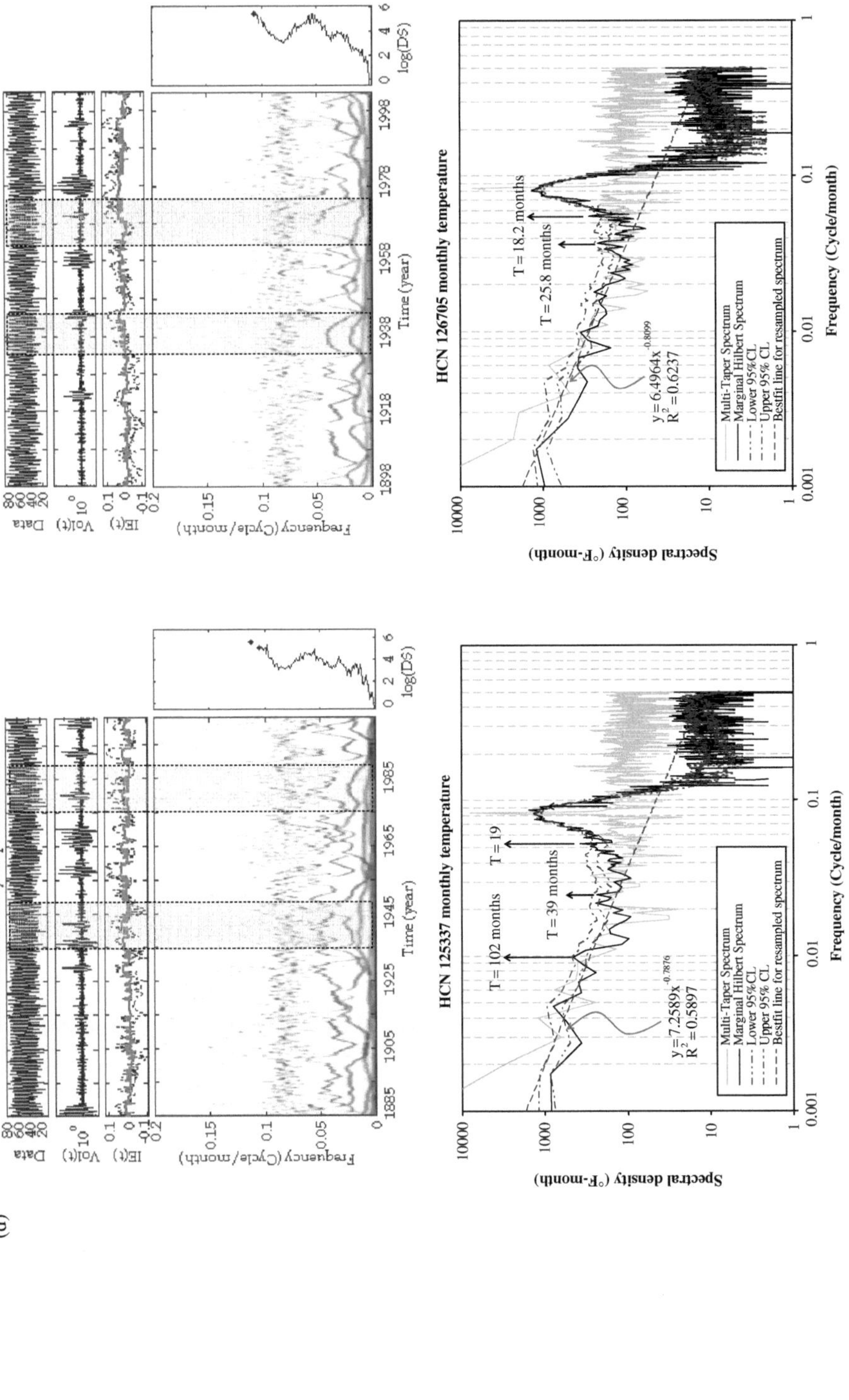

Figure 6.3.3. (*Continued*)

(e)

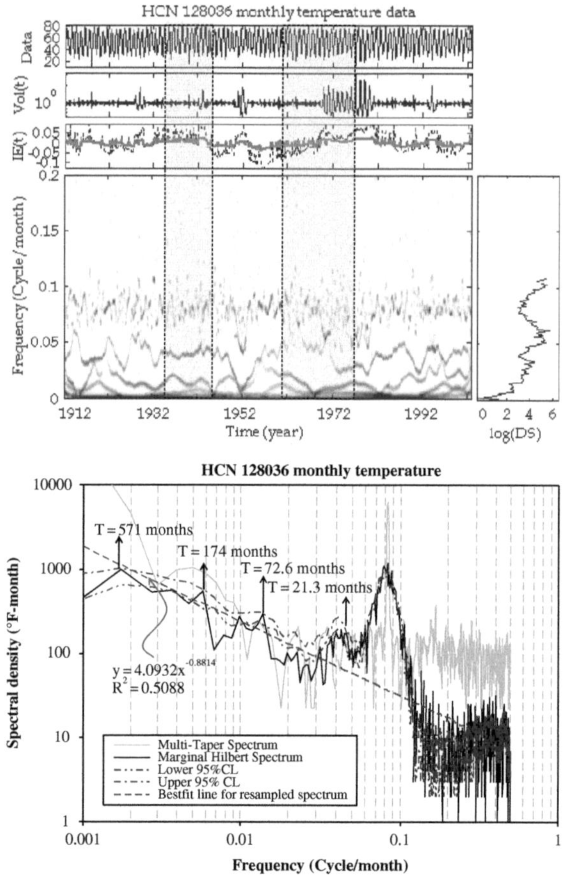

Figure 6.3.3. (*Continued*)

6.3.2 NCDC Monthly Temperature Time Series

The intrinsic mode functions of NCDC monthly temperature data are shown in Figure 6.3.7. It may be recalled that the NCDC monthly rainfall data all have an upward residual (Figure 4.3.1). The residuals of NCDC temperature data do not have such a consistent trend. It has an upward trend in division 2, 3, 6 and 9 while downward trend appeared in division 1, 4, 5, 7 and 8. HCN temperature data also do not have a consistent trend for all stations as discussed in the previous section.

The modified Mann-Kendall test is applied to the segmented data of the last IMF component. The results are shown in Table 6.3.7. The consistency between the z value and the slope of the last IMF component is examined from Figure 6.3.8. A positive slope should refer to a positive z value. The results show that 31 out of 41 segments are consistent.

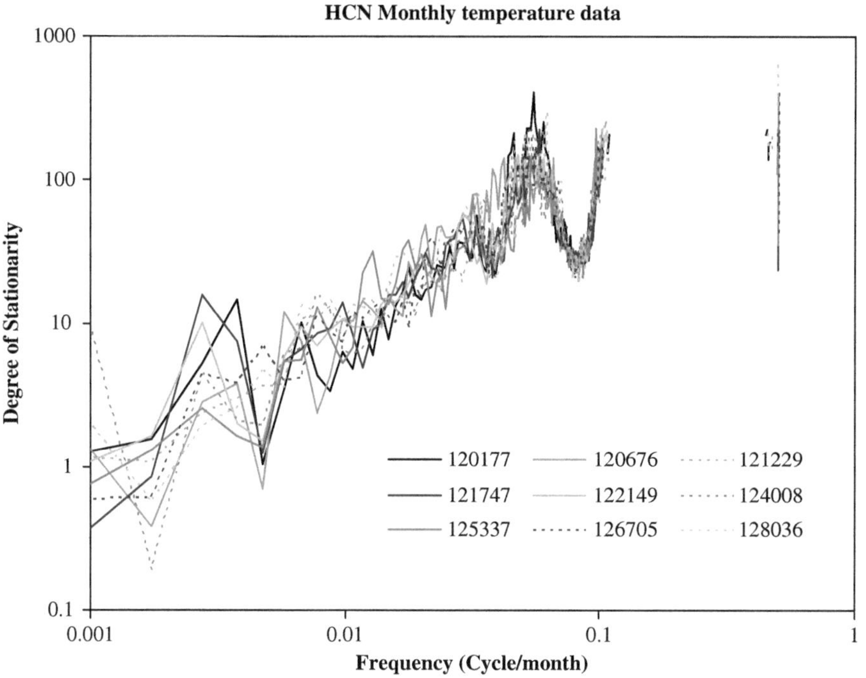

Figure 6.3.4. Degree of stationarity for HCN monthly temperature data

The time-frequency distribution is studied by using Hilbert spectra. The auxiliary time/frequency series, such as volatility, instantaneous energy, degree of stationary and marginal Hilbert spectra, are plotted in Figure 6.3.9. For the plot of the marginal Hilbert spectra, the 95 % confidence intervals, best fit line for resampled data and multi-taper spectra are also presented. The degrees of stationarity, volatility and instantaneous energy for the nine HCN stations are extracted and plotted together in Figure 6.3.10, Figure 6.3.11a, b, respectively.

In Figure 6.3.9, there are shaded areas on the top of volatility and instantaneous energy plots. These are some examples of segments with high variations. Similar to the results from HCN temperature data, the volatility and the standard deviation of

Table 6.3.5. Consistent volatilities in HCN monthly temperature time series

year	120177	120676	121229	121747	122149	124008	125337	126705	128036
1994	O	O	O	O	O	O	O	O	O
1980	X	O	O	O	X	O	O	O	O
1971	O	O	O	O	O	O	O	O	O
1958	O	O	O	O	O	O	O	O	X
1936	O	O	X	X	O	O	O	X	X

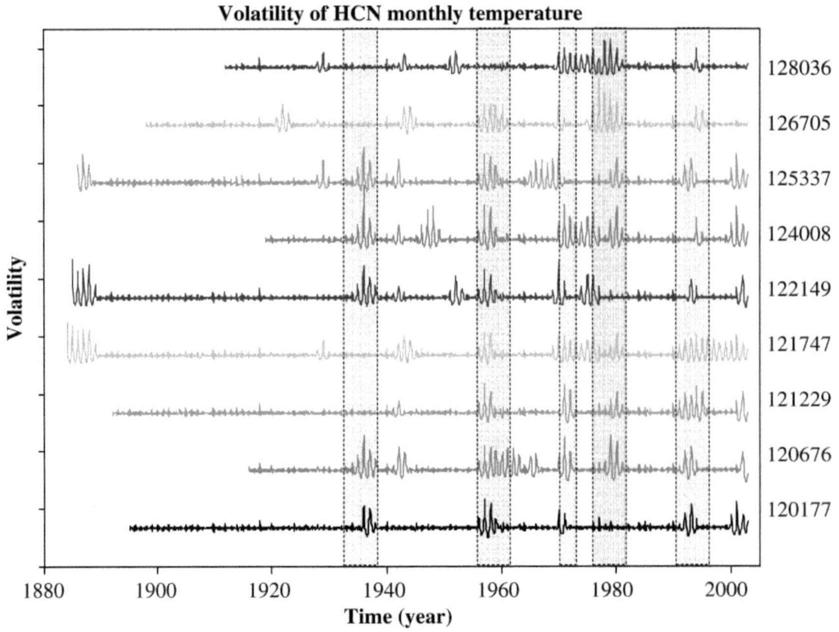

(a)

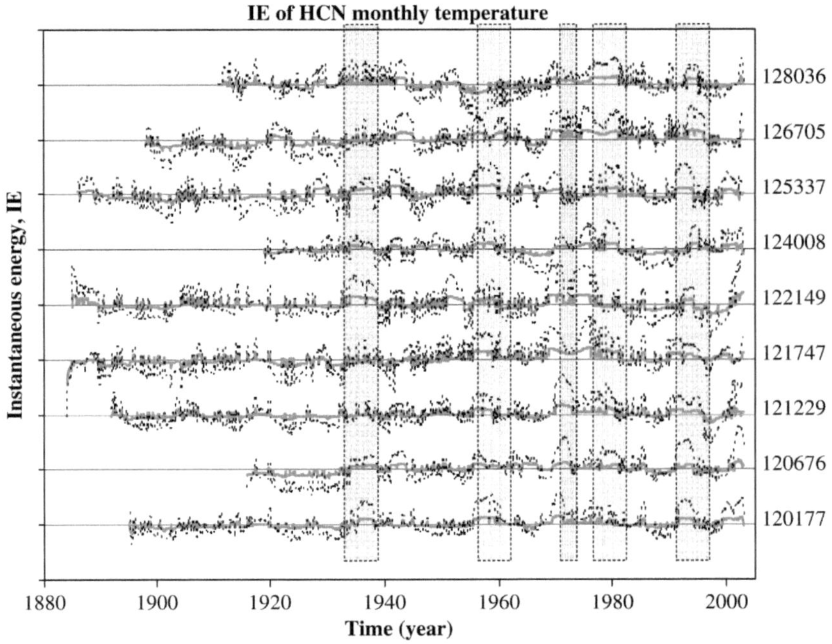

(b)

Figure 6.3.5. Volatility and instantaneous energy of Indiana HCN temperature data

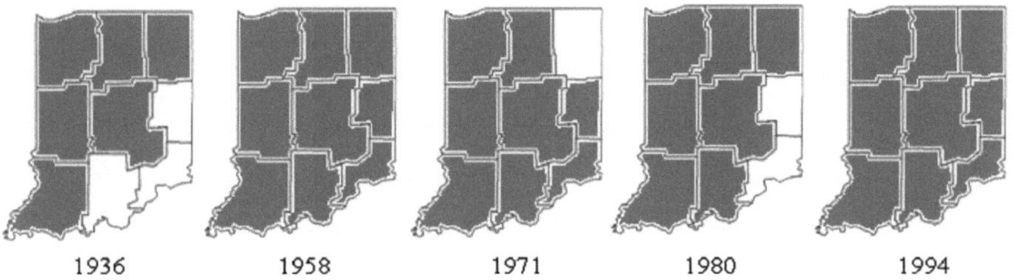

Figure 6.3.6. Significant volatilities detected for HCN monthly temperature data

instantaneous energy are quite consistent. It is not easy to distinguish the variation in the time series since the temperature data are varying in a certain range and periodically repeating. However, if more attention is paid to these segments, it is found that these peaks or dips in volatility and instantaneous energy correspond to some locally high or locally low temperature. This is an advantage of using these measures to investigate the abnormal segments or data points in a temperature series.

In Figure 6.3.10, the degree of stationarity plot indicates that the temperature data are more stationary in low frequency than in high frequency. As usual, there a dip-down of degree of stationary in annual cycle, but it is not as strong as the one in HCN monthly temperature data. Also, a few stations have a small dip-down in the frequency corresponding to 2 year period (Figure 6.3.9).

Also, from the results of degree of stationarity, it appears that some stations have high variation in the low frequency. The reason can be investigated by studying the Hilbert spectra in Figure 6.3.9. If the signal is stationary at that frequency, the energy should be close to zero at that frequency band. From the results, energy variation and discontinuity are seen in the low frequency. In Hilbert spectrum, the annual cycle is clearer in temperature data than in rainfall or runoff data according to the results shown here.

Table 6.3.6. Periodicities detected in HCN monthly temperature time series (unit: years)

No. Period	1 year	2 years		3~7 years		10 years	> 15 years
120177 (5)	1.0			7.8			17.5
120676 (3)	1.0	2.3				10.8	17.5
121229 (6)	1.0	1.7					
121747 (9)	1.0			4.7	7.1		17.5
122149 (4)	1.1	1.8					
124008 (1)	1.0	1.6	2.3	4.4			
125337 (2)	1.0	1.6		3.3		8.5	
126705 (8)	1.0	1.5	2.2				
128036 (7)	1.0	1.8		6.1		14.5	47.6

(a)

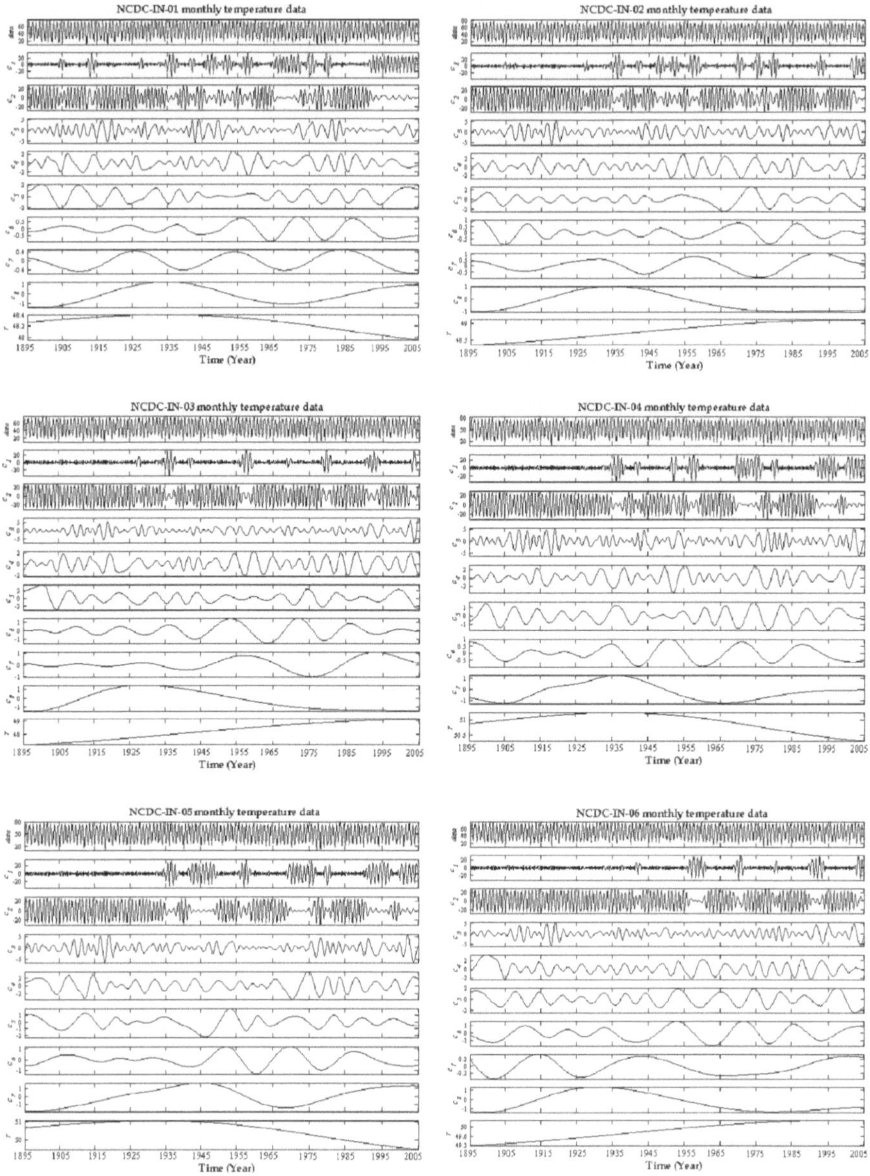

Figure 6.3.7. Intrinsic mode functions of NCDC monthly temperature data

(b)

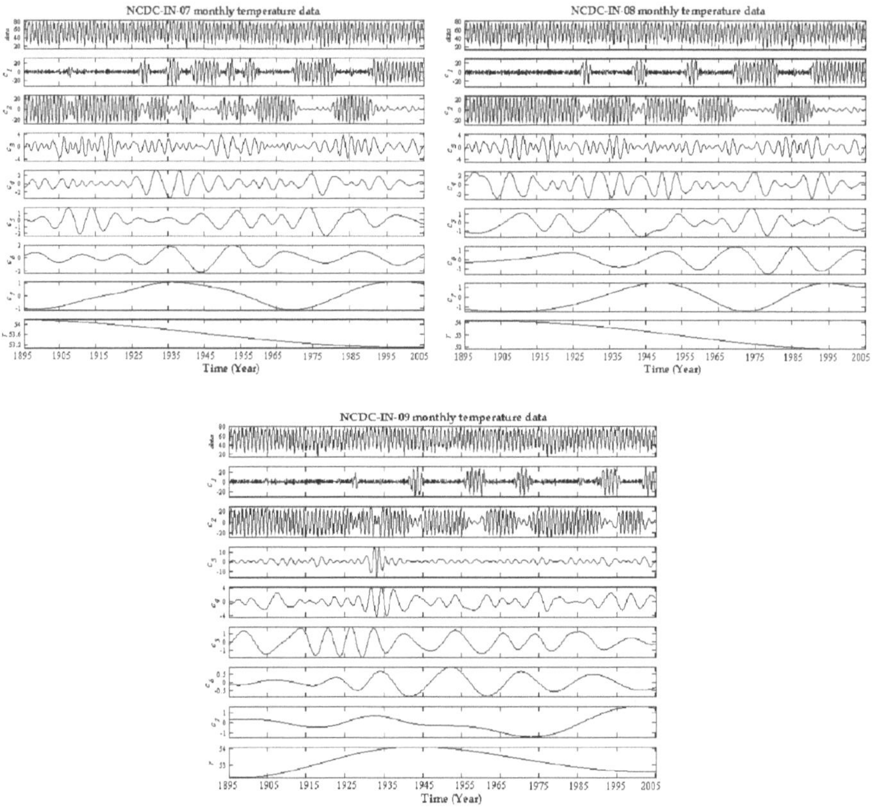

Figure 6.3.7. (*Continued*)

The spatial distribution of the significant volatilities is shown in Figure 6.3.12. For temperature data, these maps show more spatial continuity than for the rainfall data. Results from years 1936, 1958, 1971, 1980 and 1994 are used as examples to analyze the consistency of volatility. The results are shown in Table 6.3.8. If the variation of volatility is observed in a station, a circle mark ("O") is used to indicate it. Otherwise, a cross mark ("X") is used. For these 5 years, more than half of stations yield as high as variation of volatility as the others. The NCDC subdivision is shaded if the volatility is significant in a region and left blank if it is not. These are shown in Figure 6.3.12. For monthly temperature data, the volatility and standard deviation of instantaneous energy are quite consistent for most cases. Therefore, the spatial continuity study also represents the correlation of the instantaneous energy for the adjacent divisions.

Table 6.3.7. Trend test results for NCDC monthly temperature time series

Region	Segment	Modified Mann-Kendall
01	1	Upward trend (z = 1.757)
	2	Stationary (z = 1.56)
	3	Stationary (z = −0.225)
	4	Stationary (z = 0.705)
02	1	Stationary (z = 0.305)
	2	Stationary (z = 0.686)
	3	Stationary (z = −1.024)
	4	Stationary (z = −0.248)
03	1	Stationary (z = 1.273)
	2	Stationary (z = 1.269)
	3	Stationary (z = −0.046)
	4	Stationary (z = −1.25)
04	1	Stationary (z = 0.005)
	2	Stationary (z = 0.820)
	3	Stationary (z = −1.076)
	4	Stationary (z = 0.817)
05	1	Stationary (z = −0.216)
	2	Stationary (z = 1.350)
	3	Stationary (z = −0.715)
	4	Stationary (z = 1.075)
	5	Stationary (z = −1.114)
06	1	Stationary (z = −0.491)
	2	Stationary (z = 1.045)
	3	Stationary (z = −1.136)
	4	Stationary (z = −0.664)
07	1	Stationary (z = −0.141)
	2	Stationary (z = 0.632)
	3	Stationary (z = −1.091)
	4	Stationary (z = 0.150)
	5	Stationary (z = −0.164)
08	1	Stationary (z = 0.291)
	2	Stationary (z = 1.446)
	3	Stationary (z = −1.066)
	4	Stationary (z = −0.608)
	5	Stationary (z = −0.677)
09	1	Stationary (z = −0.949)
	2	Stationary (z = −0.946)
	3	Stationary (z = 1.378)
	4	Stationary (z = −0.702)
	5	Stationary (z = 0.487)
	6	Stationary (z = −1.320)

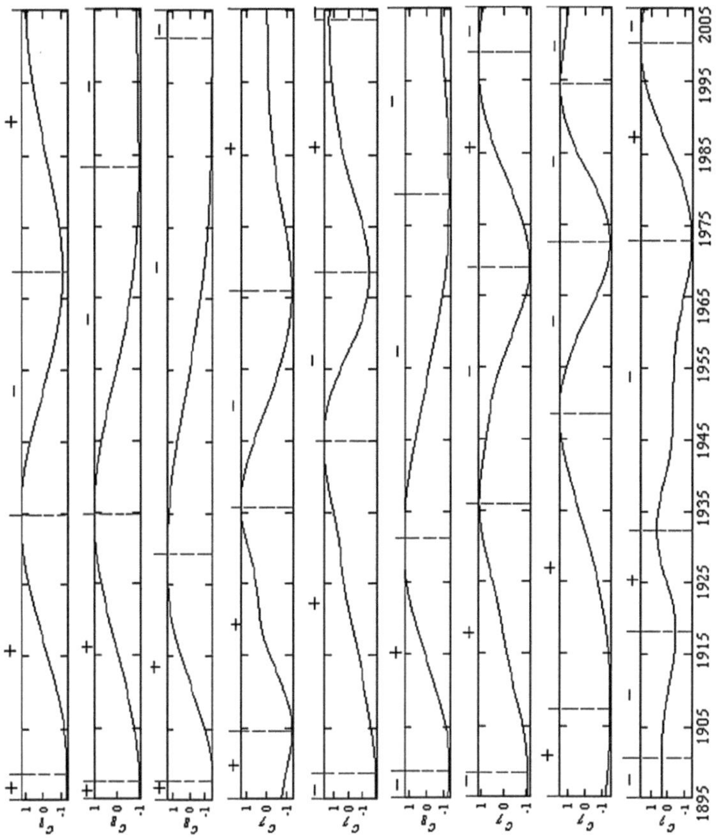

Figure 6.3.8. The last IMF component of NCDC monthly temperature data and the results of modified Mann-Kendall test for each segment

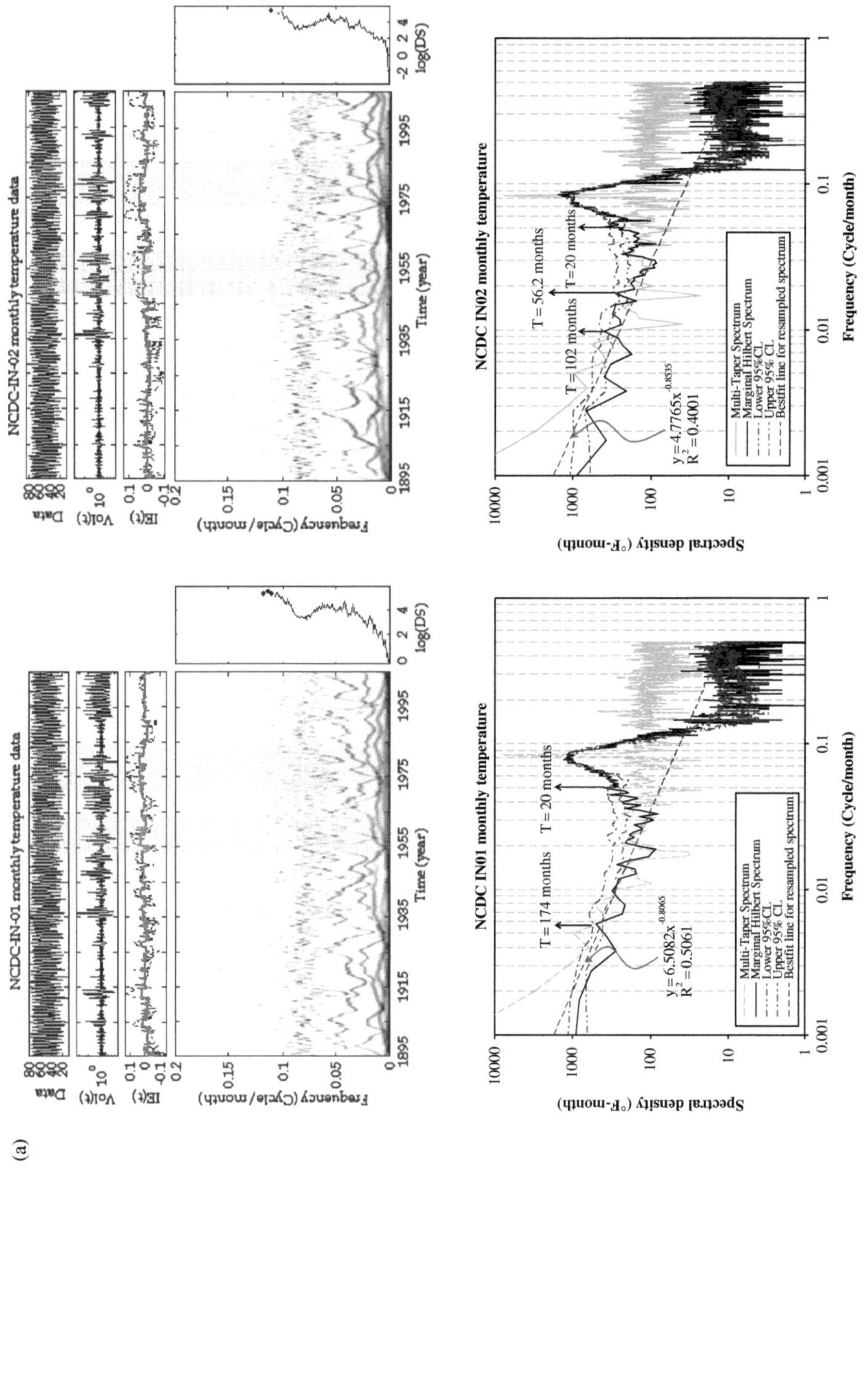

Figure 6.3.9. Time-frequency distribution and marginal Hilbert spectra of NCDC monthly temperature data

(b)

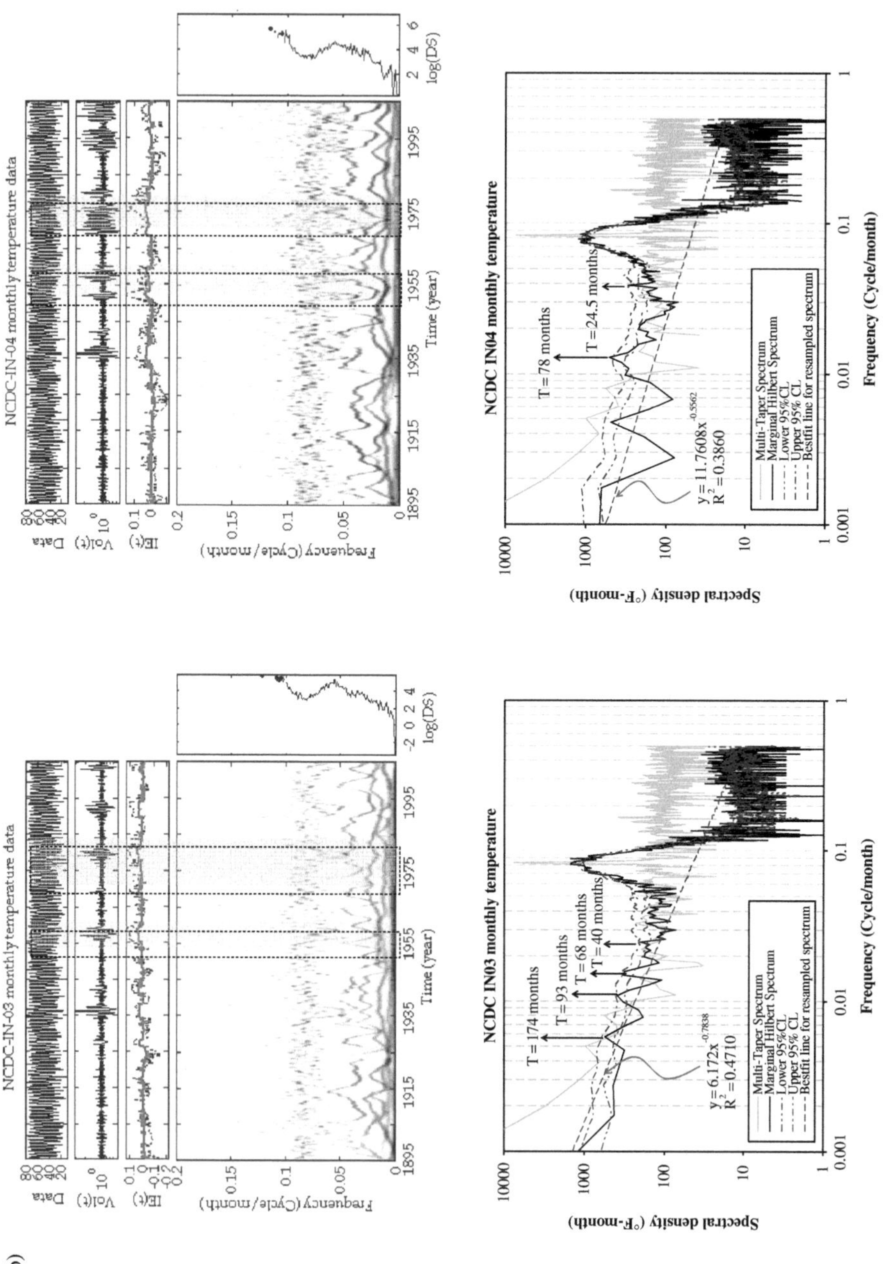

Figure 6.3.9. (Continued)

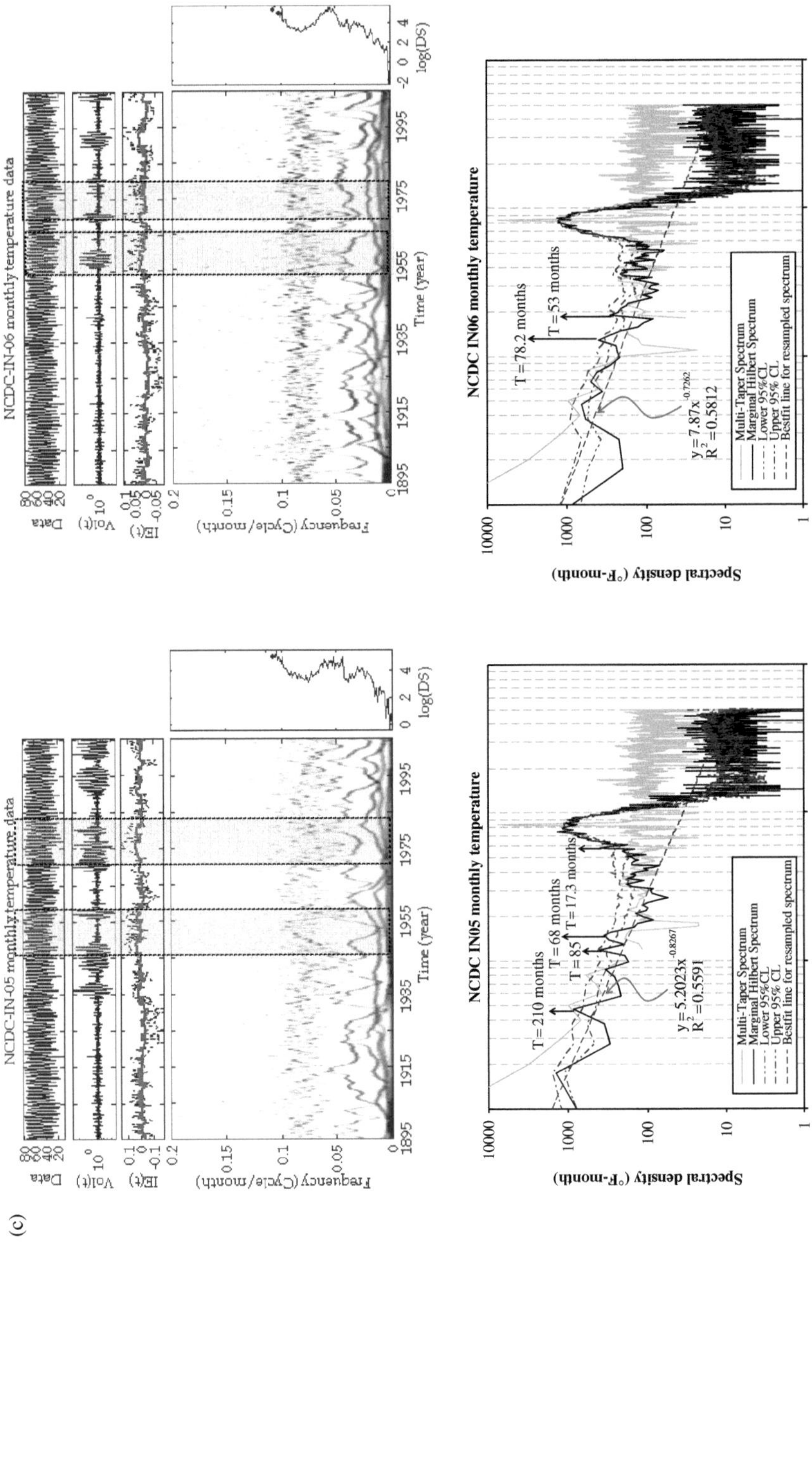

Figure 6.3.9. (Continued)

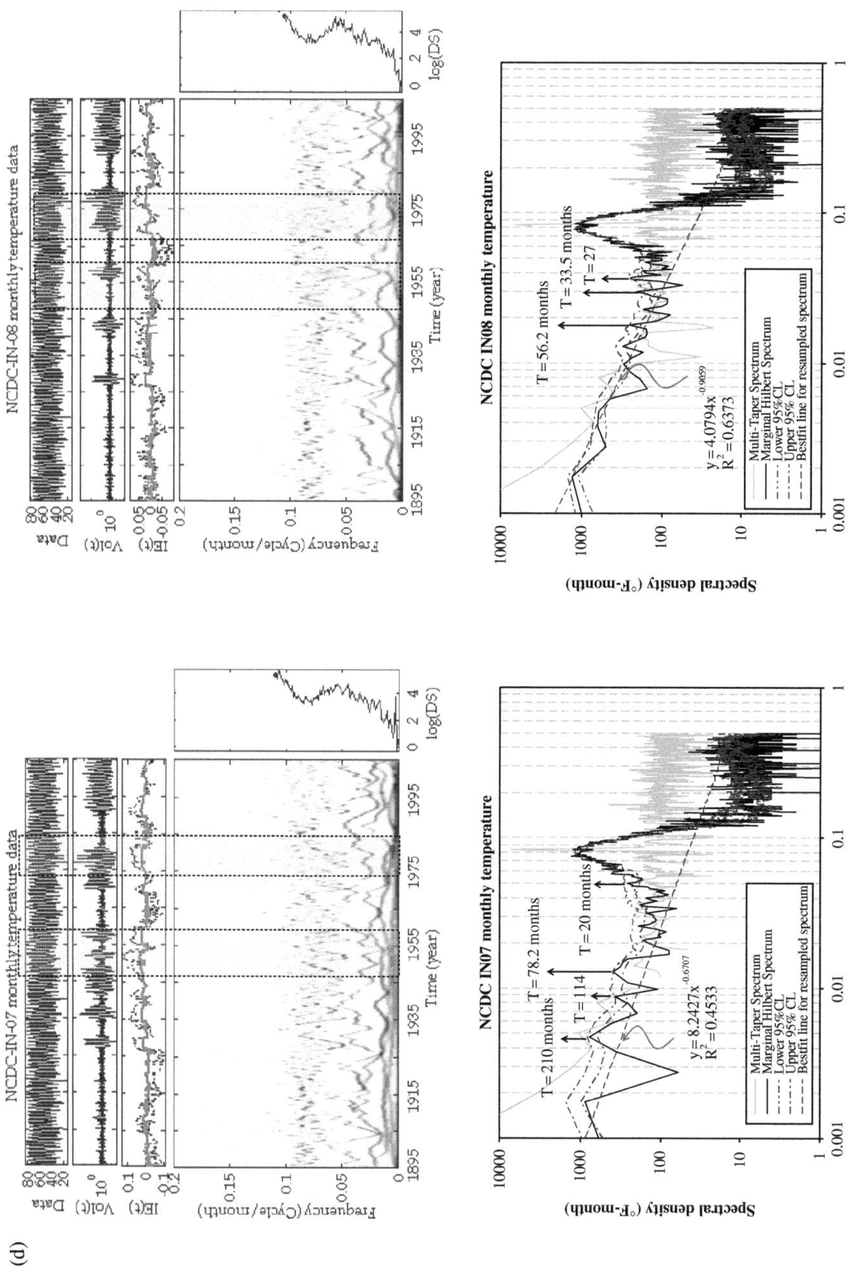

Figure 6.3.9. (Continued)

(e)

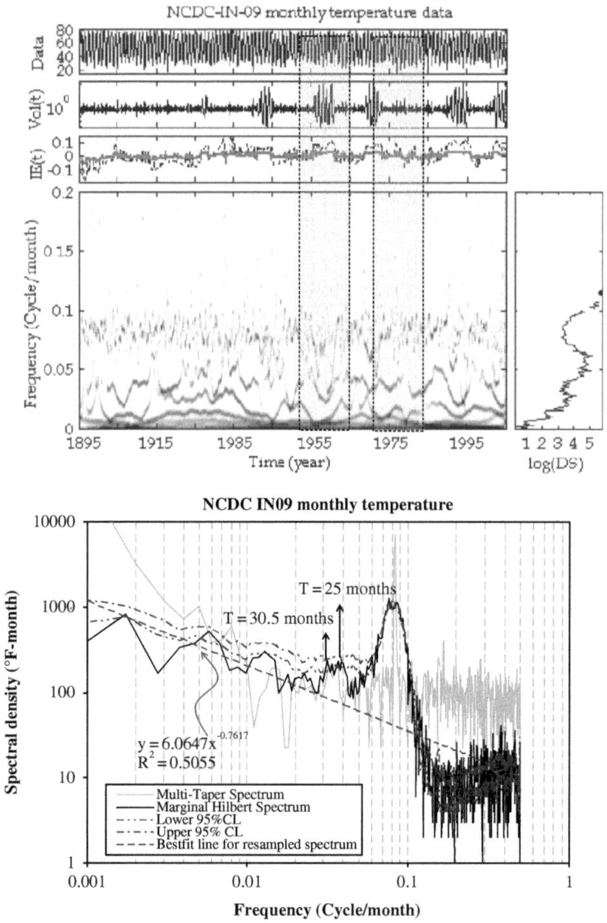

Figure 6.3.9. (Continued)

The significant periodicities are investigated by using the marginal Hilbert spectra in Figure 6.3.9. These periods are summarized in Table 6.3.9. A 2 year period is indicated in most stations. Four years, 10 years and 18 years periods are detected in some stations.

As for the best fit line for the resampled marginal Hilbert spectra, the average decay rate is −0.766 and standard deviation is 0.105. Hence, the coefficient of variation is 0.137. The fitting is poor based on the low *R*-square values. It is affected by the annual cycle and the low frequency spectra, with considerable noise with low energy. Power law is not suitable for the monthly temperature spectra.

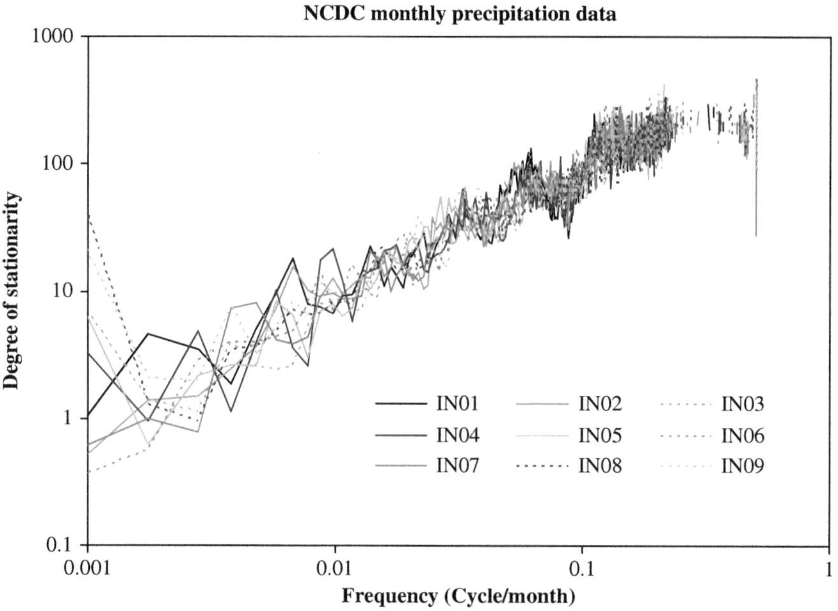

Figure 6.3.10. Degree of stationarity for NCDC monthly temperature data

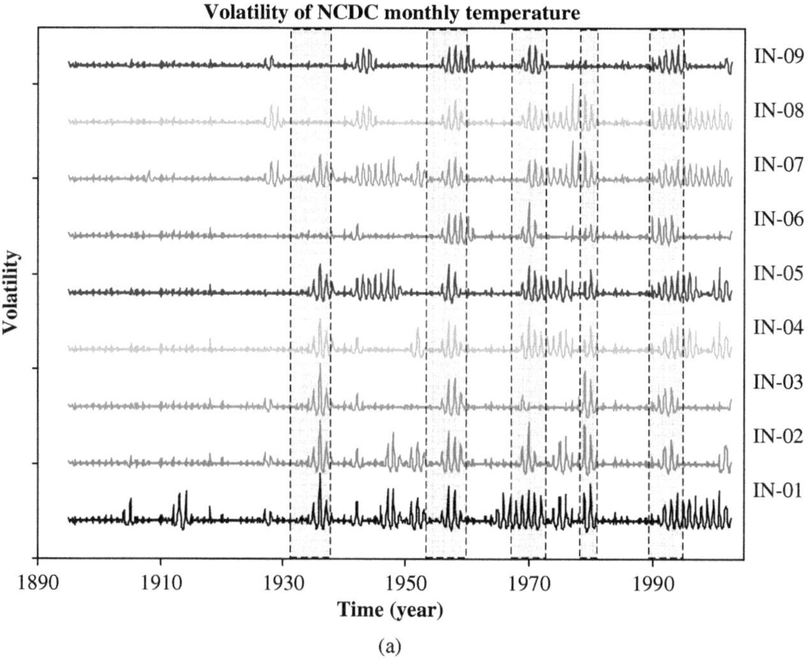

Figure 6.3.11. Volatility and instantaneous energy of Indiana NCDC temperature data

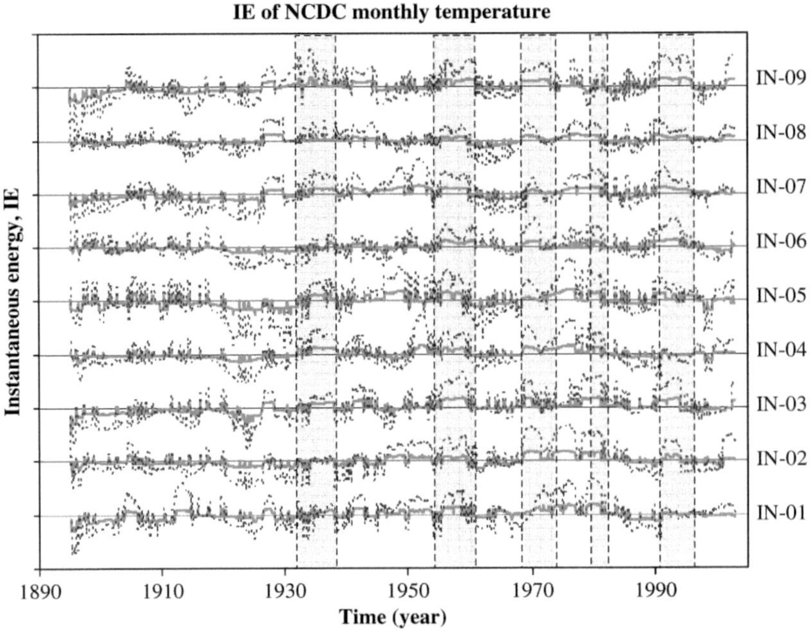

(b)

Figure 6.3.11. (*Continued*)

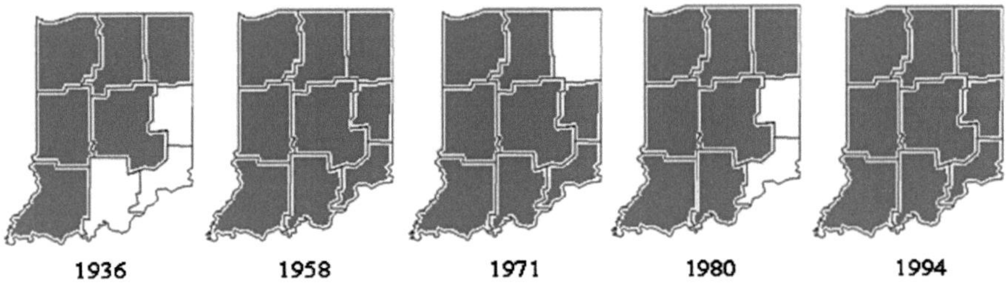

Figure 6.3.12. Significant volatilities detected for Indiana NCDC monthly temperature data

Table 6.3.8. Consistent volatilities in NCDC monthly temperature time series

year	IN-01	IN-02	IN-03	IN-04	IN-05	IN-06	IN-07	IN-08	IN-09
1994	O	O	O	O	O	O	O	O	O
1980	O	O	O	O	O	X	O	O	X
1971	O	O	X	O	O	O	O	O	O
1958	O	O	O	O	O	O	O	O	O
1936	O	O	O	O	O	X	O	X	X

Table 6.3.9. Periodicities detected in NCDC monthly temperature time series (unit: years)

No. Period	1 year	2 years		3~7 years			10 years	> 15 years
1	1.0	1.7					14.5	
2	1.0	1.7		4.7			8.5	
3	1.0			3.3	5.7	7.8	14.5	
4	1.0	2.0		6.5				
5	1.1	1.4		5.7	7.1			17.5
6	1.0			4.4	6.5			
7	1.0	1.7		6.5			9.5	17.5
8	1.0	2.3		2.8	4.7			
9	1.0	2.1	2.5					

6.4. CONCLUDING COMMENTS

The annual cycle is more clearly specified in monthly temperature data than for monthly rainfall and streamflow data in both marginal Hilbert spectra or in Hilbert spectra (time-frequency distribution). The reason may be that temperature time series have clear seasonal changes while rainfall and streamflow data are not that exact. The higher spectral energy and narrow frequency band make the 12 month periodicity much better defined. Spatial correlation yields fairly continuous regions.

The nonstationarity in these data is discussed by the analysis of the monthly temperature data in Europe. Three types of time series are studied, and they are the original data, linear-detrended data and annual-cycle removed data. These results show that Fourier based methods are very sensitive to the data with trends. The HHT has the advantage of having an instantaneous frequency. Therefore, the nonstationary behavior in time is less likely to cause distortions in the spectral behavior that make physical interpretation of the results difficult. As for the data after removing the annual cycle, the signal behaves like random noise. The features or periodicities are not easy to interpret from this signal. Consequently, the original data are used for spectral analysis.

Several periodicities were detected in the temperature series. This brings out the possibility of predicting the time series with these fundamental frequencies and the residual. The periodicities and the spectral amplitudes allow us to superpose these influences and make reasonable predictions. As long as sufficient length of data is available, the periodicities would be well estimated since they repeat themselves. However, the extrapolation of the residual is quite challenging especially when dealing with the data with trends. This aspect needs further investigation. Also, the difficulty of prediction increases when the residual sequence is not monotonic.

CHAPTER 7

WIND DATA ANALYSIS

7.1. INTRODUCTION AND DATA USED

Wind speed is an important climate-related parameter, affecting diverse activities such as the dispersal of atmospheric pollutants, the design of structures and aircraft safety. Gaining an understanding of the underlying mechanisms affecting the distribution of wind speeds, as well as any possible long term trends, could significantly influence future environmental policies and hazard mitigation strategies. Thus, proper analysis of wind speed data is considered useful and provides a notable benefit to the public. The use of the Hilbert-Huang transform has the potential to reveal new features of wind speed behavior that could not be ascertained via traditional techniques.

The wind data used in this study are hourly measurements at four National Weather Service stations in the state of Indiana. Complete data are obtained from both National Climatic Data Center (NCDC, http://cdo.ncdc.noaa.gov/ulcd/ulcd) and the Midwestern Regional Climate Center (MRCC, http://sisyphus.sws.uiuc.edu/). These agencies have the surface hourly observations for over 100 sites in the eastern half of the U.S. Parameters reported include: air temperature, dewpoint, wet-bulb temperature, pressure, relative humidity, wind speed and wind direction. The measurements are of 2 minute average wind speeds, with the averaging occurring just before the measurement is taken, Measurements are taken at a standard height of 10 m above ground using Automated Surface Observation System (ASOS). These wind speeds are measured in knots; 1 knot being equal to 1.151 miles per hour or 1.852 kilometers per hour. The time series are listed in Table 7.1.1 and plotted in Figure 7.1.1, which is averaged every 24 hours to obtain daily averages. The record length is about 14 years (1988~2002).

7.2. HOURLY WIND SPEED DATA

Four hourly wind speed series in Indiana from Evansville, Fort Wayne, Indianapolis, and South Bend are studied here. The length of record analyzed is from 1988 to 2002, hence there are 14-years of data available. The result of empirical mode decomposition of the hourly wind speed data is shown in Figure 7.2.1 and

Table 7.1.1. Stations for wind speed data

Station	Latitude	Longitude	Elevation (above sea level)
Evansville, IN	38°02'N	87°32'W	418 ft
Fort Wayne, IN	41°01'N	85°13'W	803 ft
Indianapolis, IN	39°43'N	86°16'W	794 ft
South Bend, IN	41°43'N	86°20'W	777 ft

Figure 7.2.2. Starting at the bottom of each figure, we note that the residual for three of the four stations (Fort Wayne, Indianapolis and South Bend) shows a rising trend with time, albeit at a very low rate for each of these stations. The highest order IMF for each of these stations (c_{18} for Indianapolis, c_{17} for Fort Wayne and South Bend) is a very long period oscillation of fairly low amplitude.

The modified Mann-Kendall test is applied to the hourly wind speed data based on the segments defined by the last IMF component. The results are shown in Table 7.2.1. The comparison of the segment trends and z values are shown in Figure 7.2.3. Only 12 segments out of 21 are consistent. The reason may be most of the z values are relatively small for most segments (less than 1) and this means that the trend embedded in the data is not obvious.

The residual signal from the Evansville station shows a non-monotonic behavior that varies from 5.8 to 6.6 knots. The last IMF for this station (c_{16}) does not exhibit the very long 11 year period of the other three stations but rather varies with a period of approximately 3.5 years. This leads us to conjecture that there is an unresolved

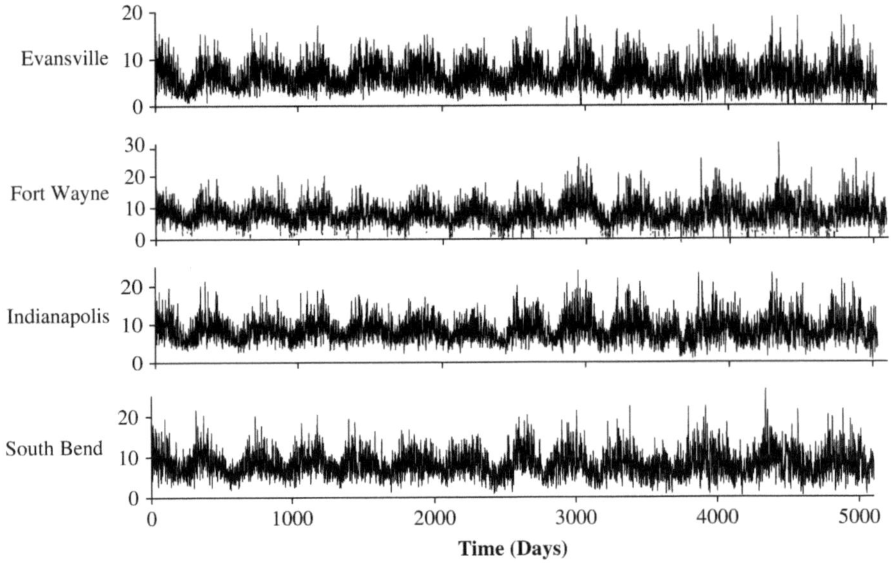

Figure 7.1.1. Daily average wind speed data (unit: knots)

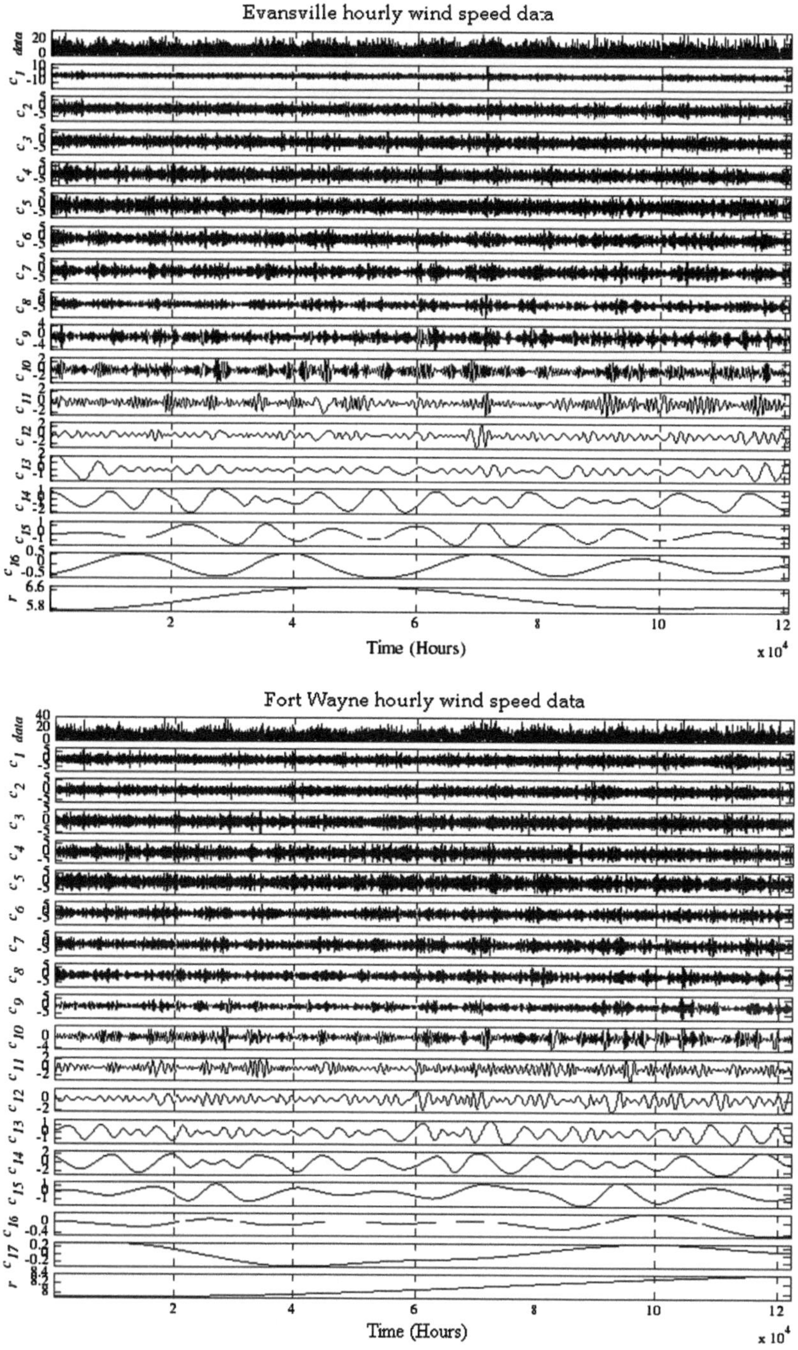

Figure 7.2.1. Intrinsic mode decomposition functions of the hourly wind speed data at Evansville and Fort Wayne

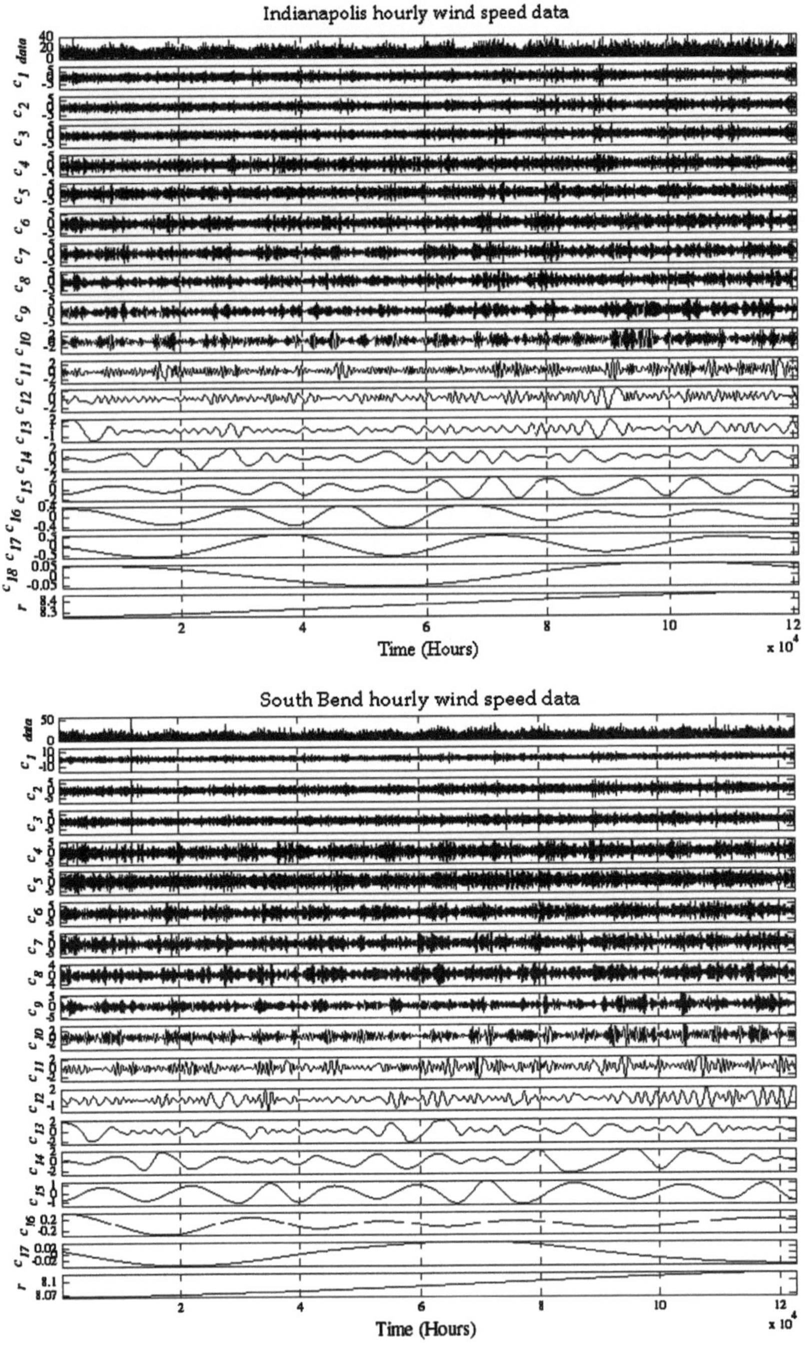

Figure 7.2.2. Intrinsic mode decomposition functions of the hourly wind speed data at Indianapolis and South Bend

Table 7.2.1. Trend test results from hourly wind speed data

Station	Segment	Modified Mann-Kendall
Evansville	1	Stationary (z = −0.536)
	2	Stationary (z = 0.136)
	3	Stationary (z = 1.204)
	4	Stationary (z = −0.635)
	5	Stationary (z = 0.529)
	6	Stationary (z = −0.146)
	7	Stationary (z = 1.426)
	8	Stationary (z = −0.083)
	9	Stationary (z = −1.202)
Fort Wayne	1	Downward trend (z = −2.351)
	2	Downward trend (z = −1.920)
	3	Stationary (z = 0.489)
	4	Stationary (z = 0.275)
Indianapolis	1	Stationary (z = 0.776)
	2	Stationary (z = 0.619)
	3	Stationary (z = 0.252)
	4	Downward trend (z = −1.756)
South Bend	1	Downward trend (z = −1.914)
	2	Stationary (z = −0.499)
	3	Stationary (z = 0.424)
	4	Downward trend (z = −2.666)

oscillation still present in the residual of the Evansville data, possibly of the same period as the 11 year oscillation observed in the other data sets. The Hilbert-Huang transform appears to break down in this situation due to the fact that there are an insufficient number of local extrema to properly resolve the presumed additional IMF. This issue is not unique to the wind speed data. It has been seen in some of the previously studied time series. This situation does not affect the spectral computation much because its amplitudes are relatively small. For the overall trend interpretation, the trend still can be indicated by investigating the slope of the non-monotonic residual.

The property of interest is the behavior of the signal in frequency domain. Taking Hilbert transform of all these intrinsic mode functions and calculating the instantaneous frequencies and amplitudes, the Hilbert spectra of each series are shown in Figure 7.2.4. The Hilbert spectrum is given as a time-frequency representation. The frequency axis scale is taken up to 0.06 cycles/hour since there are no graphically significant characteristics beyond that. It is also seen from time-frequency distribution that the southern wind speed (Evansville) is different from the northern wind speed data (South Bend).

At the lowest frequency (<0.005 cycle/hour), the amplitudes in the Hilbert spectrum are not constant in time. Spectral amplitudes are smaller at higher frequencies. This demonstrates a random characteristic of intermittent phenomena.

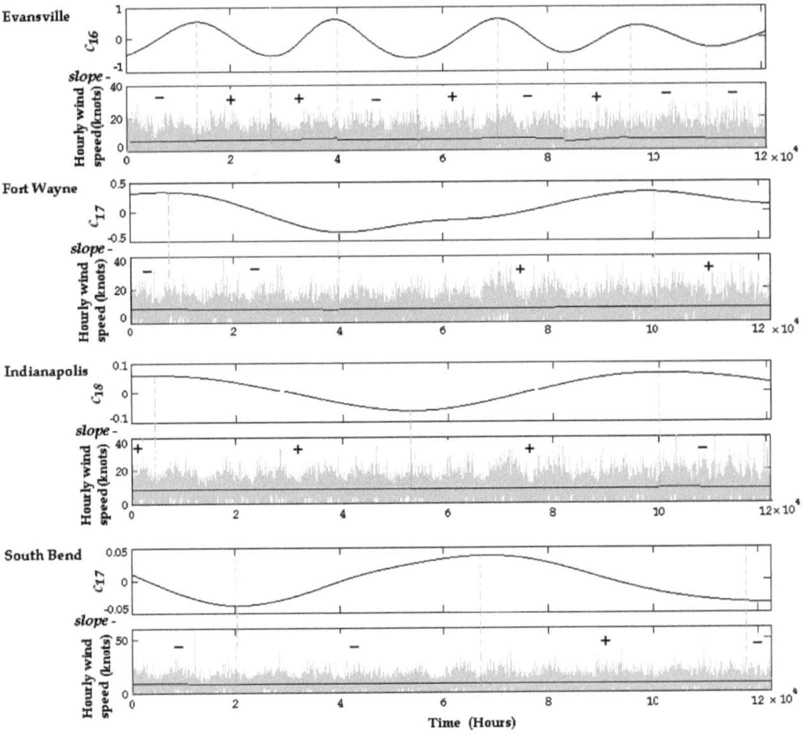

Figure 7.2.3. Investigation of the long-term oscillation of hourly wind speed data by stepwise linear regression and the last intrinsic mode function

Comparing the four stations, Evansville data has the smallest amplitudes, especially in the low frequency. This result is consistent to the lower wind speed data for Evansville. The wind speed data in Evansville ranges from 0 to 20 knots, while the other stations vary from 0 to 40 knots. It is noticeable that Evansville data has smaller wind speed values than the others. The reason is that the station located in a much lower elevation of 418 foot while the others are located at a elevation around 800 foot (Table 7.1.1).

To investigate the periodicity of the hourly wind speed data, marginal Hilbert spectrum is used. The results are shown in Figure 7.2.5. Fourier and multi-taper spectra are plotted for comparison. Also, three best fit lines for the entire spectrum, the resampled spectrum and the low frequency segment are indicated in the figure. Several interesting features are seen in the marginal Hilbert spectra. The Hilbert spectra are smoother than the corresponding Fourier and multi-taper spectra. One consequence of the smoothness of the Hilbert spectra is that it is difficult to pick out the frequencies with strong or weak energy. For the hourly data, the 24 hour period, i.e., daily cycle, is the only one can be clearly identified from the Hilbert spectra in four stations. Other than this, there are almost no other peaks or valleys

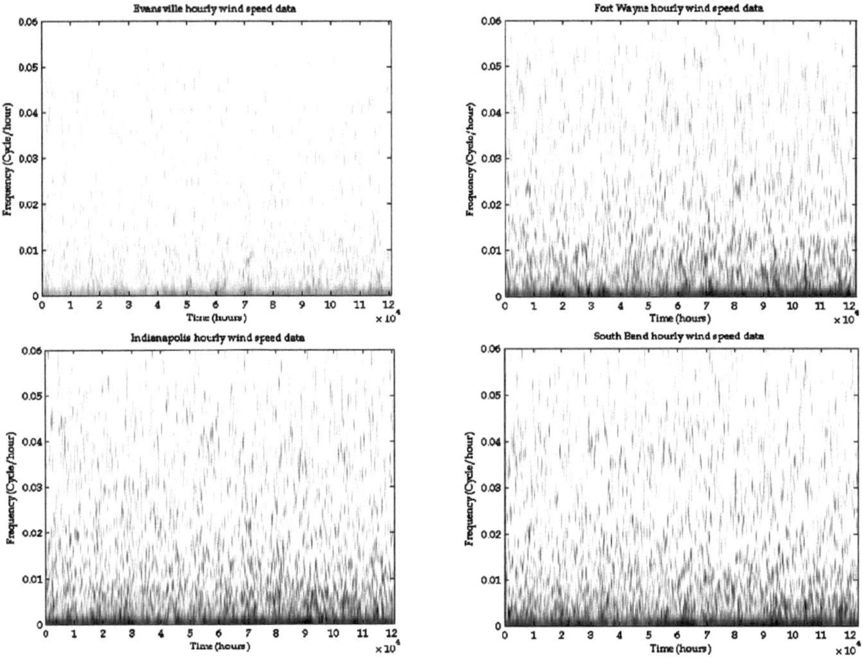

Figure 7.2.4. Time-frequency distribution (Hilbert spectra) of the hourly wind speed data

observed. This may represent a loss of frequency information, although some of the peaks found in the other spectra could simply be artifacts of transformation process.

These results are similar to the Hilbert spectrum for the daily streamflow in Warta River. The smoothness makes it easier to identify overall trends in the Hilbert spectra as a function of frequency as compared to the other techniques. In Figure 7.2.5, there are best fit lines for the various power law trends. First is a best lit line to the full spectrum, which uses all the data from the originally given spectrum. Second is a best fit line to the resampled marginal Hilbert spectrum in which data points with uniform spacing with respect to the logarithmic frequency axis are used. Use of the full spectrum weights the high frequency contributions more heavily since there are more points in that region. The third best fit line is to the low frequency segment, which ranges from 0.001 to 0.008 cycles/hour. The reason to add this analysis is that there appears to be two distinct ranges of power law-like behavior for marginal Hilbert spectra of wind speed data.

The Hilbert spectra are fitted by using the power law equation $y = af^b$, where y is the spectra, f is frequency and a, b are the coefficient pending for determined. The power law decay rate, b, and the R-square values for the fitting are summarized in Table 7.2.2. As the marginal Hilbert spectrum is an integral of Hilbert spectrum amplitudes over time, the direct physical significance of these amplitudes is debatable. Hence, the amplitude, a, is not used for discussion.

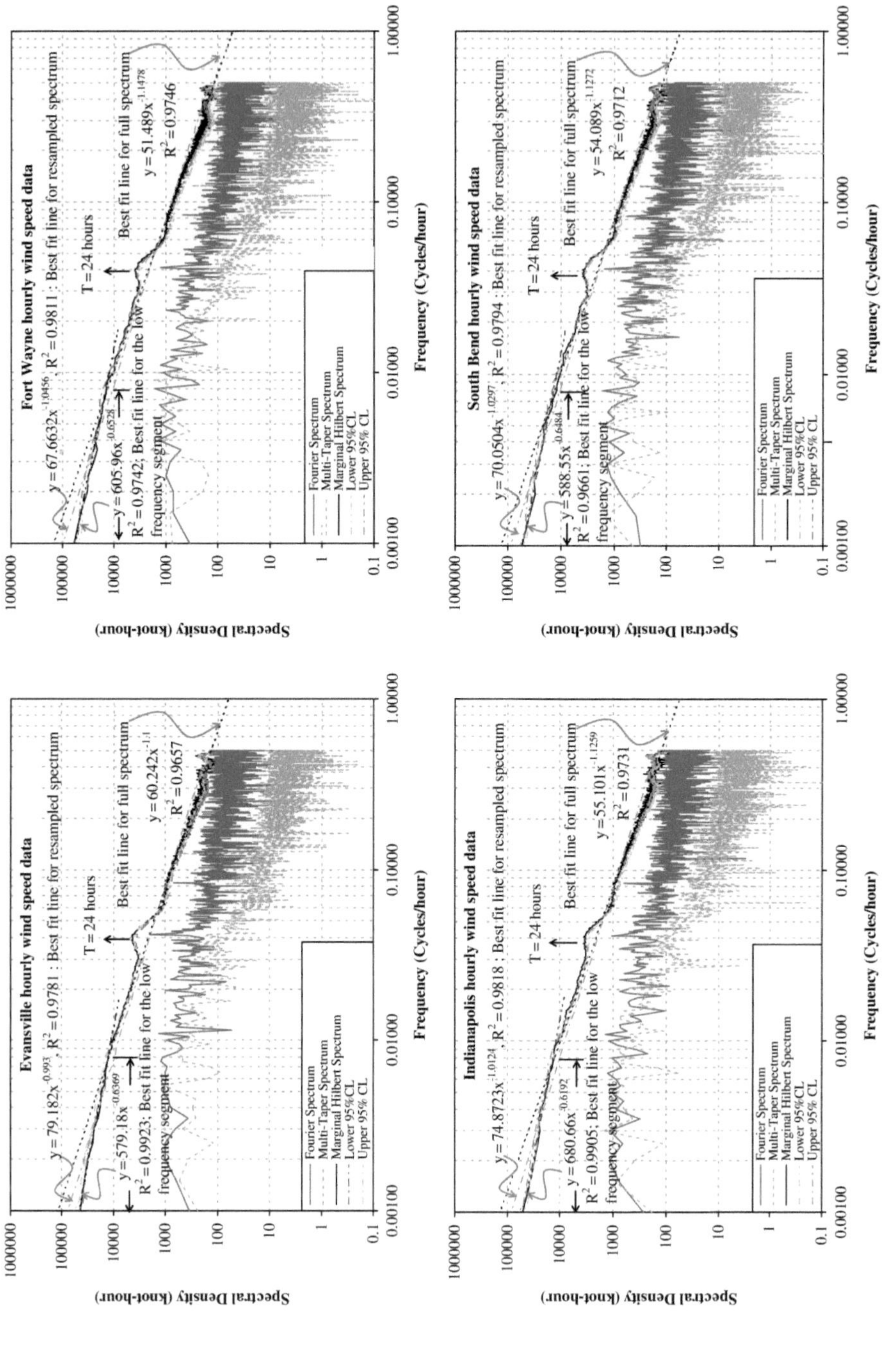

Figure 7.2.5. The marginal Hilbert spectra, Fourier spectra and MTM spectra of the hourly wind speed data

Table 7.2.2. Power law decay rates and R-square values for the best fit lines to the marginal Hilbert spectra of hourly wind speed data

Station	Full spectrum		Resampled spectrum		Low frequency segment	
	b	R^2	b	R^2	b	R^2
Evansville	−1.1000	0.9657	−0.9930	0.9781	−0.6369	0.9923
Fort Wayne	−1.1478	0.9746	−1.0456	0.9811	−0.6528	0.9742
Indianapolis	−1.1259	0.9731	−1.0124	0.9818	−0.6192	0.9905
South Bend	−1.1272	0.9712	−1.0297	0.9794	−0.6484	0.9661

All the spectra show prominently the contribution of the 24 hour cycle and they display similar fluctuation in power near the 0.5 cycle/hour cut-off. By comparing the Hilbert spectra, a great similarity is found among these results. This speaks to an overall common feature in wind data, at least for the data encountered herein. Fourier and Multi-taper spectra yield three peaks in the high frequency range. The 24-hour period has the strongest energy and is detected by all three methods. It is especially identified by HHT method. There are frequencies 0.041, 0.082 and 0.126 cycles/hour, which correspond to the periods of 24 hours, 12 hours and 8 hours, detected by Fourier and Multi-taper methods. Twelve-hour and 8-hour are the frequencies that may be embedded in the daily cycle. HHT cannot locate these two frequencies well because the oscillation of higher instantaneous frequency makes the marginal spectrum smoother and it is hard to locate the possible harmonic frequencies.

To show the similarity of Hilbert spectra, the results of these four stations are plotted in Figure 7.2.6. Only the best fit lines using the resampled marginal spectrum are compared. The power laws for the full spectra and resampled spectra are almost identical for these four stations. The four best fit lines to the full spectra have an average amplitude of 55.23 knot-hours and an average decay rate of −1.125, with relatively little variation around both of these averages. Sample standard deviations are 3.67 knot-hours for the amplitude and 0.020 for the decay rate, giving coefficients of variation of 6.64 % and 1.78 %, respectively. The quality of these fits is good; R^2 values ranged from 0.9657 to 0.9746. When the resampled spectra are used, the amplitude average increase to 72.94 knot-hours while the decay rate average dropped to −1.020. Moreover, the resampled average amplitude showed about the same level of variability as the full spectrum average (sample standard deviation = 5.13 knot-hours, c.o.v. = 7.03 %), as did the decay rates (sample standard deviation = 0.023, c.o.v. = 2.25 %). It is conjectured that the differences between the fits using the full and resampled spectra arise from the fact that the fits of the resampled spectra are influenced more strongly by the change in trend that occurs in moving from the low frequency segment to the higher frequencies. Phrased differently, the best fit lines for the full spectra effectively reflect the trends in the high frequency regime, since the preponderance of data points are outside the low frequency segment. R^2 values for these fits are still very high, ranging

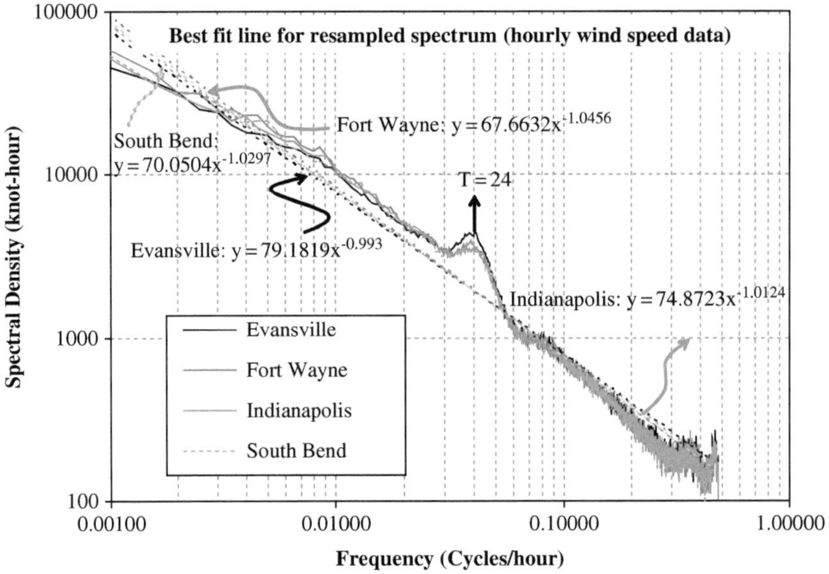

Figure 7.2.6. Comparison of marginal Hilbert spectra of hourly wind speed data in the four stations

from 0.9781 to 0.9818. In the low frequency segment, the best fit trends for the four stations reveal larger amplitudes with variability comparable to the other fits (mean = 613.6 knot-hours, sample standard deviation = 46.1 knot-hours, c.o.v. = 7.51 %) and lower decay rates, also of comparable variability (mean = −0.639, sample standard deviation = 0.015, c.o.v. = 2.35 %). The quality of these fits was good. R^2 values were above 0.9661 for all stations.

Dyrbye and Hansen (1997) cite a meteorological classification of the patterns of atmosphere motions in Figure 7.2.7a. The figure shows local weather systems and large scale variation, which encircle the entire globe and have a lifetime of several days. These phenomena are referred to as microscale, convective scale and macroscale. Figure 7.2.7b shows two different spectra. There is a great deal of variance in movements lasting approximately 4 days. Also, a clear peak at a period of 1 day is observed.

In addition, the transition in decay rate from the low frequency segment to the higher frequencies suggests a change in the underlying mechanism responsible for the spectral contributions in these two regimes. The transition frequency, roughly identified as 0.008 cycles/hour, corresponds to a period of approximately 125 hours, or about 5 days. It is the same period as that associated with the passage of extratropical cyclones and associated fronts (Dyrbye and Hansen, 1997); thus, the transition in the marginal spectrum's decay rates could be a manifestation of the importance of this large-scale meteorological phenomenon in the lower frequency ranges. The role of the daily cycle on the overall fits (particularly on the resampled spectrum) also seems to be significant and could possibly be interacting with the

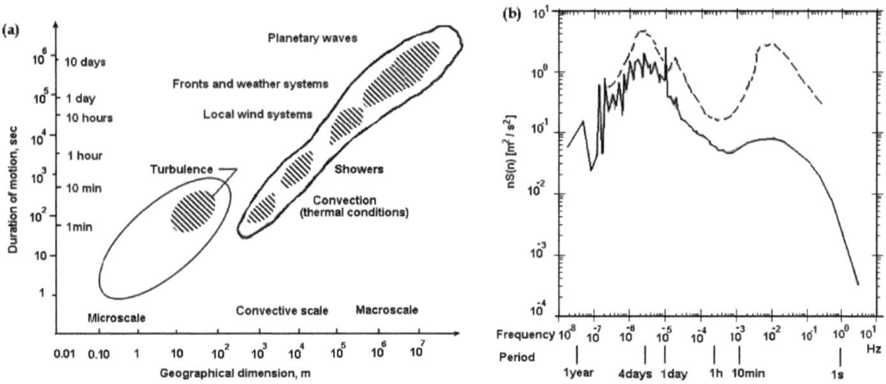

Figure 7.2.7. (a) Order of magnitude in space and time for different patterns of motions in the atmosphere (b) Autospectra for the wind velocity (Source: Dyrbye, 1997)
(The solid curve refers to the autospectrum for the wind measured at 30m height at Lammefjord, Denmark (1990); the dotted line is a spectrum for wind speed measured at 100m height in Brookhaven, NY, USA.)

front passage cycle to produce the observed transition. It would be much more difficult to extract these power law trends from either the Fourier or multi-taper spectra due to their much more erratic behavior with respect to frequency.

7.3. DAILY AVERAGE WIND SPEED DATA

The other way to investigate the low frequency characteristics is rearrange the hourly data into daily data and calculate the new spectra. Two different types of data are used in considering daily data, one is the daily average and the other is the daily peak data. Daily average data gives us an idea of the daily variation while daily peak data shows how the extreme records behave. Behavior of peak data is interesting for the risk assessment in structural design. The spectra of daily wind data are helpful in investigating the low frequency range; the other advantage to using the daily data is that it captures the long-term oscillation.

The intrinsic mode functions for the daily average wind data are shown in Figure 7.3.1 Except for Evansville, the other three stations have a slightly upward trend. The highest order IMFs (e.g., c_{10} and c_{11}) represent weaker, multi-year oscillations in the wind speed, with periods ranging from approximately 6 to 12 years. The IMFs c_9 and c_8 appear to correspond to yearly fluctuations in wind speed and are noticeably larger in amplitude. The remaining IMFs vary on a seasonal to daily time spans, which could be associated with meteorological phenomena such as the passage of storm fronts.

To investigate the long-term oscillation, the data are fitted by the stepwise trends and compared to c_{10} (Evansville), c_{11} (Fort Wayne), c_{11} (Indianapolis) and c_{10} (South Bend) as shown in Figure 7.3.2. The time series is thus segmented based on the last IMF component. The consistency between the trends of the segments and the slope

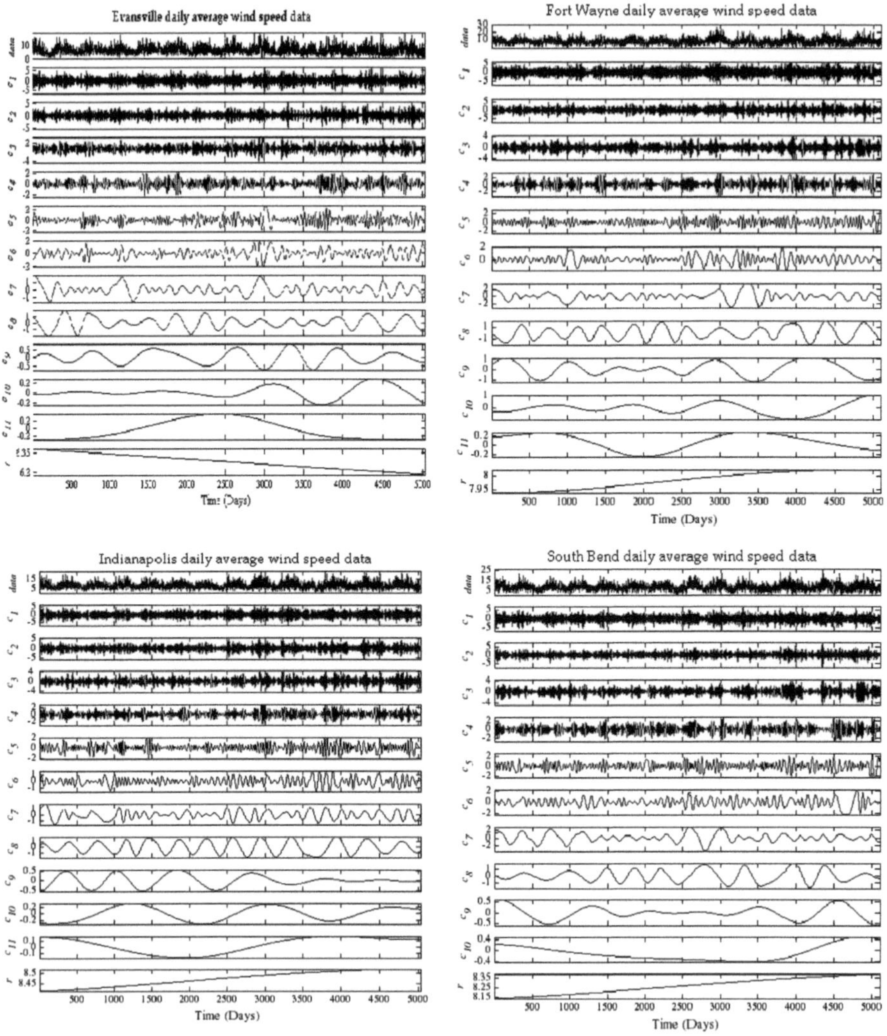

Figure 7.3.1. Intrinsic mode decomposition functions of the daily average wind speed data

of the last IMF component is investigated in Figure 7.3.2. The linear regression yields results similar to the empirical mode decomposition. This observation again emphasizes the strength of empirical mode decomposition in studying long-term oscillations. The highest order IMFs can be analyzed to provide some indication of long-term trends in the daily wind speed data. The high frequency is quite noisy and not easy to interpret from IMF directly.

To study the periodicities revealed from the IMFs, the Hilbert spectra is computed as shown in Figure 7.3.3. Hilbert-Huang transform gives a clear indication of the

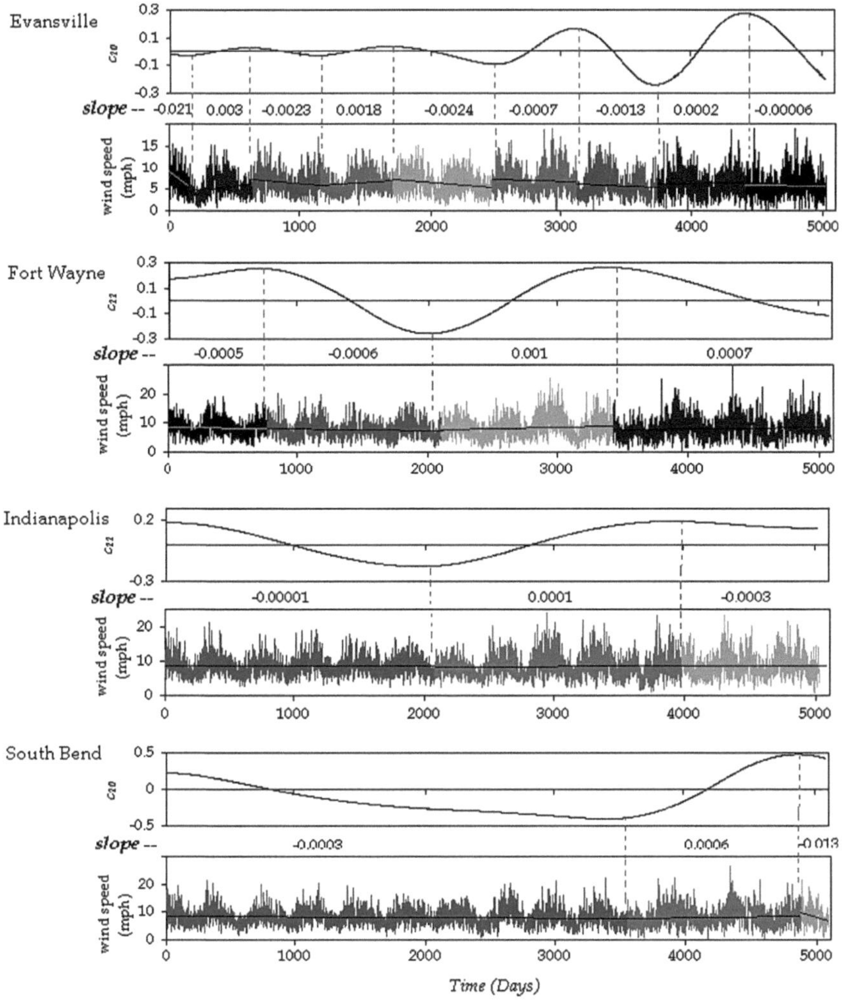

Figure 7.3.2. Investigation of the long-term oscillation by stepwise linear regression and the last intrinsic mode function

fluctuation at higher frequencies, whereas Fourier and multi-taper spectra gather most energy in the low frequency. Fourier based methods tend to blur the spectral amplitudes when the signal is nonstationary. The nonstationary feature in the low frequency is addressed well by the Hilbert-Huang transform. By looking at the frequency band from 0 to 0.003 cycles/day, the strength of energy is not constant over the time span for four stations. Some segments have much higher energy than the other adjacent segments. This indicates the daily wind speed data is nonstationary.

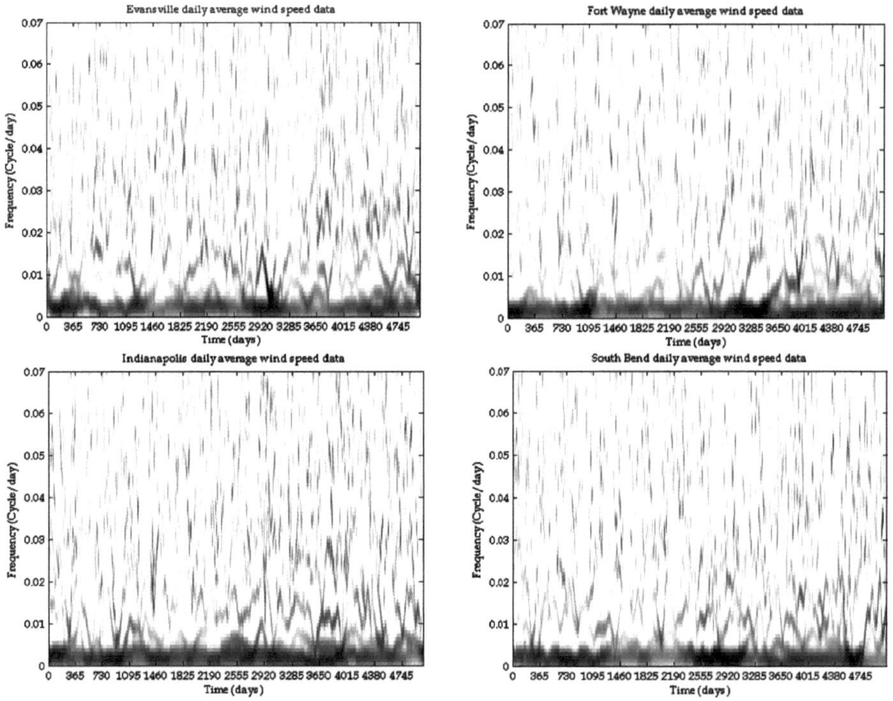

Figure 7.3.3. Time-frequency distribution (Hilbert spectra) of daily average wind speed data

The marginal Hilbert spectra for four stations are shown in Figure 7.3.4. These are useful for detecting the periodicities. Fourier and multi-taper spectra are plotted for comparison. There is no distinct separation in marginal Hilbert spectra as seen in hourly wind speed data, so herein only two best fit lines for the entire spectrum and the resampled spectrum are indicated in the figure. The Hilbert spectra are fitted by using the power law equation $y = af^b$, where y is the spectra, f is frequency and a, b are the coefficient pending for determined. The power law decay rate, b, and the R-square values for the fitting are summarized in Table 7.3.1. Again, the amplitude, a, is not used for discussion since it does not have too much universality feature.

As for the high frequency range, Fourier spectra look very spiky. There is considerable energy leakage so it is hard to tell where the peak frequency is. Multi-taper and marginal Hilbert spectra locate the peak frequency better. The significant periodicities are indicated by the 95 % confidence intervals. There is a 10 day period detected for these stations and 3~5 day period is quite strong in the high frequency domain. This is similar to the conclusion made by Dyrbye (1997). The 4-day period is relevant to passing storms or weather fronts. Hilbert-Huang spectrum extracts more detailed features of non-stationarity than the others. An annual cycle, which is around 300~400 days, is detected. Some interannual oscillations are also indicated in these spectra, such as 65 days and 180 days.

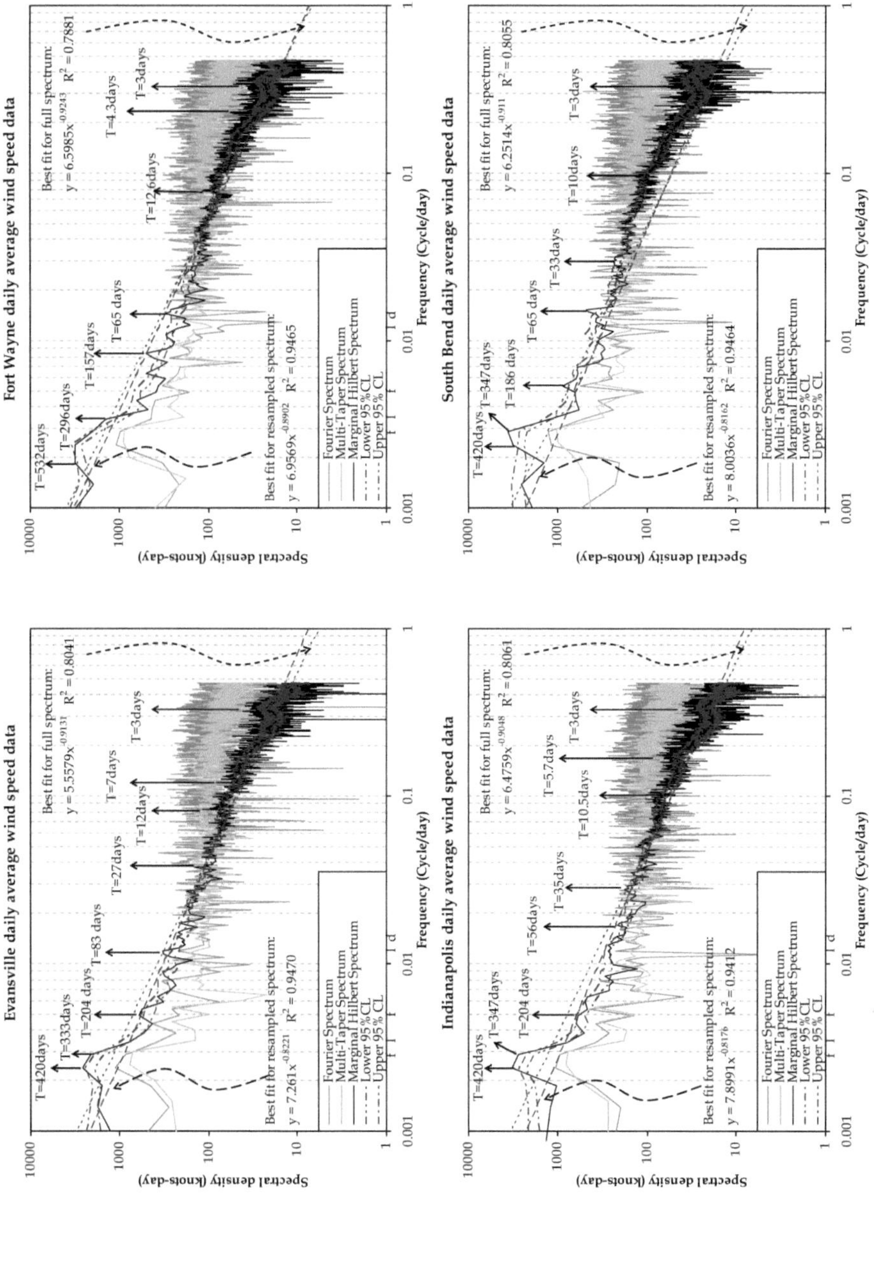

Figure 7.3.4. The marginal Hilbert spectra, Fourier spectra and MTM spectra of the daily peak wind speed data

Table 7.3.1. Power law decay rates and R-squared values for the best fit lines to the marginal Hilbert spectra of daily average wind speed data

Station	Full spectrum		Resampled spectrum	
	b	R^2	b	R^2
Evansville	−0.9131	0.8041	−0.8221	0.9470
Fort Wayne	−0.9243	0.7881	−0.8902	0.9465
Indianapolis	−0.9048	0.8061	−0.8176	0.9412
South Bend	−0.9110	0.8055	−0.8162	0.9464

In Figure 7.3.5, a power trend line is fitted for the marginal Hilbert spectrum in order to study how the slope varies for different stations. Only the best fit lines using the resampled marginal spectrum are compared. The slopes for Evansville, Fort Wayne, Indianapolis and South Bend are −0.8221, −0.8902, −0.8176 and −0.8162, respectively. The maximum difference among these slopes is 8.3 %. The degree of stationarities for these four stations are shown in Figure 7.3.6. They are quite close and have a dip down at the annual cycle.

The marginal Hilbert spectra of the daily average wind speed data show a great deal of fluctuation and randomness, particularly for high frequencies. This permits

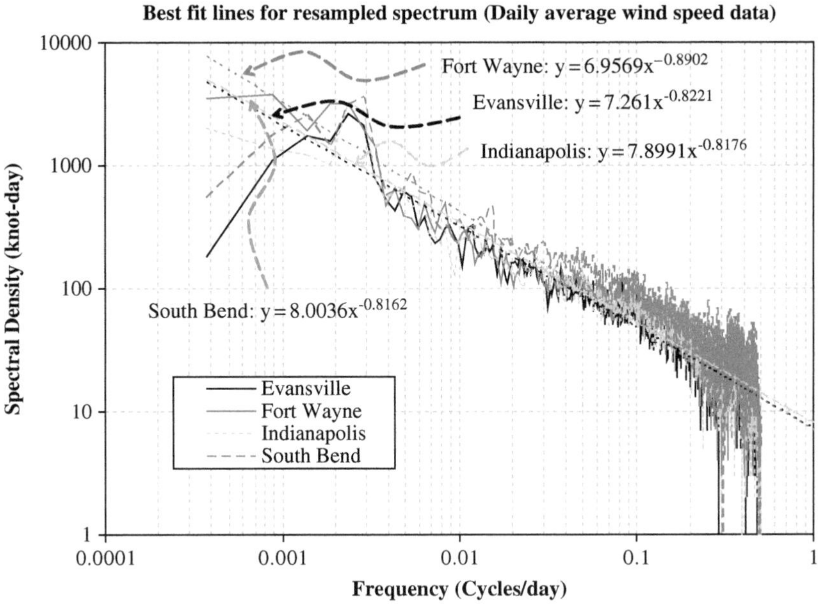

Figure 7.3.5. Comparison of marginal Hilbert spectra of daily average wind speed data in the four stations

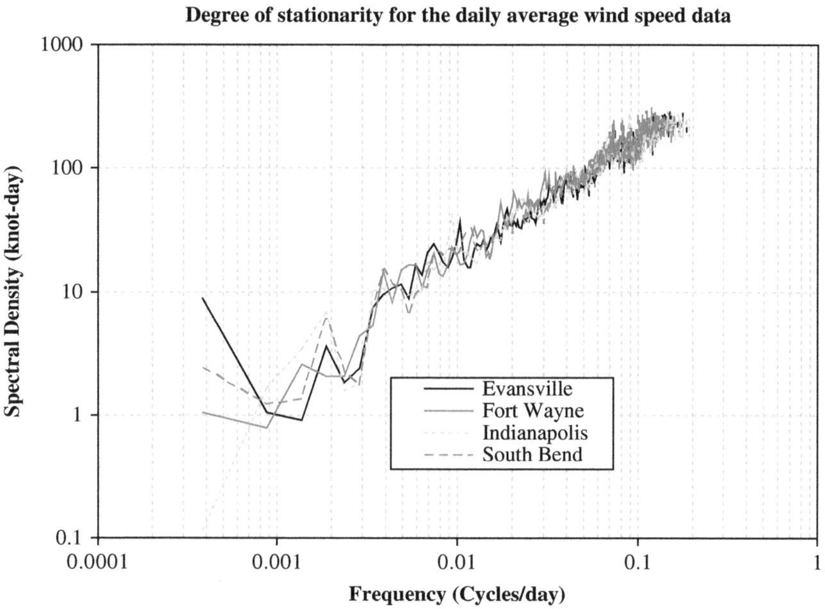

Figure 7.3.6. Degree of stationarity of the daily average wind speed data in the four stations

us to identify some stronger peaks in the marginal Hilbert spectra. In all cases, a strong annual oscillation is identified. The general behavior of the marginal Hilbert spectra again shows a great deal of commonality. The power laws corresponding to the best fit lines of the full spectra have very similar decay rates, with an average value of −0.913 knot-days and a very small variation (sample standard deviation = 0.008, c.o.v. = 0.89 %) around this value. Quality of fit was still quite good, with R^2 value around 0.8. Using a resampled version of the marginal spectra for each station results in a lower average decay rate of −0.836 with a higher sample standard deviation of 0.036 (c.o.v. = 4.29 %). Quality of the resampled fits actually improved significantly due to resampling, with R^2 values tightly clustered around 0.9.

Regardless of how the resampled spectrum is interpreted, there are no obvious changes in frequency trend for the marginal Hilbert spectra that could account for the difference in decay rate as compared to that of the full spectrum fits. It is still likely, however, that the fits from the resampled spectra conform more to a global average while the fits of the full spectra are highly influenced by the high frequency components, which may explain the lower values of the latter quantities. The presence of the strong annual cycle clearly affects the overall fits as well. However, the overall consistency of the decay rates again points toward an underlying commonality in the wind speed mechanisms that these stations share. Analysis of other data would be useful in identifying other important mechanisms.

7.4. DAILY PEAK WIND SPEED DATA

The intrinsic mode functions for the daily peak wind data are shown in Figure 7.4.1. The residual signals from these data sets behave in more or less the same manner as those of the hourly wind speed data set, with the exception that the residual signal for the daily average wind speeds in Evansville decreases slightly over time. Hence, data from these four stations have a slight upward trend.

The highest order IMFs (e.g., c_{10} and c_{11}) represent weaker, multi-year oscillations in the wind speed, with periods ranging from approximately 6 to 12 years. The IMFs c_9 and c_8 appear to correspond to yearly fluctuations in wind speed and are

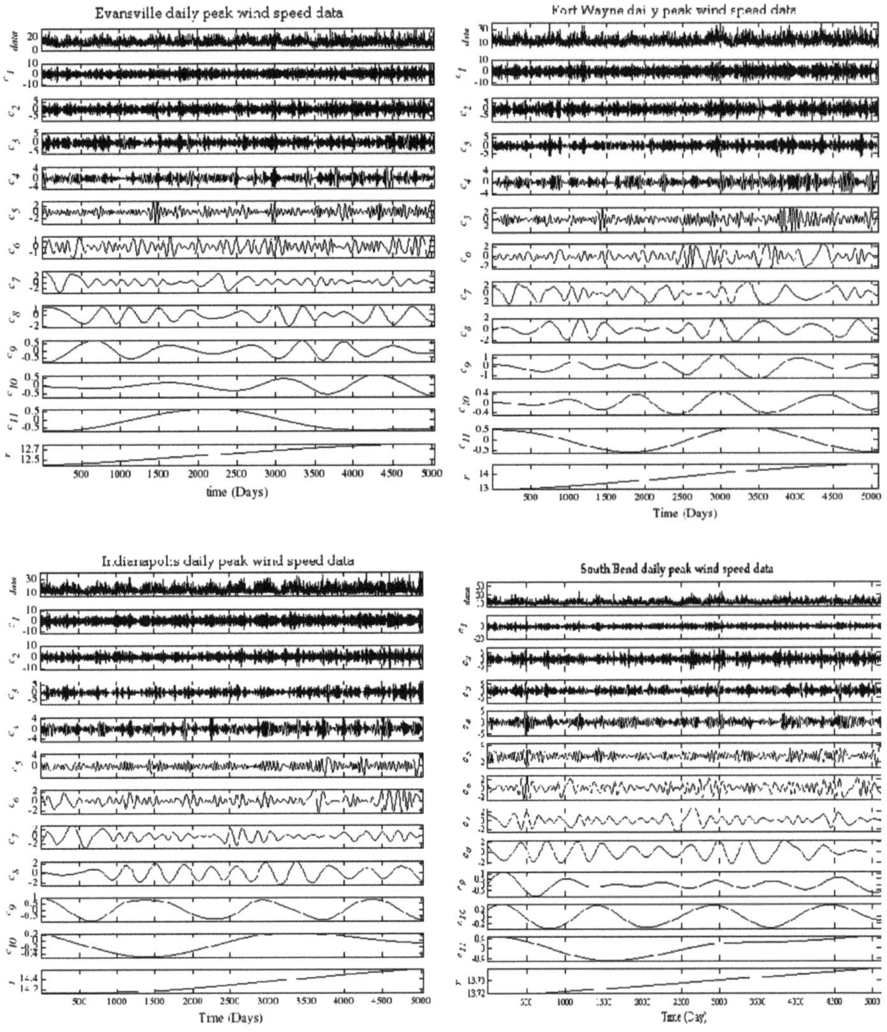

Figure 7.4.1. Intrinsic mode decomposition functions of the daily peak wind speed data

noticeably larger in amplitude. The remaining IMFs vary on a seasonal to daily time spans, which could be associated with meteorological phenomena such as the passage of storm fronts.

To study the periodicities in the IMFs, the Hilbert spectrum is computed as shown in Figure 7.4.2. The results are quite similar to the results of daily average wind speed data (Figure 7.3.3). Comparing the results of time-frequency representations to Fourier based methods, Hilbert-Huang transform gives a clear indication of the fluctuation at higher frequencies. Fourier based methods tend to blur the spectral amplitudes when the signal is nonstationary. Hence, they are not suitable for analyzing the nonstationary or nonlinear signals. The nonstationary feature in the low frequency is addressed well by the Hilbert-Huang transform. The strength of energy in low frequency (0–0.03 cycle/day) is not constant over the time span for four stations. Some segments have much higher energy than the adjacent segments. Hilbert-Huang spectra extract more detailed features of the non-stationarity than Fourier and multi-taper spectra.

The marginal Hilbert spectra for four stations are plotted in Figure 7.4.3. Each subplot represents the spectra at one station, and the results from three spectral analyses, Fourier, multi-taper and Hilbert-Huang transform, are plotted in the same log-log scale. Best fit lines for the entire spectrum, best fit line for the resampled spectrum and 95 % confidence intervals are indicated in the figure.

The marginal Hilbert spectra are fitted by using the power law equation $y = af^b$, where y is the spectra, f is frequency and a, b are the coefficients. The power law decay rate, b, and the R-square values for the fitting are summarized in Table 7.4.1. Again, the amplitude, a, is not used for discussion. These decay rates are milder than the slopes from daily average wind speed data. A comparison of marginal Hilbert spectra of daily peak wind speed data in four stations is shown in Figure 7.4.4. The slope is about $-0.816 \sim -0.890$ for daily average wind speed and $-0.723 \sim -0.855$ for daily peak wind speed. The daily average wind speed has a similar feature in frequency domain and daily peak wind speed data has more variations in spectra.

The spectra of daily wind data are helpful in investigating the low frequency range. An annual cycle, which is around 300~400 days, is detected in four stations. Other seasonal and interannual cycles are found through the spectral analysis. The period for these is not exactly consistent, but roughly speaking they are somewhere around 100 days, 150 days and 200 days. A 724-day, which is biannual cycle, is detected in Fort Wayne for the daily peak wind speed data. As for the high frequency, a 3~4 day and a 7~12 day cycle are detected. The degree of stantionarity is calculated for the daily peak wind speed data as shown in Figure 7.4.5. It is interesting to see that there are two dip downs which correspond to the 0.5 and 1 year periods.

Compared to the spectra of hourly data, the spectra of either daily average or daily peak wind speed data has clearer features in low frequency. Unless the periodicity less than 3 days is of interest, daily wind speed data are easier than hourly data to analyze. Variations occurring over hours to days are important for wind speed evaluation. Wind speed records typically show large upswings and downswings that

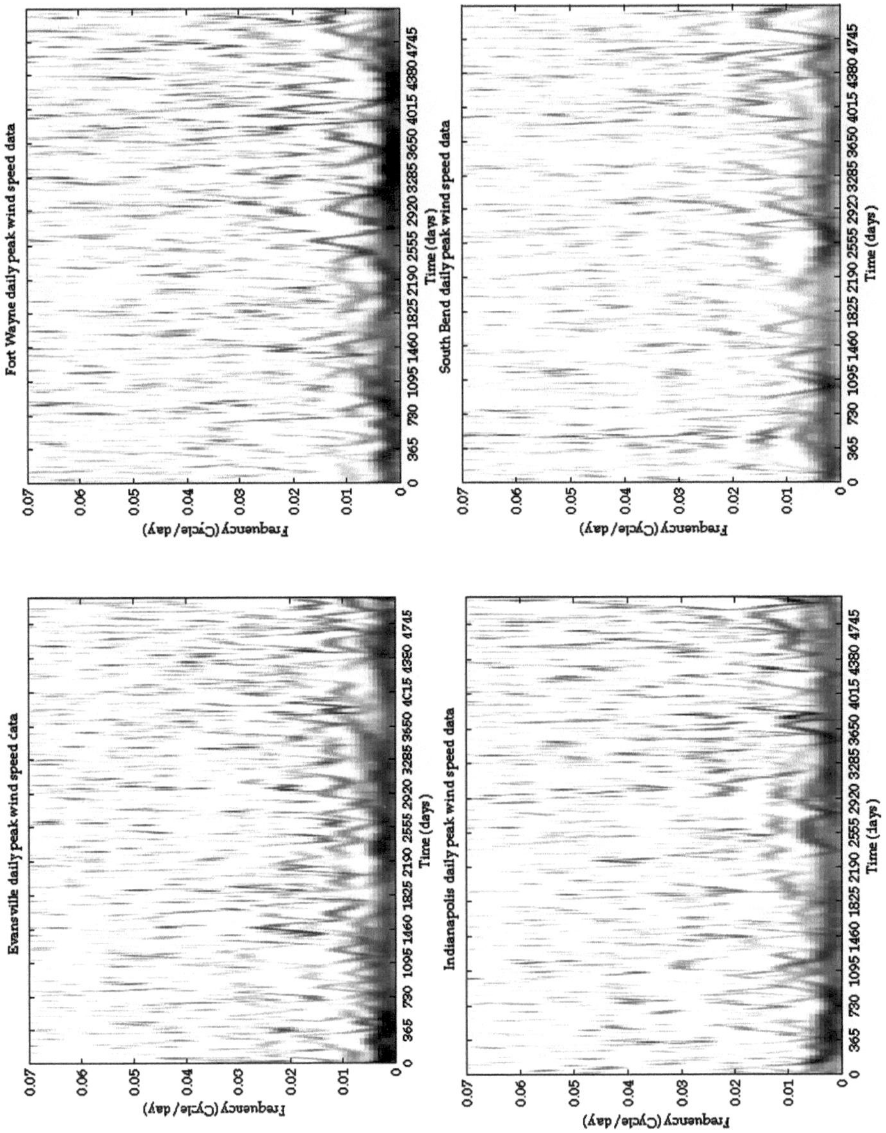

Figure 7.4.2. Time-frequency distributions (Hilbert spectra) of the daily peak wind speed data

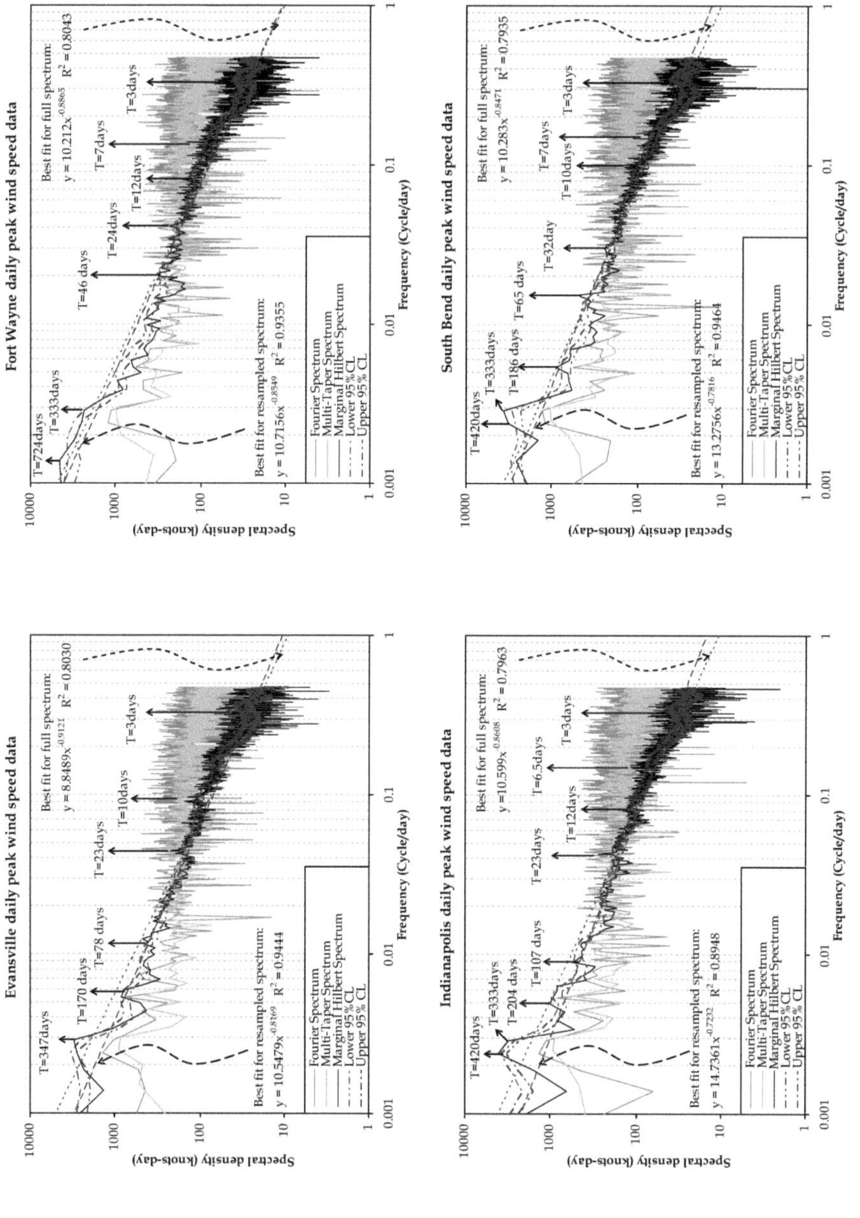

Figure 7.4.3. The marginal Hilbert spectra, Fourier spectra and MTM spectra of the daily peak wind speed data

Table 7.4.1. Power law decay rates and R-square values for the best fit lines to the marginal Hilbert spectra of daily peak wind speed data

Station	Full spectrum		Resampled spectrum	
	b	R^2	b	R^2
Evansville	−0.9121	0.8030	−0.8169	0.9444
Fort Wayne	−0.8865	0.8043	−0.8549	0.9355
Indianapolis	−0.8608	0.7963	−0.7232	0.8948
South Bend	−0.8471	0.7935	−0.7816	0.9464

persist up to several days, reflecting passing storms and weather fronts. The high frequency domain, which is affected by a small and local scale wind variation, is dominated by strong breezes caused by the temperature differences between land and sea and between mountains and valleys.

7.5. CONCLUDING COMMENTS

Comparing the results from the Hilbert-Huang transform method and other conventional methods for both hourly and daily wind speed data reveals that the Hilbert-Huang transformation provides a better picture of the time and frequency behavior. In particular, the marginal Hilbert spectrum of the hourly wind speed data shows

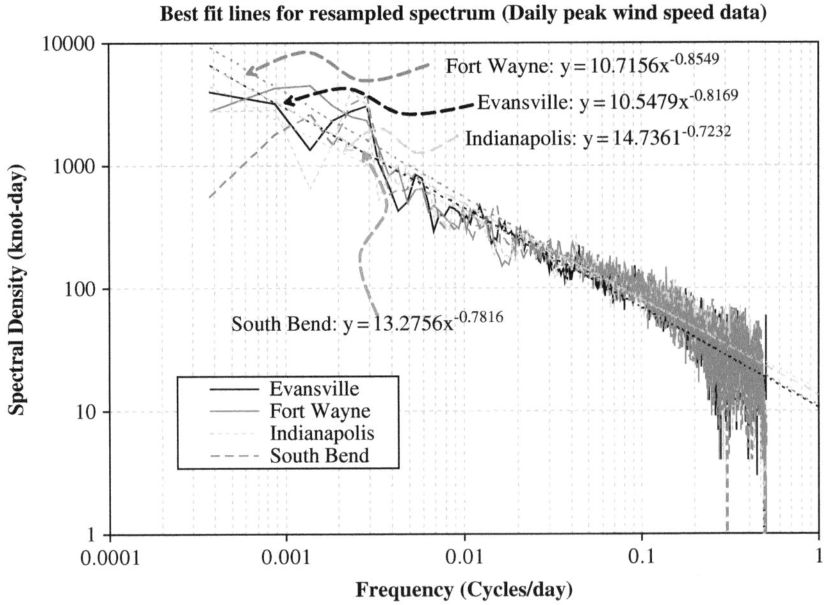

Figure 7.4.4. Comparison of marginal Hilbert spectra of daily peak wind speed data in the four stations

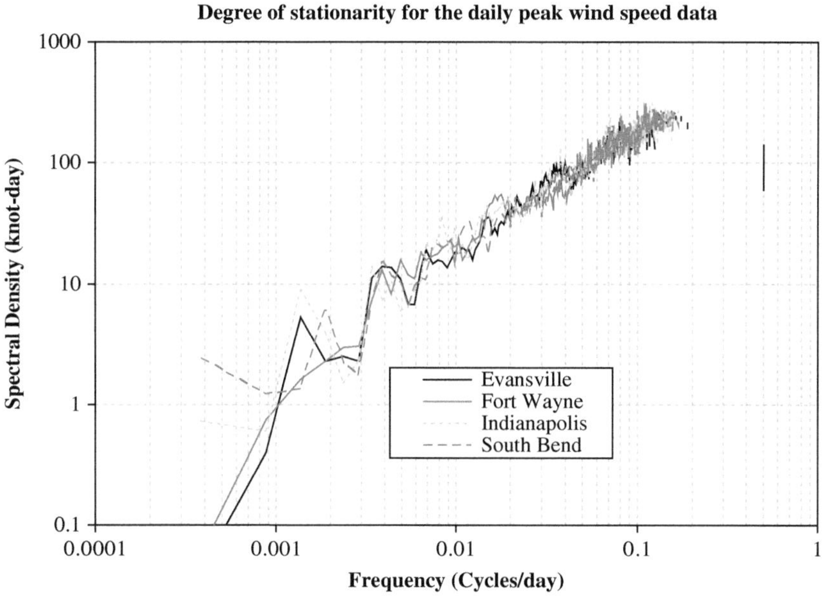

Figure 7.4.5. Degree of stationarity of the daily peak wind speed data in the four stations

a common feature across frequencies. The power law is used to characterize the relationship between frequency and spectral amplitude. All four stations displayed a distinct change in power law trend in the low frequencies, with a substantial drop in decay rate. Similarities are also observed in the daily average and daily peak wind data. The decay rates are found around -1.0 for the marginal Hilbert spectra of the hourly wind speed data. Furthermore, daily average wind speed data have slightly higher rates of decay compared to the daily peak wind speed data. Their values are about -0.8. Further research needs to be done to determine whether the differences in frequency behavior among the three data sets represent a true difference in underlying physical behavior or a consequence of the analysis procedure. These behaviors of the decay rates imply a greater sensitivity of marginal Hilbert spectra to sampling effects, as the variation in the fitted quantities from the daily peak data have increased rather significantly. These results suggest similar mechanisms responding for both peak and average wind speed data, with the Hilbert spectra showing these similarities most conclusively.

An obvious choice of future research is to extend this analysis to larger data sets in order to enhance resolution of possible long-term oscillations. Also, questions arise in the physical interpretation of these intrinsic mode functions. The identification of the nonstationary trends would be useful, as the techniques used herein represent only early attempts to examine such behavior.

CHAPTER 8

LAKE TEMPERATURE DATA ANALYSIS

8.1. INTRODUCTION AND DATA USED

In the previous chapters, characteristics of different time series are studied. The main interest was in investigating longterm oscillations and nonstationarity. When it comes to highly nonstationary spatial data, the interpretation of the results from the application of Hilbert-Huang transform is of interest. The spatial data studied here is the lake water temperature profile with respect to lake depth. The sampling site is Square Lake, which is a ground water fed lake of 38.4 km^2 located in Washington County, Minnesota, USA. The source of data is Warnnars (2005). The lake has a maximum depth of about 21.7 m. The location of the sampling site with respect to Square Lake is shown in Figure 8.1.1.

The samples are taken by the Self Contained Autonomous MicroProfiler (SCAMP), which is a portable, lightweight microstructure profiler designed by Precision Measurement Engineering, Inc. to measure extremely small scale (order 1 mm) fluctuations of temperature, light, conductivity, and oxygen concentration in lakes. It is equipped with sensors that feature high spatial resolution and fast-time response.

The SCAMP profiler is shown in Figure 8.1.2b. A SCAMP is capable of sensing small scale (1 mm) fluctuations and is a useful tool in estimating turbulent kinetic energy dissipation rates in aquatic environments. PAR sensor was available on SCAMP.

The unique drag plate and floats allow the SCAMP to ascend or descend vertically making measurement (100 Hz or 100 samples per second) through undisturbed water. The SCAMP can sample while traveling at 10 cm/sec either downward or upward through the water column. In downward mode, it can sample undisturbed water beginning somewhat below the surface and continuing all the way to the bottom. In upward mode, it can sample undisturbed water up through the surface. The two modes of travel enable the SCAMP to measure both surface mixing and benthic boundary layer mixing.

A monthly sampling regime was conducted from June to October 2004. Sampling was conducted from a boat moored near the deepest location of the lake. Water samples were collected, using a vertical Wildco student water bottle kit (Forestry suppliers, Inc), at half-meter intervals for the upper 5 m and every meter below that

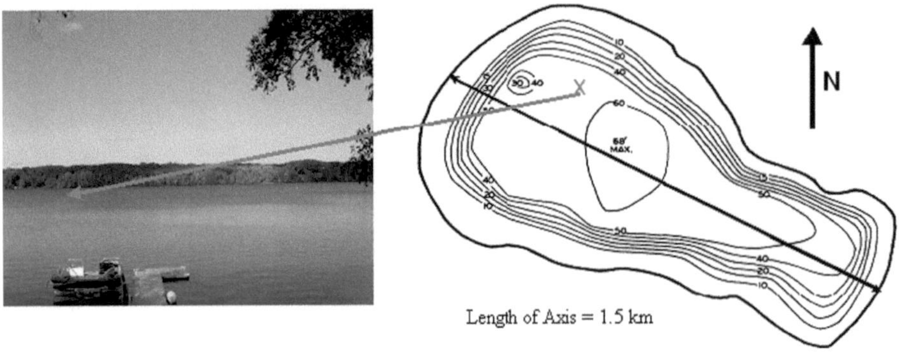

Figure 8.1.1. Square Lake with sampling site marked with an "X" (contours given in feet) (picture source: Warnnars, 2005)

(Figure 8.1.2a). Samples were placed in acid-washed one liter bottles for same-day analysis at the laboratory.

In this study, the availability of data types collected on July 8, Aug 10, Sept 17 and Oct 20 in 2004 are given in Table 8.1.1. All of them are measured in an upward direction toward water surface. In this study, the temperatures and PAR data are of interest. There are three major temperature profiles, including Fast T0, Fast T1 and accurate temperature. These profiles are shown in Figure 8.1.3. The PAR profile is shown in Figure 8.1.4. The entire profiles are shown in Figure 8.1.3a and

(a) (b)

Figure 8.1.2. Image of (a) water sampler (b) SCAMP (self contained autonomous micorprofiler) (source: Warnnars, 2005)

Table 8.1.1. The available variables for each data set (X: available; –: no data)

Data	7/8/04	8/10/04	9/17/04	10/20/04
Fast T0	X	X	X	X
Fast T1	X	X	X	X
Fast C	–	–	–	–
Dissolved oxygen	X	X	X	X
Accurate C	–	–	–	–
Accurate T	X	X	X	X
Fast T2	–	–	–	–
Fluorometer	X	X	X	X
PAR	X	X	X	X
Depth	X	X	X	X
Grad Fast T0	X	X	X	X
Grad Fast T1	X	X	X	X
Velocity	X	X	X	X
Sigma-T	X	X	X	X
Salinity	–	–	–	–
Sorted Sigma-T	X	X	X	X

PAR: photosynthetically active radiation
Fluorometer: to measure chlorophyll concentration

Figure 8.1.4a while part of these profiles are zoomed in as shown in Figure 8.1.3b and Figure 8.1.4b.

Fast T0 and Fast T1 should have similar data since they are fast sampling sensors. Accurate T is a slow sampling sensor and should not be used for spectral analysis. The difference among Fast T0, Fast T1 and accurate T is observed in Figure 8.1.3b. There is less variation in the accurate temperature. Although there are values for Fast T1, it is seen in Figure 8.1.3a that the profiles for Aug, Sep and Oct behave a vertical straight line which is not correct. Hence, in this study, Fast T0 series is the one used for Hilbert-Huang spectral analysis.

The four Fast T0 profiles eventually reach a temperature of 7°C at the 15 m depth. In July, the temperature in the top 5 m keeps a constant around 20°C and then has a transition to reach 7°C. In August, the temperature in the top 6 m keeps a constant around 20°C and then has a transition to reach 7°C. In September, the temperature in the top 8 m keeps a constant around 20°C and then has a transition to reach 7°C. In October, the temperature of the top 11 m keeps constant around 12 C and then has a transition to reach 7°C.

As for the PAR data, it designates the spectral range of solar light from 400 to 700 nm that is useful to terrestrial plants in the process of photosynthesis. The sensor has an output proportional to the number of photons received regardless of their energy levels (photon flux density: $\mu mol \cdot m^{-2} \cdot s^{-1}$). Since the energy level of each photon depends on its wavelength, the spectral response of a PAR sensor using the photovoltaic effect must be higher in the far red part of the spectrum (around 700 nm) than in the blue wavelengths (around 400 nm). This spectral region corresponds more or less to the range of visible light to the human eye. The

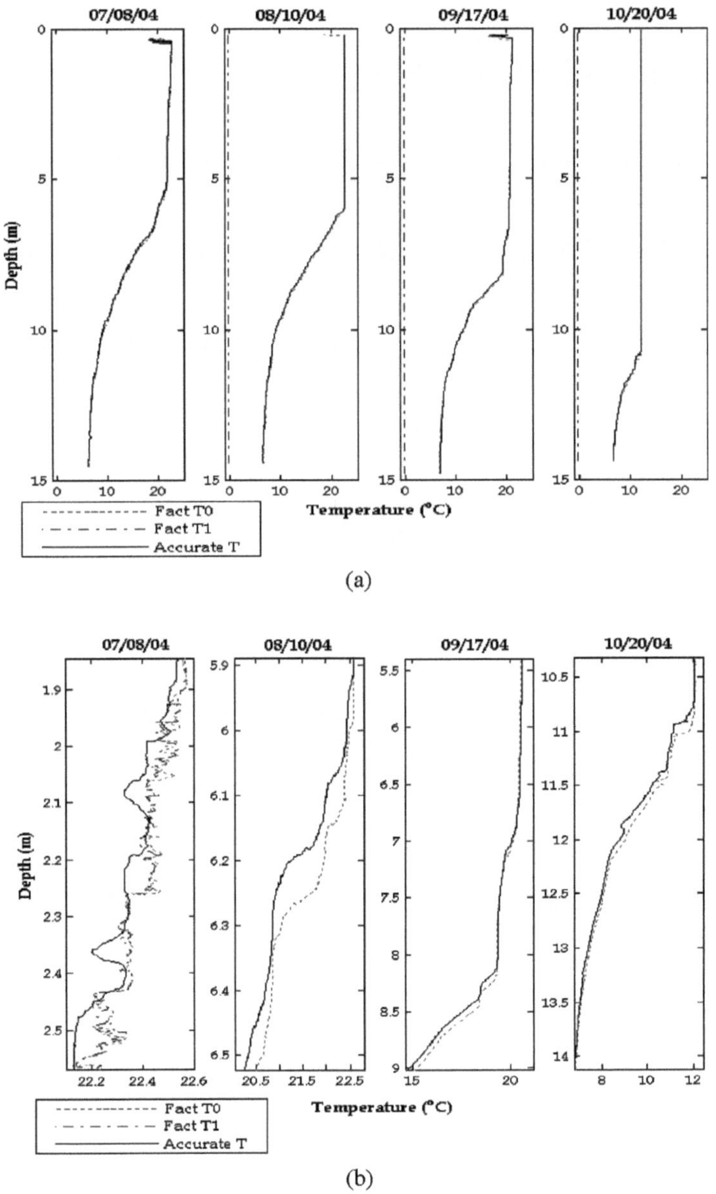

Figure 8.1.3. (a) Entire (b) zoom-in view of sampling depth profiles of temperatures

profiles from July and September have more variation in the depth of top 5 m. Other than that, the profiles behave quite linearly as the depth increases. It is noted that the original data is not evenly sampled along the depth, so each profile has been resampled with a 1-mm depth interval in this study.

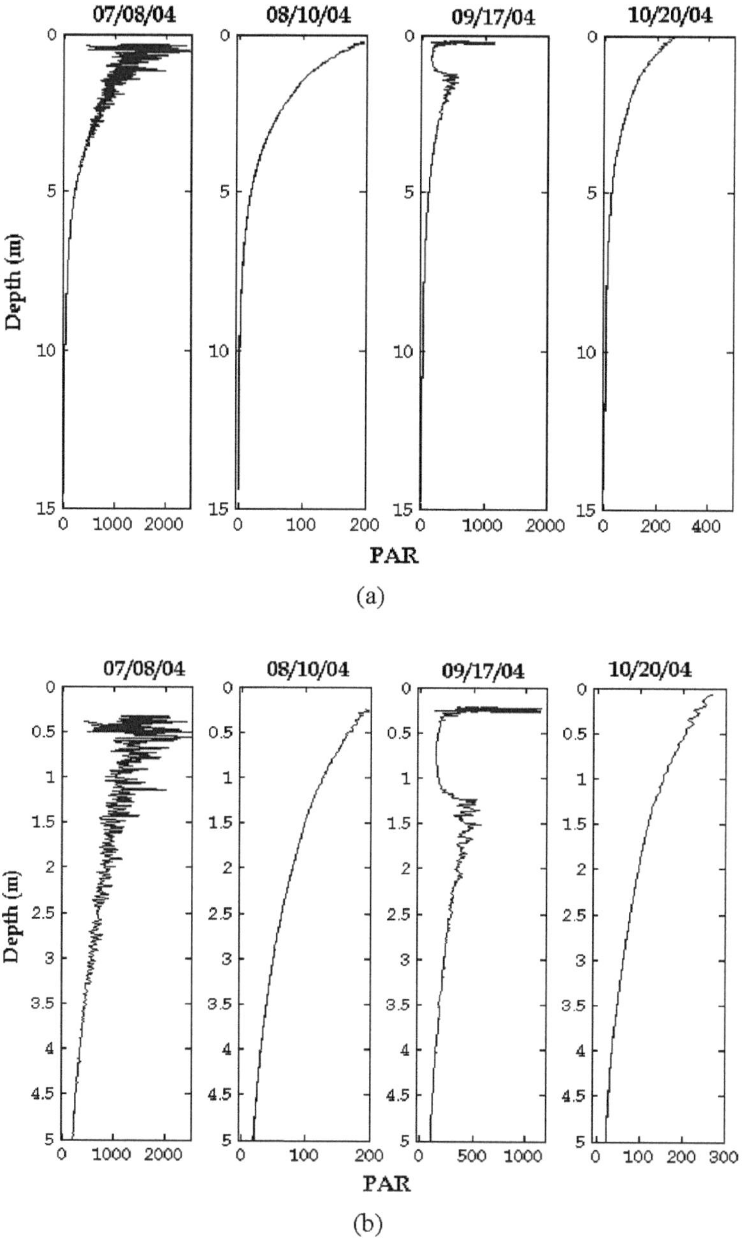

Figure 8.1.4. (a) 0–15 m (b) 0–5 m sampling depth profile of PAR

8.2. LAKE TEMPERATURE SPATIAL SERIES ANALYSIS

8.2.1 Spatial Series Analysis

The empirical mode decomposition is performed for the Fast T0 temperature and PAR data of the four samples. The decomposition is done along the depth of lake, and the intrinsic mode functions are shown in Figure 8.2.1 and Figure 8.2.2 for Fast T0 and PAR data, respectively.

From Figure 8.2.1, it is seen that all the residuals have a strongly downward trend while the intrinsic mode functions have a small fluctuations of an order less than 1°C. As for PAR data, the downward trends are observed in four series. The trend test is performed for the segments of data divided based on the last IMF components.

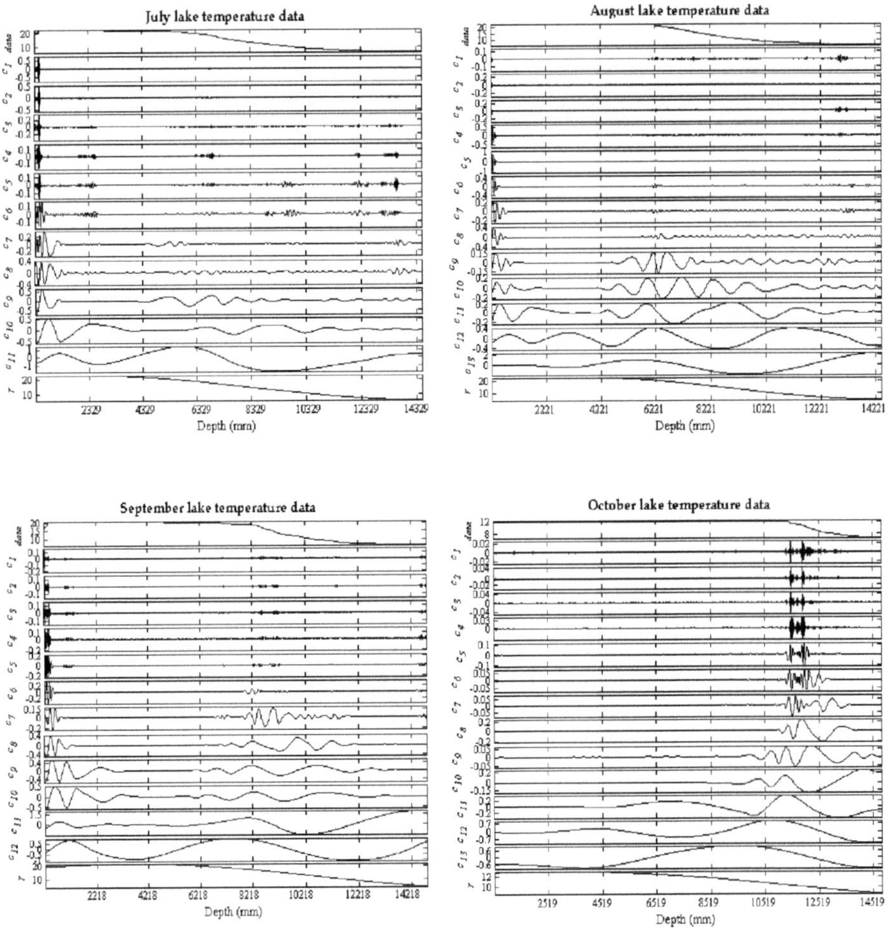

Figure 8.2.1. Intrinsic mode functions of lake temperature data

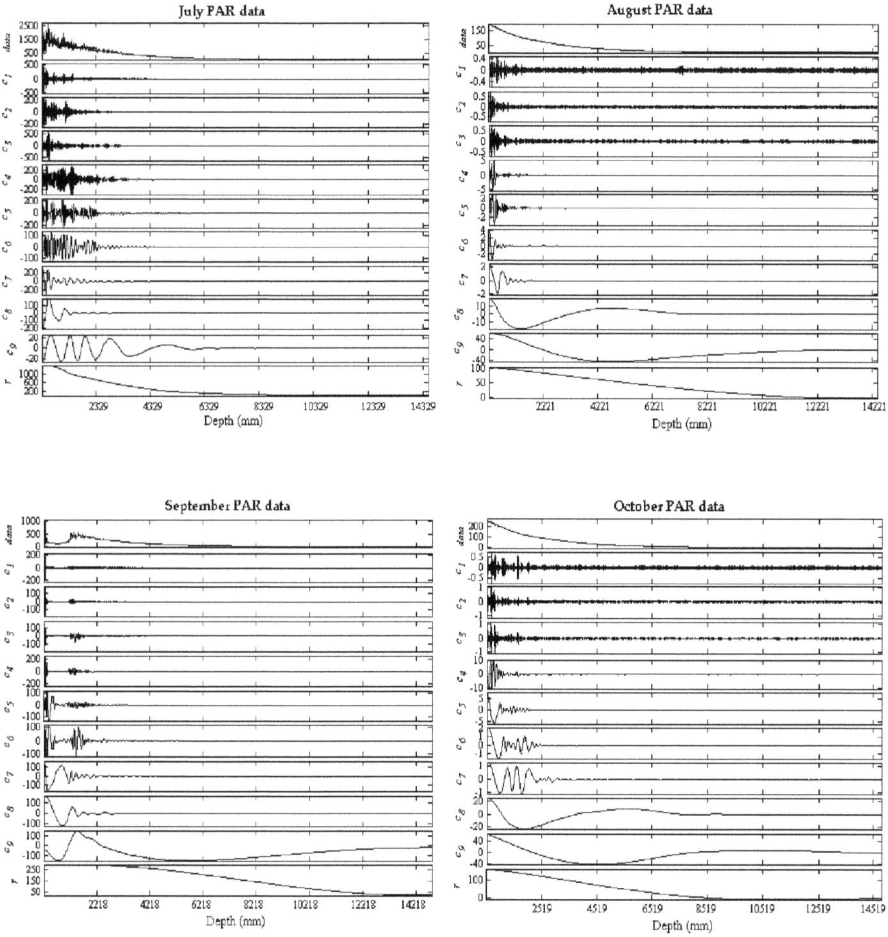

Figure 8.2.2. Intrinsic mode functions of lake PAR data

The modified Mann-Kendall test is used for testing the lake temperature and PAR data. The results are shown in Table 8.2.1. It indicates downward trends for most of the segments.

8.2.2 Time-Frequency Distribution

The Hilbert spectrum for lake temperature data is shown in Figure 8.2.3 and for the lake PAR data is shown in Figure 8.2.4. Along the time axis apart from the Hilbert spectrum, the original data, volatility and instantaneous energy (IE) are plotted. For the instantaneous energy, the bold gray line indicates the average energy at the corresponding time while the dashed black line refers to the standard deviation of the instantaneous energy. Along the frequency axis, the degree of stationarity

Table 8.2.1. Trend test results for (a) lake temperature and (b) PAR data

Time	Segment	(a) Modified Mann-Kendall
July	1	Stationary (z = −0.6302)
	2	Downward trend (z = −2.736)
	3	Downward trend (z = −3.0508)
	4	Downward trend (z = −3.0616)
	5	Downward trend (z = −3.0046)
Aug	1	Upward trend (z = 1.834)
	2	Stationary (z = 1.0395)
	3	Stationary (z = 1.3708)
	4	Downward trend (z = −2.4637)
	5	Downward trend (z = −3.0475)
	6	Downward trend (z = −2.9697)
Sep	1	Stationary (z = 0.7477)
	2	Downward trend (z = −2.9121)
	3	Downward trend (z = −3.0153)
	4	Downward trend (z = −3.0599)
	5	Downward trend (z = −2.9576)
Oct	1	Stationary (z = −1.376)
	2	Downward trend (z = −2.3167)
	3	Downward trend (z = −2.8573)
	4	Downward trend (z = −2.971)
		(b)
July	1	Upward trend (z = 1.834)
	2	Downward trend (z = −2.0836)
	3	Downward trend (z = −1.718)
	4	Downward trend (z = −2.4225)
	5	Downward trend (z = −2.2734)
	6	Downward trend (z = −2.8512)
	7	Downward trend (z = −28968)
	8	Downward trend (z = −2.978)
	9	Downward trend (z = −3.0239)
	10	Downward trend (z = −2.895)
	11	Downward trend (z = −2.335)
Aug	1	Downward trend (z = −3.0586)
	2	Downward trend (z = −3.008)
	3	Downward trend (z = −2.1501)
Sep	1	Downward trend (z = −2.811)
	2	Upward trend (z = 2.476)
	3	Downward trend (z = −2.993)
	4	Downward trend (z = −2.975)
Oct	1	Downward trend (z = −3.0496)
	2	Downward trend (z = −3.0477)
	3	Downward trend (z = −2.8519)

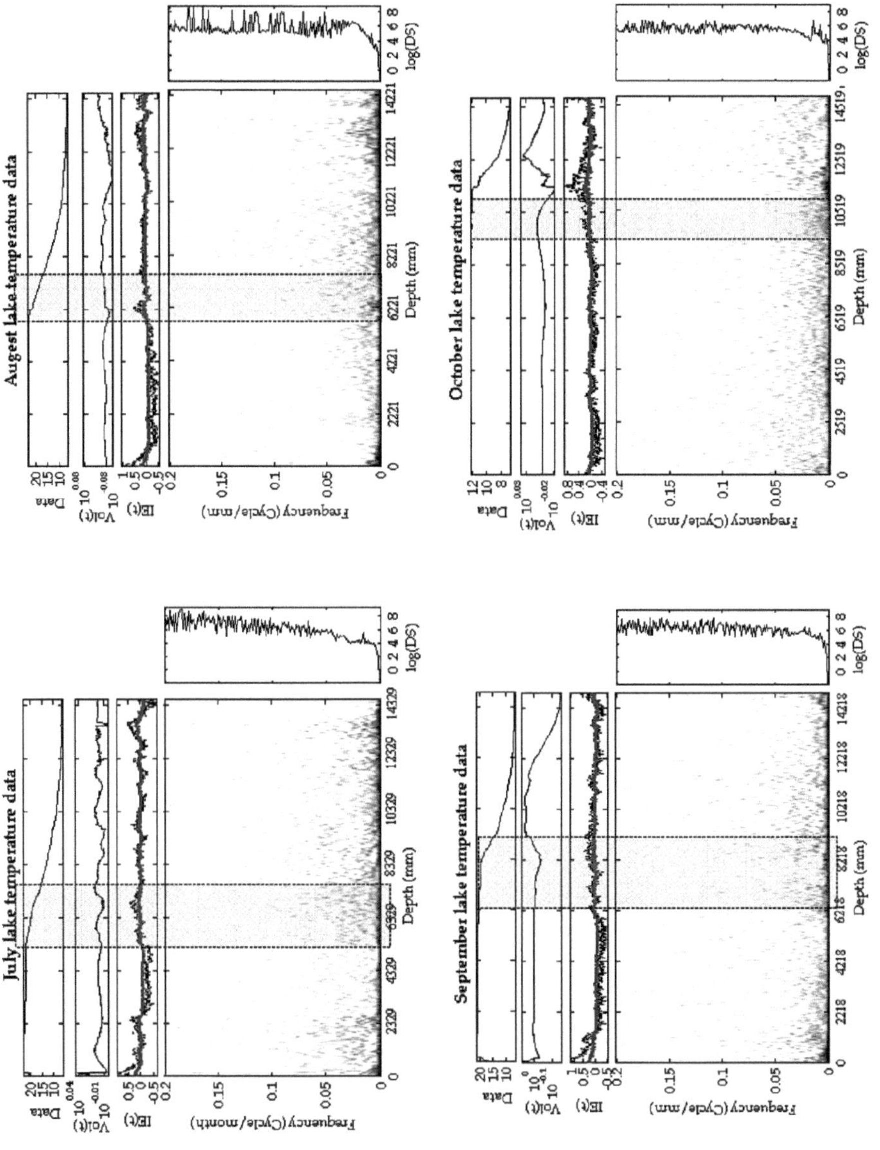

Figure 8.2.3. Time-frequency distribution of lake temperature data

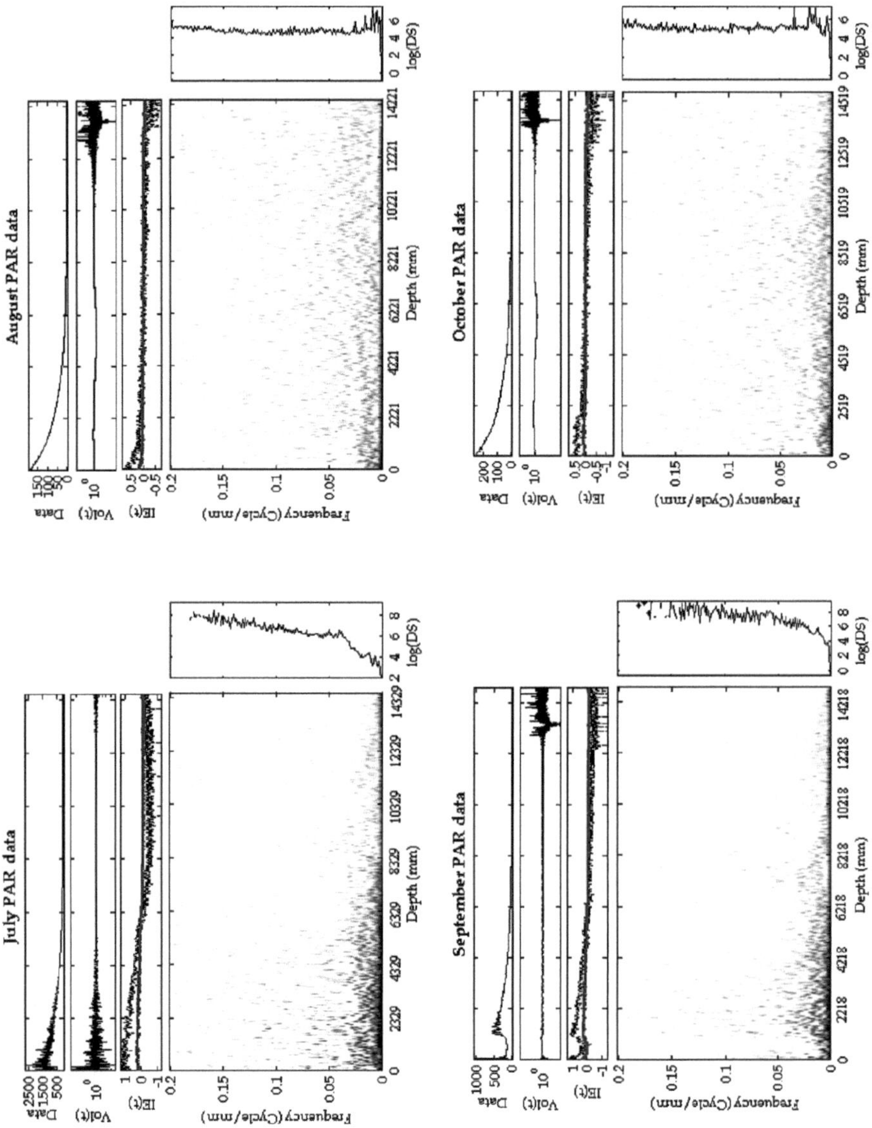

Figure 8.2.4. Time-frequency distribution of lake PAR data

corresponding to each frequency is shown. The smaller degree of stationarity means that the data is more stationary at that frequency.

In the time-frequency distribution, the strong energy occurring in temperature refers to the transition between a constant segment and a gradually varied segment. The strong energy in lake PAR data refers to the high variation segment in the signal. For example, the time-frequency distributions of PAR profile in July and September show strong spectral amplitudes in the beginning, which corresponds to the data from the top water surface. The results from standard deviation of instantaneous energy are also consistent to the variation in the data.

For the lake temperature data, there are some common characteristics of these series. First, there are some fluctuations with high frequency and small amplitude near the water surface. Then it keeps a constant for several meters (5 m, 6 m, 8 m and 11 m for the data sampled in July, Aug, Sep and October). At certain point, the temperature profile exponentially decreases and reaches another constant value. Therefore, there is a transition for each profile and it is a key component to describe these data. Through Hilbert spectral analysis (Figure 8.2.3), strong spectral amplitudes appears around these transition points (indicated by a shaded box). The advantage of Hilbert-Huang transform to investigate the nonstationary process makes it possible to locate the abrupt changes in data. Also, for the most abrupt transition in October's temperature sample, the spectral amplitude is relatively higher in the transition depth than those at other depths.

For the lake PAR data (Figure 8.2.4), there are more variations around the top 6 m. After that, it decays away and approaches zero. The volatility and standard deviation of instantaneous energy are quite consistent with the lake PAR profiles except for the depth after 13 m. It may be caused by the continuous zero values and the end effect from sifting processes. The Hilbert spectrum is of interest. These high variations of PAR data around water surface are addressed well in the corresponding spectral amplitudes. Data transitions in August and October are smooth and so are the spectra. Data in July and September are highly nonstationary in the beginning of profiles, which are also indicated in the Hilbert spectra.

The degree of stationarity behaves different from those in previous analyses (rainfall, runoff, temperature, etc.). In previous results, the degree of stationarity is low in low frequency and gradually gets higher in high frequency. However, in the cases of lake temperature and PAR data, the degrees of stationarity keep a high value for the entire frequency. The signals are highly nonstationary.

8.2.3 Frequency Domain Analysis

The marginal Hilbert spectra of the lake temperature data are shown in Figure 8.2.5. The marginal Hilbert spectra for the lake PAR data are shown in Figure 8.2.6. The 95 % confidence limits, best-fit line for the entire resampled spectrum and multi-taper spectrum are indicated as well. The multi-taper spectra are fairly flat and do not decay away in the high frequency regions.

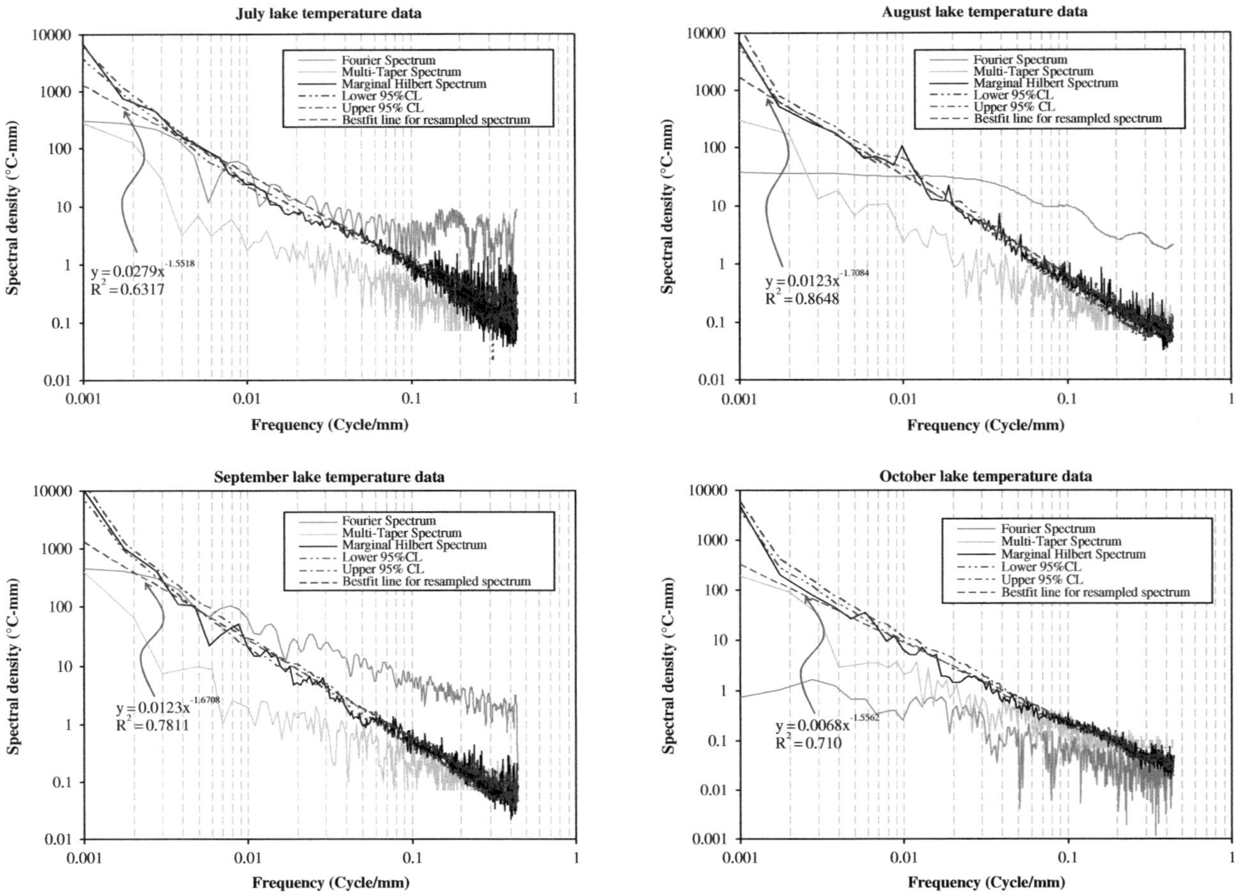

Figure 8.2.5. The marginal Hilbert spectra and confidence limits for lake temperature data

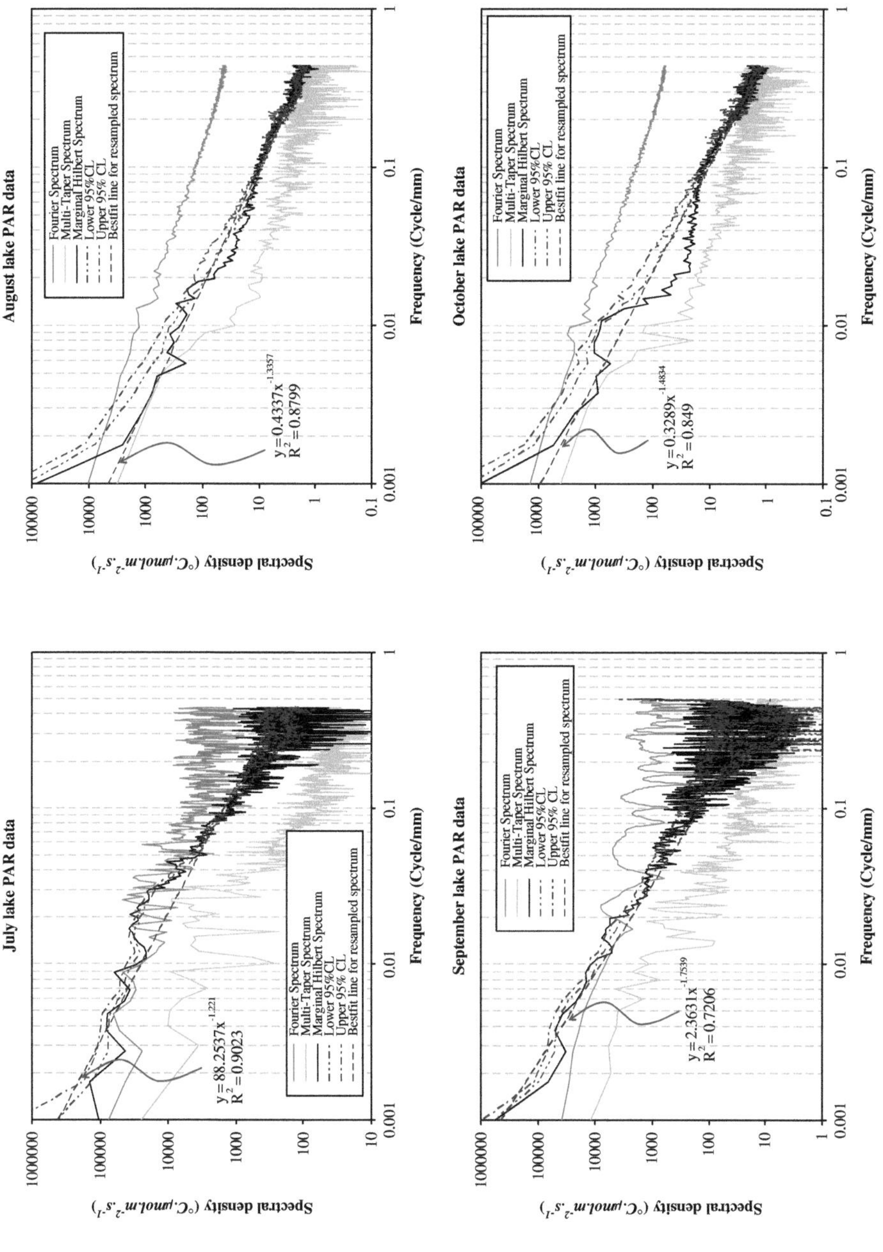

Figure 8.2.6. The marginal Hilbert spectra and confidence limits for lake PAR data

Each subplot in Figure 8.2.5 and Figure 8.2.6 represents the spectra at one station, and the results from three spectral analyses, Fourier, multi-taper and Hilbert-Huang transform, are plotted in the same log-log scale. Best fit lines for the entire spectrum and best fit line for the resampled spectrum are indicated in the figure. The marginal Hilbert spectra are fitted by using the power law equation $y = af^b$, where y is the spectra, f is frequency and a, b are the coefficient pending for determined.

The marginal Hilbert spectra are obtained by integrating the Hilbert spectrum over the entire time span. For the highly nonstationary process, the spectral power varies considerably in time, such as the results in Figure 8.2.3 and Figure 8.2.4. These Hilbert spectra make their marginal Hilbert spectra problematic. It is worthwhile to compare the Hilbert spectra to the Fourier and Multi-taper spectra. The Hilbert spectra in Figure 8.2.5 and Figure 8.2.6 are smoother than Fourier and MTM spectra, which have more spikes and valleys. However, the decay trends in spectra are fairly consistent. For the results of lake PAR data in August and October (Figure 8.2.6), the Hilbert and MTM spectra almost parallel each other.

The periodicity is not studied in this type of series since it does not make any physical sense. The power law is much of interest. The power law decay rate, b, and the R-square values are summarized in Table 8.2.2. The amplitude, a, is not used for discussion since it does not have much universality feature. Based on the coefficient of variation, temperature profile has a good fit for the power law while the power law in PAR data is not strong. The high variation in the PAR data makes the spectra have different patterns. Hence, fitting the spectra may not provide consistent results.

The results from the study of lake data reveal the strength of Hilbert-Huang Transform in investigating the transition of signals with highly nonstationary component. The periodicity is not necessary to be studied in this type of series since it does not present any physical meaning. The point of interest is in the investigation of the time variations. As the Hilbert spectra observed in lake temperature and PAR profiles, the variations of energy magnitudes and patterns represent the transition in time series. The increasing or decreasing of energy magnitudes indicates the changes of the signal frequencies and the signal amplitudes.

Table 8.2.2. Power law amplitudes, decay rates and R-square values for the lake temperature and PAR data

	Lake temperature data			Lake PAR data		
Time	a	b	R^2	a	b	R^2
July	0.0279	−1.5518	0.870	88.2537	−1.2210	0.902
August	0.0123	−1.7084	0.865	0.4337	−1.3357	0.880
September	0.0123	−1.6708	0.781	2.3631	−1.7539	0.721
October	0.0068	−1.5562	0.710	0.3289	−1.4834	0.849
Mean	0.0148	−1.6218	0.8065	22.8449	−1.4485	0.8380
Std. Dev.	0.0091	0.0798	0.0761	43.6159	0.2302	0.0812
C.V.	0.6134	−0.0492	0.0944	1.9092	−0.1589	0.0969

For these time series without strong periodicities, a smooth marginal Hilbert spectrum is revealed and that is good for power law analysis. The average decay rate obtained for the lake temperature data is around -1.6 while the rates for the lake PAR varies considerably from one to another. The identification of the nonstationary trends is useful in the analysis of the nonstationary processes. It enhances the utility of the Hilbert-Huang transform, building upon its strength in nonstationary analysis. The lake data analysis also demonstrates the idea of studying data not only in time but also in space. Therefore, for any type of data of interest, the data feature can be interpreted and described by following the procedures provided in this study. The data needs to be investigated in time domain by looking at the intrinsic mode functions, trend tests and measures, in frequency domain by looking at the marginal Hilbert spectra and power law, and in the time-frequency distribution by looking at the energy variation from time to time or from one frequency to another.

CHAPTER 9

CONCLUSIONS

With the concept of instantaneous frequency, HHT provides a better representation in frequency domain than the conventional Fourier spectra, which have fixed frequency bins; hence the HHT provides a finer resolution in time-frequency representation. It provides the local, adaptive and efficient information directly from the data without any linearity or stationary assumptions or restrictions. Therefore it can be applied to nonlinear time-variant signals. These strengths make it a powerful tool to investigate climatic and hydrological time series that are often nonstationary.

Empirical mode decomposition can be treated as a time-frequency filtering method through the representation of the intrinsic mode function (IMF) components. The empirical mode decomposition technique enables signals to be separated into several oscillation modes by extracting the characteristic scales embedded in the data. Traditionally, filtering is carried out in frequency domain only; however, frequency domain filtering is difficult when the data are either nonstationary or nonlinear, or both. The low-pass, high-pass and band-pass filters can be designed from the IMF components.

In time-frequency distribution, if the signal is nonstationary or nonlinear, the spectrum obtained from Fourier based methods blurs in the low frequency and results in some spurious oscillations. The variation of energy amplitude in time is not easy to investigate from the spectrogram obtained by Fourier based method, The high frequency variation can be seen by removing the trend; however, if the trend is not a simple one for the entire time series, this removal would lead to misleading results. Compared to these Fourier based methods, Hilbert-Huang transform has the advantage to directly analyze nonstationary data without removing trend or make additional assumptions. It is adaptive and these periods, oscillations and trends are obtained by the decomposition of the original data.

The measures used to interpret the results of Hilbert spectra are volatility, degree of stationarity, instantaneous energy and standard deviation of the instantaneous energy. Volatility is the measure obtained from summation of selected IMF components; hence it is function of time. Several modification options for computing volatility are discussed in this study. Instantaneous energy is a measure for investigating the variation of spectral energy at a frequency span. For the rainfall and streamflow data studied, standard deviation of instantaneous energy is more consistent with the peaks and valleys in the data than volatility and instantaneous energy. For temperature data, these measures are quite consistent with the variations in the time series. In addition, the spatial correlations of these volatilities are studied and the data from the state of Indiana is used as an example. The results show spatial continuity.

Degree of stationarity, which is calculated from the Hilbert-Huang spectrum, shows how uniformly the power is distributed in a frequency bin. Higher value of degree of stationarity indicates more variation in frequency through the time. It is investigated herein by using different lengths of data. The results show that while the high-frequency components are nonstationary, they can still be statistically stationary with shorter time spans.

Another advantage of Hilbert-Huang transform is that it makes possible the simulation of nonstationary time series. There are three methods investigated in this study. They are simulated by generating the random phases, generating both random phases and amplitudes, and the Wen-Yeh method. These methods are examined by using several natural time series and compared by using the spectra and the capability to reconstruct the data. Second method (simulated with random phase and amplitudes) is not preferred because it varies too much from the data characteristics. The results from the first (simulated with random phase) and the third method (Wen-Yeh) are quite close. But, in this study, the first method is recommended to use. The reason is that it is easy to implement and does not involve too many assumptions. Therefore, the simulations done in this study are based on the method 1.

The success of nonstationary time series simulation also brought out the idea of identifying the confidence limits by utilizing the simulated data. The introduction of the confidence limits for the marginal Hilbert spectra is discussed and used in this study. Once the confidence limits are available, the significant periods are detected in a systematic way. This procedure is much more elegant than guessing the significance of peaks in spectra.

For large amount of data points, such as daily Warta River data and hourly wind speed data, a smooth marginal Hilbert spectrum results. A consequence of the smoothness of the Hilbert spectra is that it is difficult to pick out strong or weak contributions at individual frequencies. Except for the strongest periodicity, there are almost no other peaks or valleys. This may represent a loss of frequency information in the Hilbert spectra, although some of the peaks found in the other spectra could simply be artifacts of the transformation process used to obtain them.

This smoothness also brought out the idea of investigating the power law. It would be much more difficult to extract these power law trends from either the Fourier or multi-taper spectra due to their much more erratic behavior with respect to frequency. Thus, the Hilbert-Huang transform does appear to be capable of providing information on frequency behavior that other methods cannot provide. For most of spectra discussed in this study, one or two best fit lines are used to the full or resampled Hilbert spectra. For these smooth spectra, the decay rates are fairly consistent for the same type of data. This may lead to universality conclusions while applying the Hilbert spectral analysis to these data.

Physical interpretation is the most challenging part in HHT analysis. From the empirical mode decomposition and spectral analysis, the information embedded in the signal are extracted to find out the possible oscillations, such as annual cycle, quasi-biennial oscillation, ENSO-like mode, solar cycle, trends, and others. Through the analysis, the annual cycle is clearly seen in rainfall, runoff, temperature and wind speed data. The indication of 1 year period from time-frequency distribution is better seen in temperature data than the others. For wind speed data, daily and annual cycles are clearly detected. A 4~5 day period from fronts and weather system is detected as well.

The overall advantages of Hilbert-Huang transform spectra are summarized as follows. (1) Conventional spectral analysis methods are very sensitive to nonstationarity in the data. They produce spurious low frequency harmonic waves when handling nonstationary signals. The interpretation of the results in the low frequency domain is affected and the long-term oscillations cannot be accurately interpreted. (2) The empirical mode decomposition reduces the signal into several intrinsic mode function components without any assumption for frequency resolution. Each IMF component represents the information embedded in the data from high to low frequency. The residual indicates the overall trend in the data. The higher order IMF components represent the long-term oscillations as well. This is a powerful advantage in time series analysis. (3) The power leakage of HHT spectra is smaller than in other methods. (4) The frequency defined in HHT is the instantaneous frequency, which is a function of time. Unlike other methods in which frequencies are assigned and limited by the trade-off with time resolution, HHT precisely presents the instantaneous frequency at that point of time. This results in the nonstationary situation to have smaller impact on the location of frequency.

REFERENCES

Balocchi R, Menicucci D, Santarcangelo E, Sebastiani L, Gemignani A, Ghelarducci B, Varanini, M (April 2004) Deriving the respiratory sinus arrhythmia from the heartbeat time series using empirical mode decomposition. Chaos, Solitons & Fractals 20(1).

Box GEP, Jenkins G (1976) Time series analysis: Forecasting and control. Holden-Day.

Bretthorst GL (1998) Bayesian spectrum analysis and parameter estimation. Lecture Notes in Statistics, 48. Springer-Verlag, New York..

Brockwell PJ, Davis RA (1991) Time series: Theory and methods. Springer, New York,.

Camuffo D, Jones P (2002) Improved understanding of past climate variability from early daily European instrumental sources. Kluwer Academic Publishers, Dordrecht.

Chen J, Xu YL, Zhang RC (2004) Modal parameter identification of Tsing Ma suspension bridge under Typhoon Victor: EMD-HT method. Journal of Wind Engineering & Industrial Aerodynamics 92:805–827.

Douka E, Hadjileontiadis LJ (2005) Time-frequency analysis of the free vibration response of a beam with a breathing crack. NDT&E International 38:3–10.

Dyrbye C, Hansen SO (1997) Wind loads on structures. John Wiley & Sons Publishers, USA.

Flandrin P (1999) Time-frequency/ time-scale analysis. Academic Press, San Diego, CA.

Flandrin P, Gonçalvès P April 5, (2004) Empirical mode decompositions as data-driven wavelet-like expansions. International Journal of Wavelets, Multiresolution and Information Processing.

Flandrin P, Rilling G, Goncalves P (2003) Empirical mode decomposition as a filter bank. IEEE Signal Processing Letters.

Gai G (2006) The processing of rotor startup signals based on empirical mode decomposition. Mechanical Systems and Signal Processing 20:222–235.

Gloersen P, Huang N (2003) Comparison of interannual intrinsic modes in Hemispheric sea ice covers and other geophysical parameters. IEEE Transactions on Geoscience and Remote Sensing 41(5).

Gu P and Wen YK (2005). Simulation of Nonstationary Random Processes Using Instantaneous Frequency and Amplitude from Hilbert-Huang Transform. In Huang, NE, Attoh-Okine NO. The Hilbert-Huang Transform in Engineering. CRC Press, Taylor & Francis Group, 2005.

Hamed KH, Rao AR (1998). A modified Mann-Kendall trend test for autocorrelated data. Journal of Hydrology 204:182–196.

Hsu E, Rao AR (2005). Bayesian spectral analysis of climatic and hydrologic time series. World Water & Environmental Resources Congress, Anchorage, Alaska,.

Huang NE, Attoh-Okine NO (2005). The Hilbert-Huang transform in engineering. CRC Press, Taylor & Francis Group.

Huang NE, Chern CC, Huang K, Salvino LW, Long SR, Fan KL (2001) A new spectral analysis of station TCU129, Chi-Chi, Taiwan, 21 September 1999. Bulletin of the Seismological of America 91(5):1310–1338.

Huang NE, Shen SSP (2005) Hilbert-Huang Transform and its Applications. Interdisciplinary Mathematical Sciences, Vol. 5, World Scientific Publication Co. Pte. Ltd.

Huang NE, Shen Z, Long SR, Wu MC, Shih HH, Zheng Q, Yen N, Tung C., Liu HH (1998) The empirical mode decomposition and the Hilbert spectrum for nonlinear and non-stationary time series analysis. Proceedings of the Royal Society London 1998. A454:903–995.

Huang NE, Wu MC, Long SR, Shen SSP, Qu W, Gloersen P, Fan KL (2003a) A confidence limit for the empirical mode decomposition and Hilbert spectral analysis. Proceedings of the Royal Society London 459:2317–2345.

Huang NE, Wu M, Qu W, Long SR, Shen SSP (2003b) Applications of Hilbert-Huang transform to non-stationary financial time series analysis. Applied Stochastic Models in Business and Industry 19:245–268.

Huang W, Shen Z, Huang NE, Fung YC (1999a) Engineering analysis of biological variables: An example of blood pressure over 1 day. Proc. Natl. Acad. Sci. USA 95:4816–4821.

Huang W, Shen Z, Huang NE, Fung YC (1999b) Nonlinear indicial response of complex nonstationary oscillations as pulmonary hypertension responding to step hypoxia. Proc. Natl. Acad. Sci. USA 96.

Hwang PA, Kaihatu JM, Wang DW (2002) A comparison of the energy flux computation of shoaling waves using Hilbert and wavelet spectral analysis techniques. 7th Int. Workshop on Wave Hindcasting and Forecasting. 10:21–25.

Karl TR, Williams Jr CN, Young PJ, National Climatic Data Center, Wayne M. Wendland, Illinois State Water Survey (1986). A model to estimate the time of observation bias associated with monthly mean maximum, minimum and mean temperatures for the United States. Journal of Climate and Applied Meteorology 25:145–160, American Meteorological Society, Boston, MA.

Kendall MG (1975) Rank correlation methods. Charles Griffin, London.

Liang H, Bressler SL, Desimone R, Fries P (2005). Empirical mode decomposition: A method for analyzing neural data. Neurocomputing 65–66:801–807.

Loutridis SJ (2005). Instantaneous energy density as a feature for gear fault detection. Mechanical Systems and Signal Processing 20(5):239–1253.

Mann HB (1945) Non-parametric tests against trend. Econometrica 13:245–259.

Montesinos ME, Munoz-Cobo JL, Perez C (2003) Hilbert-Huang analysis of BWR neutron detector signals: application to DR calculation and to corrupted signal analysis. Annuals of Nuclear Energy 30:715–727.

Olejnik K (1991) Przeplywy Warty W Poznaniu. Poznan.

Pan J, Yan X, Zheng Q, Liu WT, Klemas VV (2002) Interpretation of scatterometer ocean surface wind vector EOFs over the Northwestern Pacific. Remote Sensing of Environment 84:53–68.

Percival DB, Walden AT (1993) Spectral analysis for physical applications; Multitaper and conventional univariate techniques. Cambridge University Press.

Peng ZK, Tse P.W, Chu FL (2005). A comparison study of improved Hilbert-Huang transform and wavelet transform: Application to fault diagnosis for rolling bearing. Mechanical Systems and Signal Processing 19(5):974–988.

Ponomarenko VI, Prokhorov MD, Bespyatov AB, Bodrov MB,. Gridnev VI (2005) Deriving main rhythms of the human cardiovascular system from the heartbeat time series and detecting their synchronization. Chaos. Solitons and Fractals 23:1429–1428.

Priestley MB (1965) Evolutionary spectra and non-stationary processes. Journal of Royal Statistics Society, Ser. B27:204–237.

Quek ST, Tua PS, Wang Q (2003) Detecting anomalies in beams and plate based on Hilbert-Huang transform of real signals. Smart Materials and Structures 12:447–460.

Quinlan FT, Karl TR, Williams Jr CN (1987) United States Historical Climatology Network (HCN) serial temperature and rainfall data. NDP019. Carbon Dioxide Information Analysis Center. Oak Ridge National Laboratory.

Thomson DJ (1982) Spectrum estimation and harmonic analysis. Proc. IEEE 70:1055–1096.

Tirtotjondro WW (1992) Bayesian methods in hydrology. Ph.D. thesis, Purdue University, School of Civil Engineering, p 248.

Veltcheva AD (2002) Wave and group transformation by a Hilbert Spectrum. Coastal Engineering Journal 44(4): 283–300.

Vörösmarty CJ, Fekete B, Tucker BA (1996) River Discharge Database, Version 1.0 (RivDIS v1.0), Volumes 0 through 6. A contribution to IHP-V Theme 1. Technical Documents in Hydrology Series. UNESCO, Paris.

Warnnars TA (2005) The influence of fluid motion on freshwater algae: A biophysical investigation, Ph.D. thesis, University of Minnesota, Department of Civil Engineering, p 124.

Wen YW, Gu P (2004) Description and simulation of nonstationary processes based on Hilbert Spectra. ASCE Journal of Engineering Mechanics 942–951.

Yang JN, Lei Y, Pan S, Huang NE (2003). System identification of linear structures based on Hilbert-Huang spectral analysis. Part 2: Complex modes. Earthquake Engineering and Structural Dynamics 32:1533–1554.

Yu D, Cheng J, Yang Y (2005) Application of EMD method and Hilbert spectrum to the fault diagnosis of roller bearings. Mechanical Systems and Signal Processing 19:259–270.

Zhang RR, King R, Olson L, Xu Y (2005) Dynamic response of the Trinity River Relief Bridge to controlled pile damage: Modeling and experimental data analysis comparing Fourier and Hilbert-Huang techniques. Journal of Sound and Vibration 285(4–5):1049–1070.

Zhang RR, VanDemark L, Liang J, Hu Y (2004) On estimating site damping with soil non-linear from earthquake recordings. International Journal of Non-linear Mechanics 39: 1502–1517.

INDEX

Water Science and Technology Library

1. A.S. Eikum and R.W. Seabloom (eds.): *Alternative Wastewater Treatment.* Low-Cost Small Systems, Research and Development. Proceedings of the Conference held in Oslo, Norway (7–10 September 1981). 1982 ISBN 90-277-1430-4
2. W. Brutsaert and G.H. Jirka (eds.): *Gas Transfer at Water Surfaces.* 1984 ISBN 90-277-1697-8
3. D.A. Kraijenhoff and J.R. Moll (eds.): *River Flow Modelling and Forecasting.* 1986 ISBN 90-277-2082-7
4. World Meteorological Organization (ed.): *Microprocessors in Operational Hydrology.* Proceedings of a Conference held in Geneva (4–5 September 1984). 1986 ISBN 90-277-2156-4
5. J. N.ěmec: *Hydrological Forecasting.* Design and Operation of Hydrological Forecasting Systems. 1986 ISBN 90-277-2259-5
6. V.K. Gupta, I. Rodríguez-Iturbe and E.F. Wood (eds.): *Scale Problems in Hydrology.* Runoff Generation and Basin Response. 1986 ISBN 90-277-2258-7
7. D.C. Major and H.E. Schwarz: *Large-Scale Regional Water Resources Planning.* The North Atlantic Regional Study. 1990 ISBN 0-7923-0711-9
8. W.H. Hager: *Energy Dissipators and Hydraulic Jump.* 1992 ISBN 0-7923-1508-1
9. V.P. Singh and M. Fiorentino (eds.): *Entropy and Energy Dissipation in Water Resources.* 1992 ISBN 0-7923-1696-7
10. K.W. Hipel (ed.): *Stochastic and Statistical Methods in Hydrology and Environmental Engineering.* A Four Volume Work Resulting from the International Conference in Honour of Professor T.E. Unny (21–23 June 1993). 1994
    10/1: Extreme values: floods and droughts ISBN 0-7923-2756-X
    10/2: Stochastic and statistical modelling with groundwater and surface water applications ISBN 0-7923-2757-8
    10/3: Time series analysis in hydrology and environmental engineering ISBN 0-7923-2758-6
    10/4: Effective environmental management for sustainable development ISBN 0-7923-2759-4
    Set 10/1–10/4: ISBN 0-7923-2760-8
11. S.N. Rodionov: *Global and Regional Climate Interaction: The Caspian Sea Experience.* 1994 ISBN 0-7923-2784-5
12. A. Peters, G. Wittum, B. Herrling, U. Meissner, C.A. Brebbia, W.G. Gray and G.F. Pinder (eds.): *Computational Methods in Water Resources X.* 1994 Set 12/1–12/2: ISBN 0-7923-2937-6
13. C.B. Vreugdenhil: *Numerical Methods for Shallow-Water Flow.* 1994 ISBN 0-7923-3164-8
14. E. Cabrera and A.F. Vela (eds.): *Improving Efficiency and Reliability in Water Distribution Systems.* 1995 ISBN 0-7923-3536-8
15. V.P. Singh (ed.): *Environmental Hydrology.* 1995 ISBN 0-7923-3549-X
16. V.P. Singh and B. Kumar (eds.): *Proceedings of the International Conference on Hydrology and Water Resources* (New Delhi, 1993). 1996
    16/1: Surface-water hydrology ISBN 0-7923-3650-X
    16/2: Subsurface-water hydrology ISBN 0-7923-3651-8

16/3: Water-quality hydrology ISBN 0-7923-3652-6
16/4: Water resources planning and management ISBN 0-7923-3653-4
Set 16/1–16/4 ISBN 0-7923-3654-2

17. V.P. Singh: *Dam Breach Modeling Technology*. 1996 ISBN 0-7923-3925-8
18. Z. Kaczmarek, K.M. Strzepek, L. Somlyódy and V. Priazhinskaya (eds.): *Water Resources Management in the Face of Climatic/Hydrologic Uncertainties*. 1996 ISBN 0-7923-3927-4
19. V.P. Singh and W.H. Hager (eds.): *Environmental Hydraulics*. 1996 ISBN 0-7923-3983-5
20. G.B. Engelen and F.H. Kloosterman: *Hydrological Systems Analysis*. Methods and Applications. 1996 ISBN 0-7923-3986-X
21. A.S. Issar and S.D. Resnick (eds.): *Runoff, Infiltration and Subsurface Flow of Water in Arid and Semi-Arid Regions*. 1996 ISBN 0-7923-4034-5
22. M.B. Abbott and J.C. Refsgaard (eds.): *Distributed Hydrological Modelling*. 1996 ISBN 0-7923-4042-6
23. J. Gottlieb and P. DuChateau (eds.): *Parameter Identification and Inverse Problems in Hydrology, Geology and Ecology*. 1996 ISBN 0-7923-4089-2
24. V.P. Singh (ed.): *Hydrology of Disasters*. 1996 ISBN 0-7923-4092-2
25. A. Gianguzza, E. Pelizzetti and S. Sammartano (eds.): *Marine Chemistry*. An Environmental Analytical Chemistry Approach. 1997 ISBN 0-7923-4622-X
26. V.P. Singh and M. Fiorentino (eds.): *Geographical Information Systems in Hydrology*. 1996 ISBN 0-7923-4226-7
27. N.B. Harmancioglu, V.P. Singh and M.N. Alpaslan (eds.): *Environmental Data Management*. 1998 ISBN 0-7923-4857-5
28. G. Gambolati (ed.): *CENAS. Coastline Evolution of the Upper Adriatic Sea Due to Sea Level Rise and Natural and Anthropogenic Land Subsidence*. 1998 ISBN 0-7923-5119-3
29. D. Stephenson: *Water Supply Management*. 1998 ISBN 0-7923-5136-3
30. V.P. Singh: *Entropy-Based Parameter Estimation in Hydrology*. 1998 ISBN 0-7923-5224-6
31. A.S. Issar and N. Brown (eds.): *Water, Environment and Society in Times of Climatic Change*. 1998 ISBN 0-7923-5282-3
32. E. Cabrera and J. García-Serra (eds.): *Drought Management Planning in Water Supply Systems*. 1999 ISBN 0-7923-5294-7
33. N.B. Harmancioglu, O. Fistikoglu, S.D. Ozkul, V.P. Singh and M.N. Alpaslan: *Water Quality Monitoring Network Design*. 1999 ISBN 0-7923-5506-7
34. I. Stober and K. Bucher (eds): *Hydrogeology of Crystalline Rocks*. 2000 ISBN 0-7923-6082-6
35. J.S. Whitmore: *Drought Management on Farmland*. 2000 ISBN 0-7923-5998-4
36. R.S. Govindaraju and A. Ramachandra Rao (eds.): *Artificial Neural Networks in Hydrology*. 2000 ISBN 0-7923-6226-8
37. P. Singh and V.P. Singh: *Snow and Glacier Hydrology*. 2001 ISBN 0-7923-6767-7
38. B.E. Vieux: *Distributed Hydrologic Modeling Using GIS*. 2001 ISBN 0-7923-7002-3
39. I.V. Nagy, K. Asante-Duah and I. Zsuffa: *Hydrological Dimensioning and Operation of Reservoirs*. Practical Design Concepts and Principles. 2002 ISBN 1-4020-0438-9

Water Science and Technology Library

40. I. Stober and K. Bucher (eds.): *Water-Rock Interaction.* 2002 ISBN 1-4020-0497-4
41. M. Shahin: *Hydrology and Water Resources of Africa.* 2002 ISBN 1-4020-0866-X
42. S.K. Mishra and V.P. Singh: *Soil Conservation Service Curve Number (SCS-CN) Methodology.* 2003 ISBN 1-4020-1132-6
43. C. Ray, G. Melin and R.B. Linsky (eds.): *Riverbank Filtration.* Improving Source-Water Quality. 2003 ISBN 1-4020-1133-4
44. G. Rossi, A. Cancelliere, L.S. Pereira, T. Oweis, M. Shatanawi and A. Zairi (eds.): *Tools for Drought Mitigation in Mediterranean Regions.* 2003 ISBN 1-4020-1140-7
45. A. Ramachandra Rao, K.H. Hamed and H.-L. Chen: *Nonstationarities in Hydrologic and Environmental Time Series.* 2003 ISBN 1-4020-1297-7
46. D.E. Agthe, R.B. Billings and N. Buras (eds.): *Managing Urban Water Supply.* 2003 ISBN 1-4020-1720-0
47. V.P. Singh, N. Sharma and C.S.P. Ojha (eds.): *The Brahmaputra Basin Water Resources.* 2004 ISBN 1-4020-1737-5
48. B.E. Vieux: *Distributed Hydrologic Modeling Using GIS. Second Edition.* 2004 ISBN 1-4020-2459-2
49. M. Monirul Qader Mirza (ed.): *The Ganges Water Diversion: Environmental Effects and Implications.* 2004 ISBN 1-4020-2479-7
50. Y. Rubin and S.S. Hubbard (eds.): *Hydrogeophysics.* 2005 ISBN 1-4020-3101-7
51. K.H. Johannesson (ed.): *Rare Earth Elements in Groundwater Flow Systems.* 2005 ISBN 1-4020-3233-1
52. R.S. Harmon (ed.): *The Río Chagres, Panama.* A Multidisciplinary Profile of a Tropical Watershed. 2005 ISBN 1-4020-3298-6
53. To be published.
54. V. Badescu, R.S. Cathcart and R.D. Schuiling (eds): Macro-Engineering: A Challenge for the Future. 2006 ISBN 1-4020-3739-2
55. K.-P. Seiler and J.R. Gat: *Groundwater Recharge from Run-off, Infiltration and Percolation.* 2008 ISBN 978-1-4020-5305-4
56. G. Salvadori, C. De Michele, N.T. Kottegoda and R. Renzo: *Extremes in Nature.* An Approach Using Copulas. 2007 ISBN 1-4020-4414-3
57. S.K. Jain, R.K. Agarwal and V.P. Singh: *Hydrology and Water Resources of India.* 2007 ISBN 1-4020-4414-4
58. To be published.
59. M. Shahin: *Water Resources and Hydrometeorology of the Arab Region.* 2007 ISBN 1-4020-4577-8
60. A.R. Rao and E-C. Hsu: *Hilbert-Huang Transform Analysis of Hydrological and Environmental Time Series.* 2008 ISBN 978-1-4020-6453-1
61. R.S. Govindaraju and B.S. Das: *Moment Analysis for Subsurface Hydrologic Applications.* 2007 ISBN 978-1-4020-5751-9
62. Giuseppe Rossi, Teodoro Vega and Brunella Bonaccorso (eds.): *Methods and Tools for Drought Analysis and Management.* 2007 ISBN 978-1-4020-5923-0

MIX
Papier aus verantwortungsvollen Quellen
Paper from responsible sources
FSC® C105338
FSC
www.fsc.org

If you have any concerns about our products,
you can contact us on
ProductSafety@springernature.com

In case Publisher is established outside the EU,
the EU authorized representative is:
Springer Nature Customer Service Center GmbH
Europaplatz 3, 69115 Heidelberg, Germany

Printed by Libri Plureos GmbH
in Hamburg, Germany